Practical Waste Management

Practical Waste Management

Edited by

John R. Holmes
*Technical Director, Grandmet Waste
Services Limited, Aylesbury, UK*

A Wiley–Interscience Publication

JOHN WILEY & SONS
Chichester · New York · Brisbane · Toronto · Singapore

Library of Congress Cataloging in Publication Data:
Main entry under title:

Practical waste managment.

  'A Wiley–Interscience publication.'
  Includes index.
  1. Refuse and refuse disposal. 2. Sanitary
landfills. I. Holmes, John R.
TD791.P65 1983   628.4'45   82-8633

ISBN 0 471 10491 4   AACR2

British Library Cataloguing in Publication Data:

Practical waste management.
  1. Refuse and refuse disposal. 2. Factory
  and trade waste   3. Sewage disposal
  I. Holmes, John R.
  628.4'4   TD791

ISBN 0 471 10491 4

Typeset by Photo-Graphics, Honiton, Devon

and printed by Pitman Press Ltd., Bath, Avon.

Acknowledgments

The editor and contributors gratefully acknowledge the support given by the following organizations, institutions, manufacturers, and waste management companies whose systems, equipment, processes, and plant have been mentioned in this book.

INSTITUTIONS AND LEARNED BODIES
Institute of Solid Wastes Management
Institute of Public Finance and Accountancy
Oxfam
Keep Britain Tidy Group

RESEARCH ORGANIZATIONS
Harwell Laboratory Environmental Support Group

TRADE ASSOCIATIONS
National Association of Waste Disposal Contractors
Glass Manufacturers Federation

TRADE JOURNALS AND PERIODICALS
Surveyor Magazine
Municipal Engineering
NAWDC News

CONSULTANTS AND MANAGEMENT COMPANIES
Environmental Resources Ltd
Powell Duffryn Pollution Control Ltd

PUBLIC AUTHORITIES AND ORGANIZATIONS
South Yorkshire County Council
West Yorkshire County Council
West Midlands County Council
Anglian Water Authority
County Surveyors Society
Leeds City Council

Ashford Borough Council
London Borough of Bromley
London Borough of Waltham Forest
The City of Copenhagen
The City of Hamburg
The Department of the Environment
Environmental Protection Agency USA

PROCESS PLANT
Powell Dufryn Engineering Ltd
Peabody Holmes Ltd
Murphy Solid Waste Systems Ltd
Motherwell Bridge Tacol Ltd
Flakt, Sweden
Bruunt Sorensen AS, Denmark
Blue Circle Industries Ltd
Imperial Metal Industries
Thames Board Ltd
Newell Dunford Engineering Ltd
Buhler Brothers Ltd
NCC Rexco Ltd

VEHICLES AND EQUIPMENT
Macpactor Ltd
Caterpillar Company Ltd
Scapa Ltd
Cope Alkman Plastics Ltd
The Boughton Group Ltd
Columbus Dixon Inc.
Anchorpac Ltd
SSI Plastics Ltd
Shelvoke & Drury Ltd
Jack Allen Car Bodies Ltd
Hestair Dennis Ltd
Mechanical Products Supply Co Ltd
Acrowpactor Ltd
Cabac Ltd
Johnson Brothers (Engineering) Ltd
Applied Cleaning Economics Ltd
Glasdon Ltd
Concrete Products Ltd
Reed Medway Waste Handling Systems Ltd
Bomag (Great Britain) Ltd

International Harvester
Whale Tankers Ltd
Vacall Ltd
Bradley Municipal Vehicles Ltd
Hestair Eagle Ltd
Smiths Electric Vehicles Ltd
JCB Sales Ltd
White Cross Equipment Ltd

WASTE MANAGEMENT COMPANIES
Cleanaway Ltd
Haul Waste Ltd
Waste Management Ltd
London Brick Landfill Ltd
W. Hemmings Waste Disposal
Biffa Waste Services
Hales Containers Ltd
Thomas Black Ltd
Wimpey Waste Management Ltd
Grundons Ltd
AVG, Hamburg
Kali and Salz AG
Weston Tank Cleaning Ltd
Polymeric Treatments Ltd
Packington Estate Enterprises Ltd
West Wales Tank Cleaning Ltd
Croda Chemicals Ltd
PLM Sweden
Hargreaves Ltd
ReChem International Ltd
Rigbys Ltd
Safeway Ltd
Ocean Combustion Services BV
T.R. International Ltd
Leigh Analytical Services Ltd
Kommunekemi AS, Denmark
Lancashire Tar Distillers Ltd
Surface Control Ltd

List of contributors

J.A. AMBROSE — *Peabody Holmes Limited, Turnbridge, Huddersfield, West Yorkshire, HD1 6RB*

K.J. BRATLEY — *Deputy County Waste Disposal Officer, West Midlands County Council, 1 Lancaster Circus, Queensway, Birmingham B4 7DJ*

W.A. CLENNELL — *Managing Director, Motherwell Bridge Tacol Limited, Green Dragon House, 64/70 High Street, Croydon, Surrey, CR0 1NA*

A. COAD — *WEDC Unit, University of Technology, Loughborough, Leicestershire LE11 3TU*

J.M. COTTERIDGE — *London Borough of Waltham Forest, Works Department Depot, Markhouse Avenue, London E17 8BS*

P.N.O. CRICK — *Director of Landfill Development, Cleanaway Limited, Claydons Lane, Rayleigh, Essex SS6 7UW*

C.H.C. HALEY — *Blue Circle Technical, Research Division, London Road, Greenhithe, Kent DA9 9JO*

M.C. HARRISON — *Managing Director, Haul Waste Limited, The Old Silk Mills, Taunton, Devon TA2 6QB*

R.G.P. HAWKINS — *Barrister, Cleanaway Limited, Claydons Lane, Rayleigh, Essex SS6 7UW*

A.E. HIGGINSON, MBE — *Consultant, 4 Grange Park, Cranleigh, Surrey GU6 7HY*

J.R. HOLMES — *Technical Director, Grandmet Waste Services Limited, Buckingham Street, Aylesbury, Bucks HP21 2LA*

A.Q. KHAN — *Toxic Wastes Officer, Environment Department, South Yorkshire County Council, Regent Street, Barnsley, South Yorkshsire S70 2HF*

H.W. LUTHER — *Waste Management Limited, Rixton Old Hall, Manchester Road, Rixton, Warrington, Cheshire WA3 6EW*

B. McCartney

Managing Director, Macpactor Limited, 30A Cook Street, Hazel Grove, Stockport, Cheshire SK7 4EG

P.A. Oluwande

Professor of Civil Engineering, Faculty of Technology, University of Ibadan, Nigeria

R.J. Owens

John Taylor and Sons, Consulting Engineers, Artillery House, Artillery Row, Westminster, London SW1P 12Y

A. Parker

Hazardous Materials Service, Building 146.3, Harwell Laboratory. AERE, Harwell, Oxfordshire OX11 0RA

P.K. Patrick

Consultant, Friars Lea, 10 The Promenade, Peacehaven, Sussex BN9 8QF

J. Pickford

Group Leader, Waste and Environment Developing Countries (WEDC), Department of Civil Engineering, University of Technology, Loughborough, Leicestershire LE11 3TU

A. Sowerby

Murphy Solid Waste Systems Limited, Highbury House, Highbury Corner, London N5 1RL

P.R. Spencer

Operations Director, London Brick Landfill Limited, Stewartby, Bedfordshire MK43 9LZ

F.W. Stokes

Consultant, Adeney, Llanvaches, Newport, Gwent NP6 3AY

C.I. Tunaley

Waste Disposal Officer, Environment Department, South Yorkshire County Council, Regent Street, Barnsley, South Yorkshire S70 2HF

G.L. Vizard

County Surveyor, Dorset County Council, County Hall, Dorchester, Dorset DT1 1XJ

S.L. Willetts

Steetley Construction Materials Limited, Gibbetts Lane, Shawell, Lutterworth, Leicestershire LE17 6AA

Contents

Foreword

A.E. Higginson, MBE, FISWM, FCIT
Past President, Institute of Solid Wastes Management

During the last decade management services have witnessed a great upheaval in their history in terms of legislative controls, new codes of practice, elevation of standards, and developments in technological approach to all waste arising and disposal. Even a glossary of technical terms has been produced for international agreement on exact meaning in discussing waste management expressions for accurate documentation. Considerable expense has been incurred on research into the behaviour of wastes in landfill sites tracking leachate generation, composition, and its treatment. In the USA enormous amounts of government funding have been allocated into studies of resource recovery, energy from waste, incineration practices, and recoverable heat to meet the challenge of dwindling oil supplies.

A summary of these modern developments was needed, to guide responsible officials both in the public and private sector on trends, latest methods, and market forces, not only applying to the sciences but in business management and financial accountability. John Holmes has succeeded in drawing together experts in their respective fields to discuss, inform, guide, and instruct on many of these developments affecting waste management services. He has marshalled a whole range of important subjects in one book which will be invaluable to all who seek reliable data, up-to-date information, information on latest techniques, and on the effect of wastes and their environmental impact upon society.

Foreword

W. STAPLETON, OBE, FBIM, MISWM
National Association of Waste Disposal Contractors

The National Association of Waste Disposal Contractors is proud to have contributed much towards the greater professionalism and expertise that now prevails in our industry. We value our close links with the Institute of Solid Wastes Management and our sister associations in other parts of Europe and the world. NAWDC's Training Committee has been central to the spread and exchange of knowledge in waste management and our annual courses are well attended from every sector of the industry.

John Holmes' collection of essays includes many contributions by executives of our member companies, and I am happy as NAWDC's Director to endorse the publication of this book. With topics of options, legislation, decision-making, treatment methods, the carriage and transfer of wastes, refuse collection, and street sweeping, the essays are an important aid to those who wish to be better informed in this vital industry.

Practical Waste Management
Edited by J.R. Holmes
©1983 John Wiley & Sons Ltd

1

Waste management options and decisions

JOHN R. HOLMES*
Technical Director, Grandmet Waste Services Limited

ABSTRACT

A review of refuse collection and disposal in the United Kingdom
and Western Europe, this chapter describes the financial and
economic background to the service and explains the methods by
which management decisions are taken between one treatment
system and another. The size and scope of the private sector of the
industry, the growth and viability of waste reclamation systems, and
the future development of refuse collection and disposal in the
United Kingdom are explored. The differences in legislation background between the United Kingdom and some other European
countries are described.

THE ENVIRONMENTAL PRESSURES

With the current interest in environmental matters, waste collection, disposal, and reclamation, it is necessary for environmentalists, manufacturers,
developers of processes, and other entrepreneurs to understand how the
public authorities see their duties and to understand the economic and
management realities that force them to adopt one course or another. To
many committed people the public authorities and the private sector of the
waste disposal industry may seem to adopt philistine and insensitive attitudes
to waste recycling and reclamation, or persist in pursuing cheap and seemingly irresponsible courses of action. Before they are held to account for these
real and imagined failings, it is important to understand their point of view.
Faced with pressures from the general public and elected counsellors,
bombarded by the process industry with a plethora of schemes capable of
separating and converting waste into various forms, yet constrained to
operate a vital public service in an efficient and economic way, the operation-

---

* Formerly Waste Disposal Officer, South Yorkshire County Council.

al decisions are not easily made and this chapter sets out to show the policies and attitudes held by the majority of public and private sector waste managers performing a vital public service against a background of very limited financial resources.

THE BROAD ALTERNATIVES

All methods of waste disposal are in the final analysis landfill, and even the most sophisticated processes produce residues that require to be landfilled. The choice lies in selecting the level of capital cost, operating cost, pollution risk and environmental impact of the various options open to the WDAs and private operators. The spectrum of options ranges from the simple sanitary

Figure 1 Waste—reclaim or bury? The dilemma facing the developed world. (Courtesy of The Caterpillar Company, USA)

landfill of domestic and commercial wastes through various transfer station and bulk haul alternatives on to the high capital cost systems like composting and incineration. Even the new generation of experimental waste reclamation processes still discharge significant tonnages of residue and these are land-filled. As far as liquids and sludges are concerned, much the same range of alternatives exist with an overwhelming bias towards landfill. This, as we all know, occurs in mixtures with domestic and commercial wastes and occasionally in the deep mine shafts and mineral workings such as those owned by the Leigh Interests Group in the Midlands. Every one of these alternatives requires land for its residues and, in general, operational costs and capital investment are inversely proportional to the land or space consumed by the system.

The United Kingdom annually generates about 18 million tonnes of domestic and commercial wastes and at least 20 million tonnes of industrial waste including about 3 million tonnes of building debris. Only a small proportion of industrial wastes are toxic and for these waste chemical neutralization, incineration, or carefully controlled landfill are the only safe

Table 1 Domestic and commercial waste disposal patterns in England and Wales—1980/81 estimates (percentages)

	Direct landfill	Incineration	Other methods
Metropolitan	77	18.6	4.4
Non-metropolitan	92.7	6.6	0.7
GLC	88.4	11.6	0
England and Wales overall	89	9.6	1.4

Table 2 Domestic and commercial waste treatent costs per tonne (£) in England—1980/81 estimates

	Direct landfill	Pulverization landfill	Remote haulage and landfill (contractors and agents)	Incineration	Other methods
Metropolitan	1.98	8.46	6.64	12.09	9.80
Non-metropolitan	2.65	8.19	6.71	8.89	14.15
GLC	5.08	7.50	14.60	8.74	—
England and Wales overall	2.41	8.09	10.26	10.41	11.02

Data extracted from the Chartered Institute of Public Finance & Accountancy (CIPFA)

Table 3   Capital cost of some mechanical engineering disposal plants (1980 values)
(£)

System	Investment capital/tonne/hour
Heat recovery incineration	250,000
Non-heat recovery incineration	200,000
Reclamation transfer station	80,000
Pulverization transfer station	30,000
Simple transfer station	20,000

answers. The treated residues are inert and suitable for landfill. In addition to these wastes there is about 10 million tonnes of power station waste, 60 million tonnes of colliery spoil and about 20 million tonnes of quarrying waste. Most of these vast quantities of waste are inert and disposal by any means other than return to the land, is impractical. Indeed, these wastes are often useful reclamation agents and, if sensitively and skilfully used, can serve a valuable purpose in the recovery of derelict land. In the case of domestic and commercial wastes a wide variety of alternatives is available, and an indication of the capital cost and operating costs of each system is given in Tables 1–3.

THE MUNICIPAL REFUSE COLLECTION SERVICE

No study of the finances of the problem can be considered complete without an appreciation of the municipal refuse collection service. By far the bigger spender—both in revenue costs, manpower, and capital investments in vehicles, workshops, and plant—it is without doubt the senior service overshadowed to some degree by the thrustful attitudes of the recently created Waste Disposal Authorities in the English counties. Recent estimates prepared by the Department of the Environment suggest that the refuse collection costs of the United Kingdom are of the order of £414 million, which, for an estimated 18 million tonnes of domestic and collected commercial wastes, gives a mean collection cost of £23/tonne. The net revenue expenditure of refuse disposal in England, on the 1980/81 estimates prepared by the Chartered Institute of Public Finance and Accountancy was given as £150 million. The figures for Scotland and Wales are less easy to identify but on a strict ratio of population it is likely that the municipal waste disposal cost for the whole of the United Kingdom is of the order of £160 million.

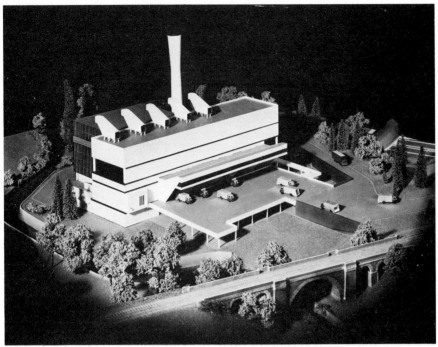

Figure 2   A model of a modern refuse incineration plant—36 tonnes/hour capacity.
(Courtesy of the City of Coventry)

Table 4   Costs of cleansing services

Authority	Population	Total cleansing service (£)	Refuse collection alone (£)	Refuse tonnage (£)	Collection cost per tonne (£)
A	600,000	4,519,000	3,663,000	200,000	18.31
B	224,000	1,303,000	1,118,000	65,000	17.19
C	305,000	1,708,000	1,667,000	100,000	16.66
D	119,000	821,000	626,000	40,000	15.64
E	744,000	5,525,000	4,286,000	223,000	19.22
F	544,000	6,618,000	4,895,000	181,000	27.04

Table 5   Collection and disposal costs

Authority	Population	Collection cost/tonne (£)	Disposal cost per tonne (£)	Collection and disposal cost/tonne (£)
A	600,000	18.31	3.07	21.38
B	224,000	17.19	3.07	20.26
C	305,000	16.66	4.09	20.75
D	119,000	15.64	1.92	17.56
E	744,000	19.22	4.09	23.31
F	544,000	27.04	2.34	29.38

Table 6   Analysis of costs (%)

Salaries, wages, and allowances	56
Repairs, maintenance, and depot charges	7
Transport costs	29
Plastic sacks	4
Special collection services	4
Total	100

Estimated Collection and Disposal Cost—United Kingdom
Collection service:	£414 million
Disposal service:	£160 million
Total:	£574 million

A selection of English District Councils in their 1980/81 budget estimates gave figures for the cost of cleansing services, including street sweeping, gully emptying, and other miscellaneous cleansing activities. These are reproduced in Table 4. Extending these figures to include the disposal costs in the County Councils in which these authorities are located, the total cost of the collection and disposal service emerges. The disposal costs shown in Table 5 are the average for each county and take account of differing combinations of landfill, pulverization, and incineration modes. A typical London Borough analyses its refuse collection costs as shown in Table 6.

THE CITY OF LEEDS

A good example of the financial and operational skeleton of city refuse collection, street sweeping, and other cleansing services is that given for the City of Leeds in the 2,092,000 population West Yorkshire Metropolitan County. Leeds, with a population of 734,000 and an area of 56,000 ha, much of which is heavily developed and industrialized, offers a contrast to those following the example of a smaller district in the south of England. The probable results for the financial year 1980/81 give a nett cost of £7,182,000, giving refuse collection and cleansing services a 4% share of the City's nett revenue requirement of £173,372,540.

Refuse collection

The major service is the annual collection of 210,000 tonnes of waste from 265,000 premises. This is done on the normal weekly cycle by over 110 vehicles and teams of collectors. In addition to the collection of 36,000 tonnes of commercial waste for which the City makes a charge, the cleansing department provides a free collection for awkward household waste including furniture and unwanted electrical appliances. The use of the service by the public averages about 600 calls per week. The cleansing department (on the order of the Regional Water Authority) is responsible for the emptying of all domestic cesspools and septic tanks. Another responsibility is the cleansing and maintenance of over 100 public toilets. In general City Centre toilets are manned by full-time attendants, while a fleet of vans and cleaners provides a mobile service to other sites.

Street sweeping and gully emptying

These services, although the responsibility of the West Yorkshire Metropolitan County Council, are still operated by the City's cleansing department and recharged to the County. The financial arrangements are such that the county is responsible for those charges that relate to the accepted highway safety standards, while the City Council pays that part that is related to amenity and aesthetic appearances. Needless to say, these arrangements require a considerable negotiating tolerance between the parties. The service is executed by mechanical sweeping of roadway channels supplemented by manual footpath sweeping using pedestrian-controlled electric trucks and hand-pushed orderly trolleys.

Gully emptying, which is also recharged to the County Council, is conducted on an ongoing programme, the optimum frequencies of which are stated below in Tables 8–10. These frequencies properly relate to earlier financial years as the Council's statement of accounts makes clear that in

Table 7   Waste collection statistics, 1978–79 actuals (summary)

(a) Non-financial data

	London Boroughs (inc. City) (31)	Metropolitan Districts (35)	Non-Met. Districts England (230)	Non-Met. Districts Wales (32)	Total England and Wales (328)
Population	6,549,900	11,210,000	22,509,390	2,498,900	42,768,190
Area (hectares)	153,735	671,809	8,990,944	1,695,405	11,511,893
Hereditaments					
Domestic	2,464,034	4,030,885	8,281,433	943,008	15,719,360
Commercial	552,233	655,928	1,344,492	132,058	2,684,711
Mixed	32,014	31,801	76,271	10,415	150,501
Industrial	14,600	36,795	54,383	4,886	110,664
Waste collected (tonnes)	2,067,216	3,204,583	7,101,474	990,057	13,363,330
Disposable sacks used (000's)					
Paper	2,170	4,840	8,843	781	16,634
Plastic	22,950	60,649	176,920	33,236	293,755
Bulk storage containers	4,651,167	3,715,205	5,011,906	377,545	13,755,823
Number of dustbins provided	13,890	268,795	135,921	4,668	423,274
Waste reclaimed					
Paper (tonnes)	18,397	23,508	64,752	759	107,416
Vehicles (number)	4,898	2,596	6,546	922	14,962
Other (tonnes)	137	797	5,592	603	7,129
Collection vehicles					
With compaction	1,485	2,115	4,035	480	8,115
Without compaction	142	270	277	55	744
General purpose	113	178	410	63	764
Total	1,740	2,563	4,722	598	9,623
Staff employed at 31 March 1979					
Drivers	930	1,593	2.064	397	4,984
Loaders	3,829	6,643	10,701	1,280	22,453
Driver/loaders	842	970	2,293	129	4,234
Other manual staff	107	164	326	43	640
Vehicle maintenance staff	364	540	801	128	1,833
Technical and administrative staff	363	642	928	142	2,075
Total	6,435	10,552	17,113	2,119	36,219
Estimate of total waste collected (tonnes)	*2,180,000*	*3,280,000*	*8,830,000*	*1,100,000*	*15,390,000*

(b) Financial data (£'000)

Expenditure					
Employees	31,387	42,918	73,023	7,933	155,261
Provision of disposable sacks	1,150	2,165	5,767	1,051	10,133
Provision of dustbins	92	937	529	35	1,593
Transport and moveable plant	13,100	17,200	31,513	4,485	66,298
Establishment expenses	3,375	4,581	6,987	597	15,540
Other running expenses	3,261	2,706	3,193	238	9,398
Agency services					
Other local authorities	21	148	259	35	463
Contractors	419	—	427	27	873
Leasing charges	338	545	1,387	236	2,506
Debt charges					
Principal	322	379	715	59	1,475
Interest	212	328	419	53	1,012
Gross expenditure	53,719	72,000	124,936	14,929	265,584
Income					
Collection charges					
Commercial waste	2,687	3,846	5,358	329	12,220
Bulky household waste	193	197	418	13	821
Other	997	453	1,002	118	2,570
Sales of reclaimed waste					
Paper	418	621	1,770	12	2,821
Vehicles	12	6	14	1	33
Other	20	21	71	9	121
Contributions from other authorities					
Waste disposal authority	938	580	684	—	2,202
Other local authority	82	7	164	37	290
Total Income	5,347	5,731	9,481	519	21,078
NET EXPENDITURE	48,372	66,269	115,455	14,410	244,506
Estimate of Total Expenditure	*51,090*	*69,540*	*143,480*	*15,510*	*279,620*

Table 8   Staff employed

	Refuse	Street sweeping	Totals
Managers and supervisors	54	13	67
Manual workers	789	219	1008
Totals	843	232	1975

1980/81 some reductions in the standards of service may have to be imposed. The County Council is also responsible for the treatment of roads when snow and ice occur, but the City's cleansing department is able to provide, and does provide, men and vehicles to supplement the County's resources.

The nature of the service

Refuse yield	Domestic waste	174,000 t
	Commercial waste	36,000 t
		210,000 t

Base data	No. of premises	265,000	
	Refuse vehicle fleet	110	
	Public conveniences	100	
	Bulky waste collections	600	
	Motorways	25	km
	Trunk roads	280	km
	Principal roads	554	km
	Secondary roads	674	km
	Other roads	4,755	km
	Totals for all Roads	*6,288 km*	
	Footpath lengths	4,068	km
	Channels	4,520	km
	Gullies	128,000	

ASHFORD BOROUGH COUNCIL

In contrast to Leeds' size and metropolitan problems, the refuse service in Ashford Borough Council operating in very different circumstances has the following shape and form. Ashford is the biggest of Kent County Council's 14

Table 9   Frequencies of services

Street sweeping	
City centre footpaths	Weekly
Industrial and commercial footpaths	6-weekly
Rural area footpaths	2-monthly
Carriageway sweeping services	
Heavy traffic roads	Weekly
Industrial and residential roads	Monthly
Rural areas	2-monthly
Other roads	6-monthly
Gully emptying services	
All roads	6-monthly
Rural areas	12-monthly

Table 10   Expenditure, refuse collection and cleansing services, financial year 1980/81

Employees	5,685,830
Premises and fixed plant	162,850
Supplies and services	317,900
Transport and moveable plant	2,581,640
Establishment expenses	337,970
Miscellaneous expenditure	3,040
	9,089,230
Debt charges	54,070
Extraordinary expenditure	6,240
	9,149,540
Revenue re-allocations (−)	8,180
	9,141,360
Income	1,959,280
	7,182,080

District Councils. It is a predominantly rural area with Ashford as the main centre of population. Of the total 83,000 population, about 45,000 (54%) live in urban areas while the rural element is 38,000 (46%).

An efficient and economic Authority, its refuse collection vehicle fleet consists of 14 vehicles, three of which are deemed to be spare with an effective operational fleet of 11 vehicles. The vehicles are mainly 10 m³ nominal capacity rear-end loaders, and one side-loader is used for remote rural areas.

The service is based on weekly kerbside collection; rear door collections are only given to disabled people and old age pensioners. There is a limited use of plastic sacks for special collections. The refuse rounds are organized as follows:

Urban domestic rounds	3
Urban trade waste round	1
Rural rounds	7

Householders are responsible for providing their own dustbins but the Council replaces bins damaged by the collection crews. In a few council house areas bins are provided by the Council. In addition to this, commercial and hospital wastes are handled by 1.5 m^3 Paladin bulk containers.

Table 11   Revenue expenditure

	Domestic collections	Bulk refuse collections
Wages and supervision	276,000	14,000
Equipment	1,900	—
Refuse sacks	2,000	—
Protective clothing	3,200	—
Vehicle running expenses	127,000	4,300
Central administration charges	62,300	2,000
Gross costs	472,400	20,300
Income	46,600	20,300
Nett costs	425,800	Nil
Nett costs of all refuse collection	*£425,800*	

Table 12   Budget for other services

	Expenditure	Income
Cesspool emptying	96,400	76,000
Slaughterhouse	15,400	15,000
Pest control and drain cleaning	37,200	4,000
Street cleaning and litter prevention	103,150	21,800
Clearances	3,500	4,500
Gross expenditure and income	255,650	121,300
Nett cost of these services	*£134,350*	

The Council also operates a bulky household waste collection service, and the County Council as the Waste Disposal Authority provides a number of civic amenity sites for the population. Refuse crew sizes vary with three or four collectors per vehicle for the urban rounds and down to one man for the most rural rounds operated by a side-loading vehicle. In addition to this the Council also handles the usual range of other services such as street sweeping, litter prevention and gully emptying. The urban population density is of the order of 5,625 people to the km^2 while the rural density is 109 per km^2.

Disposal facilities

As with all other counties in England, refuse disposal is controlled by the borough and not the district council. Ashford Borough Council takes its refuse to a Kent County Council-nominated landfill site such as Chambers Green village about 19 km west of Ashford town. The journey time from the centre of Ashford to Chambers Green is about 20 minutes each way. The landfill site is privately owned and on lease to Kent County Council.

Budget estimates

The Council's 1981/82 budget estimates for Refuse Collection are shown in Table 11. A range of several other street sweeping, cleansing, and hygiene services are operated by the Council. The budgets for these are shown in Table 12.

THE SCALE OF PRIVATE SECTOR ACTIVITY

The private sector of the waste disposal industry plays a vital role, particularly in handling the more difficult disposal of industrial waste liquids and sludges where almost the whole activity of collection, treatment, and disposal is in the hands of the private companies, many of them members of the National Association of Waste Disposal Contractors. Indeed, even in the disposal of domestic and commercial wastes collected by the municipal authorities the private sector plays a considerable part. Reference to the annual statistics published by the Society of County Treasurers and the County Surveyors Society indicate that, acting on behalf of the Waste Disposal Authorities, private contractors dispose of about 14% of collected municipal wastes in England. An assessment made by NAWDC in 1979 estimated the size of the private sector in the terms shown in Table 13.

At the lower end of the scale our business goes on with innumerable small companies equipped with tippers and skip lift vehicles. Collecting small amounts of commercial and industrial wastes, these operators make use of local authority landfills (and pay for this service). They may also perform

Table 13   Private sector share ('000 tonnes)

Liquid and sludge tanker disposal	3,800
Dry bulk waste	3,500
Skips—dry inert waste	7,500
Domestic waste WDA contracts	1,800
Totals	16,600

cesspit, gully emptying and other similar services. At the other end of the scale are the bigger operators, very often subsidiaries of major British public companies and having access to substantial financial and corporate back-up from their parent group; they are truly professional waste disposal experts.

Through the aegis of NAWDC the industry closely collaborates with Government ministries, professional bodies like the Institution of Municipal Engineers and Institute of Solid Wastes Management, and Municipal Chief Officer groupings such as the County Surveyors Society. NAWDC plays its part in all of these and this includes advising on legislation, safety, vehicle regulations, training, and standards. It is as well to note that NAWDC brought in voluntary codes of conduct and practice before they were enshrined in the legislation of 1972 and 1974.

The biggest and best known of the private sector companies own substantial assets in existing and potential landfill sites including such major facilities as Pitsea, owned and operated by Cleanaway Limited. In terms of tonnage of waste handled, this company alone deals with 1,300,000 tonnes of waste. This tonnage is greater than that handled by several county WDAs in England.

The more sophisticated capital plants such as incinerators for industrial liquids and sludges are nearly all owned and operated by private sector companies and this also applies to deep-sea disposal or more specialized wastes.

Table 14   Who handles the waste controlled by the WDAs (%)

	WDA direct	Agent authorities	Contractors	Other WDAs
Non-metropolitan counties	78	12	8	2
Metropolitan counties	99.4	—	0.6	—
Greater London Council	21	—	78	1
Totals for England	34	7.5	14	1.5

Table 15　Origins of waste handled by the WDAs

	Collection authorities (%)	Other sources (%)	Total tonnages ('000s)
Non-metropolitan counties	66	34	15,285
Metropolitan counties	56	44	7,002
Greater London Council	85	15	3,000
Totals for England	66	34	25,287

Data from Chartered Institute of Public Finance & Accountancy Waste Disposal Statistics, 1979/80.

Figure 3　The bulk transfer of domestic and commercial wastes. A Dumpmaster transfer trailer leaving the Copenhagen refuse transfer station

Municipal refuse collection by private contract is another venture area and as yet, excepting the sixteen (January, '83) awarded contracts for refuse collection and street cleaning services, the whole execution of the service is operated by the collection authorities.

A number of commercial groups are now interested in participating in the municipal waste market. These activities are well established on the disposal service. This activity will grow as the local landfills become exhausted and the larger cities must rely on remote landfills (owned by the mineral contractors). These sites will increasingly require the intervention of bulk transfer station systems by road, rail, and river. A number of long-term stable disposal contracts will be able to be negotiated and these will involve the contractor in three key activities:

(1) owning or having access to the mineral extraction voids;
(2) owning and operating the transport system;
(3) building and operating the transfer station facilities.

As far as collection services are concerned the position is quite different. Private sector interest is intense and all eyes are focused on the first contract in Southend-on-Sea. Here marketing opportunities will be swayed by political factors and without doubt the running will be made by the Tory-controlled shire towns. The larger metropolitan towns, particularly in the North of England, are the bastions of the Labour Party. Political persuasion and professional entrenchment will mean that the potentially more lucrative contracts—Leeds £7 million a year versus Ashford's £500,000—will be much harder to secure. In a way this is no bad thing as both sides to these new ventures will want to learn the ropes. An Ashford that fields a fleet of 11 vehicles, rather than a Leeds that fields 100 vehicles, is a much more easily handled proposition.

THE RANGE OF INDUSTRIAL WASTE TREATMENT FACILITIES IN THE UNITED KINGDOM

The scope and professionalism of the private sector of the waste disposal industry in the United Kingdom can be gauged from Table 16. In these main groups of treatment facilities the services offered by the industry cover chemical treatment, incineration, and various solidification systems. In this field the private sector deals with almost 100% of waste arising in this country. The information is taken from evidence given in 1980 to the House of Lords by the National Association of Waste Disposal Contractors.

Table 16   Private sector scope

(a) Chemical treatment (mainly for inorganic wastes)

Operator	Location	Type of wastes processed	Capacity ('000 tonnes/year)
DP Effluents	Runcorn, Cheshire	Acid treatment and liquid wastes	30
Polymeric Treatments Ltd	Killamarsh, North Derbyshire	Dewatering of sludges	5
Hargreaves Clearwaste Services Ltd	Wakefield, West Yorkshire	Most inorganic wastes	10
Re-Chem International Ltd	Fawley, Hampshire	Most inorganic wastes	20
Re-Chem International Ltd	Pontypool, Gwent	Most inorganic wastes	20
Re-Chem International Ltd	Roughmute, Stirlingshire	Most inorganic wastes	20
Safeway Sludge Disposal	Garretts Green, Birmingham	Acid treatment and sludge dewatering	30
Totals			135

(b) Incineration (mainly for treatment of organic wastes)

Operator	Location	Type of wastes processed	Capacity ('000 tonnes/year)
Berridge Incinerators Ltd	Hucknall, Notts	Liquids including halogenated	9
Croda Synthetic Chemicals	Four Ashes, Wolverhampton	Non-halogenated liquids	5
Hargreaves Clearwaste Services Ltd	Wakefield, West Yorkshire	Non-halogenated liquids	15
Re-Chem International Ltd	Fawley, Hampshire	Solid and liquid wastes—virtually all types	15
Re-Chem International Ltd	Pontypool	Solid and liquid wastes—virtually all types	20
Re-Chem International Ltd	Roughmute, Stirlingshire	Solid and liquid wastes—virtually all types	20
Cleanaway Ltd (formerly Redland Purle Ltd)	Ellesmere Port, Cheshire	Bulk liquids, including halogenated	20
Cleanaway Ltd (formerly Redland Purle Ltd)	Rainham, Essex	Non-halogenated	13
Totals			117

Source: NAWDC evidence to House of Lords Select Committee 1980

Table 16   *continued*

(c) 'Solidification' plants (mainly for treatment of inorganic wastes)

Operator	Location	Capacity ('000 tonnes/year)
Polymeric treatments Ltd	Aldridge, West Midlands	250
Stablex	Thurrock, Essex	300
Totals		550

BEHAVIOUR OF WASTE IN LANDFILLS

There is a wide variation (at least initially) in the land consumed by the various systems. Incineration clearly has the most marked effect on the nature of refuse and is most economical in its use of land. Other systems, like pulverization and high-density baling, are next in scale and the deposit of untreated refuse undoubtedly requires the most space. Research now going forward in the United Kingdom indicates that domestic refuse compacts and settles over several years and, after a period of 6 years or so, the densities of refuse, irrespective of the initial treatment used in its disposal, tend to converge to a common level. Incineration is still in front, but the differences in space demand are much less than the theoretical differences in density between incinerated residue and compacted and covered refuse might suggest. Operational factors, the availability of mechanical plant, the economic catchment areas of the treatment plants, and their residue landfill sites also affect the amount of space consumed, and in practice the differences in space consumption by various systems are rather less than the theoretical ratios might suggest. While there is an undoubted shortage of landfill resources close to the urban areas, mineral extraction in the United Kingdom is always creating more voids than there is waste to fill them. These holes are not always in the right places, the geology is often unsuitable, and other environmental factors preclude landfill, but by and large adequate landfill reserves exist and are available for the disposal of domestic and industrial waste. This being so, it seems likely that, in the main, landfill will continue to be the predominant disposal solution in the United Kingdom, but increasingly this will occur at larger, more remote mineral extraction sites served by bulk transfer stations, via road and rail links. This increasing use of transfer stations presents special opportunities for reclamation both by separate collection and by various separation techniques.

THE GROUND RULES FOR THE PROCESSING OF DOMESTIC AND COMMERCIAL WASTES

The United States Environmental Protection Agency, in its comprehensive reports to the United States Congress, has attributed the limited response to the use of recycled materials to the following primary causes:

(1) Natural resources occur in concentrated form whereas recycled secondary materials from waste are dispersed and have attendant high collection cost.

(2) Virgin materials, even unprocessed, tend to be more homogeneous in composition than wastes, are of higher quality, and are less contaminated. Product quality and specifications are thus easier to control.

(3) The principal process technologies are designed to use virgin materials, while waste processing requires different attitudes and technology.

(4) The use of synthetic material in combination with natural resources makes economic sorting difficult.

A later report went on to say:

(5) The use of recycled materials appears to result in a reduction of energy consumption and pollution compared with the use of virgin materials.

Figure 4   Waste paper baling in the United Kingdom. (Courtesy of Scapa Ltd)

(6) The recovery of materials from wastes is very dependent on economic factors. Manufacturing costs from secondary materials are as high or often higher than those for the use of virgin materials. Consequently only high-quality material can find a ready market and, often, artificial economic factors favour the use of virgin materials.

(7) While technology exists to separate useful materials from municipal refuse, recovery costs are high and recovery is feasible only in areas where circumstances force a high cost disposal pattern combined with suitable local markets.

THE POTENTIAL IN REFUSE

A typical analysis of municipal waste in the United Kingdom shows a composition by weight of:

	Percentage
Dust and cinder	22.9
Large cinders	4.5
Paper	32.5
Vegetable matter	19.3
Metals	7.1
Glass	7.9
Rags	2.2
Plastics	1.0
Unclassified debris	2.6
	100.0

This waste is generated at a rate of about 0.33 tonne/person per year or 12.6 kg/household per week. Its density, currently at about 150 kg/m^3 (when tipped), is falling while volume is increasing. The combination of these two factors gives a gradual increase in weight generated per person of about 1% per year. Even a superficial study of the composition of the waste shows that it contains a number of things which can usefully be recycled or, through thermal processes, provide heat energy for industrial, commercial, and domestic uses. Again, pioneering processes have been developed by which fuel oil, gas, and combustible char can be produced by the pyrolysis of refuse and other wastes. Even more advanced fundamental research has been carried out to produce protein and ethyl alcohol by the hydrolysis of domestic refuse. Many of these research projects are the realities of the future but they point the way to an exciting new concept in conserving the raw material we use.

RECOVERY TECHNIQUES

Recovery of useful materials in refuse can occur at the collection or at the disposal point.

Collection-point recovery, usually of paper but occasionally of other materials, is—with the exception of occasional operations by charitable groups—the responsibility of the District Councils or the Greater London Council Boroughs. Materials are separated at source and collected separately by refuse vehicle trailers or separate vehicles. Transport and collection costs are significant factors in the economics of these schemes and experience has shown that the long-term public response is no higher than about 1:3.

Disposal-point recovery can occur both on direct sanitary landfill sites and at mechanical treatment plants such as incinerators, pulverizers, transfer stations, and composting plants. In modern direct incineration plants materials recovery is usually confined to ferrous metals extraction from the incinerated residues but in composting plants and other treatment techniques this extraction takes place prior to treatment. Many of the early generation of cell type incinerators had extensive manual and mechanical 'front-end' separation of metals, glass, and textiles.

Figure 5   Typical salvaged waste paper bales. (Courtesy of Thomas Board Mills Ltd)

Practical waste management

Figure 6   Refuse being handled by mechanical shovel in a modern transfer station.
(Courtesy of South Yorkshire County Council)

Table 17   Examples of energetic local authority paper salvage schemes*

Area	Population	Refuse (tonnes)	Paper content at 32% (tonnes)	Actual recovery (tonnes)	Percentage of potential	Domestic premises (%)	Trade premises (%)
1	13,420	4,428	1,417	468	33	66	34
2	35,890	11,843	3,790	1,560	41	70	30
3	40,860	13,483	4,315	1,716	39	72	28
4	41,680	13,754	4,401	1,300	30	70	30
5	64,280	21,212	6,788	1,768	26	71	29
6	89,090	29,399	9,408	4,056	43	71	29
7	163,380	53,915	17,253	7,020	41	71	29
8	215,280	71,042	22,733	7,488	33	72	28

*First quarter 1973

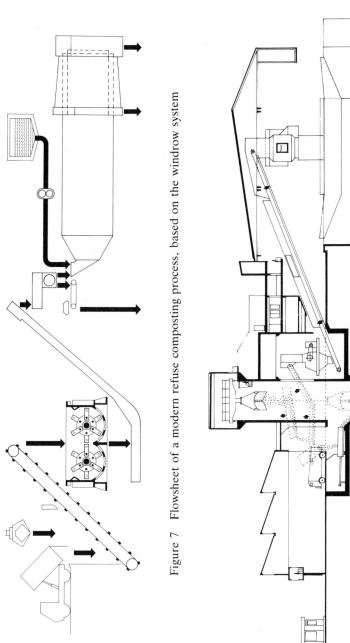

Figure 7   Flowsheet of a modern refuse composting process, based on the windrow system

Figure 8   Cross-section through a modern refuse pulverization river barge transfer station

Table 18   Analysis of public response to glass reclamation in York (1974)*

Period	Sacks issued	Sacks returned	Return of sacks (%)	No. of bottles recovered†
Week 1	864	113	13	3,211
Week 2	1,178	498	24	7,560
Week 3	864	—	—	
Week 4	1,178	302	26	2,992
Week 5	864	401	47	3,760
Week 6	1,178	284	24	3,275
Week 7	864	275	32	3,267
Week 8	1,178	286	24	2,509

*Number of households in test area: 1,021.
†Theoretical glass content: 7.5% by weight.

Table 19   Some characteristics of waste-derived fuel plant

Typical fuel generation rate	0.35 tonne/tonne
Typical metal extraction rate	0.05 tonne/tonne
Pellet dimensions	40 × 15 mm
Typical efficiency of system	60%
Tonnage of waste eliminated by process	40%
Volume of waste removed	70%

Parameters

Gross calorific value	7,250 MJ/kg
Net calorific value	6,500 MJ/kg
Moisture content	10–12%
Density of fuel	600 kg/m^3
Ash fusion temperature	1,200 °C

METHODS OF RECOVERY

Collection point separation and recovery

(a) Local authorities' refuse collection service.
(b) Private charities and entrepreneurs.

Disposal point techniques

(a) Energy recovery.
(b) Materials recovery.
(c) Pyrolysis process.
(d) Compost processes.
(e) Chemical processes.

FUTURE TRENDS

The limited long-term public response rate to separation and salvage at the collection point has caused a number of research and development authorities to reconsider separation and sorting of municipal waste at the point of disposal. This discarded idea, given new impetus by advances in materials handling technology, is again coming into favour. Pioneer research work at the Warren Spring Laboratory at Stevenage and in the USA has produced pilot plants designed to use the physical properties of the constituents in waste—density, size, mass, shape, and colour—to achieve separation by flotation, centrifugal, and ballistic methods. These developments, operating together with collection-point recovery of appropriate materials and the ultimate disposal of only the useless elements of waste, may be the way to effect practical and economic recovery by municipal authorities. In considering these new techniques it should be remembered that, in most cases, a waste reclamation plant will not pay its own way in conventional income and expenditure terms. Such plants as now operate are designed to perform a waste disposal function and operate as reclamation units to economize on their operational costs. In every case a contribution towards the treatment cost is called for by the waste-producing or disposing authority. In this country these ideas have reached fruition in the two DoE-aided projects in Tyne & Wear and South Yorkshire County Councils. The former aims primarily at waste-derived fuels and ferrous metals, while the South Yorkshire plant develops more fully the separation techniques of the Warren Spring Laboratory.

HEAT AND ENERGY

Domestic and commercial waste has a significant calorific value and this is increasing with the changes in paper and plastic content. Many modern incineration plants, particularly in continental Europe, are steam or high-pressure hot water recovery installations. Power generation, district, and process heating are common applications although there is no reason why, in steam-raising plants, turbine prime movers cannot be used in association with refrigeration bulk cold stores, air compressors, or driving the incineration plant's own fans and pumps as in the City of Coventry waste reduction unit.

As a general guide 1 tonne of domestic waste has the gross heat value of about 0.50 tonne of coal, or 0.25 tonne of fuel oil. In modern incineration plants 1 tonne of this waste can produce 2.5 tonnes of steam or 40–50 units of gas (therms). In terms of mechanical power output it can be shown that a 10 tonne/hour steam-raising refuse furnace can raise steam to support turbine prime movers of the order of 1,000 kW rating. The internal fans and pumps of the furnace will require about 450 kW of power, leaving a surplus of 550 kW to be applied to external loads. This available surplus of mechanical power can be used in a variety of applications requiring rotating drives, including air compressors or refrigeration plant applications.

Throughout Europe there are numerous fine examples of domestic waste incineration plants, most with heat recovery to steam or high-pressure hot water, utilized in the generation of electricity or district heating. In all these installations the various designers and manufacturers have contributed to plants of generous design and elegant concept in their engineering and architecture.

WASTE-DERIVED FUELS: THE STATE OF ART

As energy costs continue to rise the potential savings from a wider use of waste-derived fuels (WDFs) becomes increasingly attractive. Before these fuels can be more generally adopted, particular attention needs to be given not only to the techniques of producing WDFs but to the scale of capital investment required and the marketing of the product. The Waste Disposal Authorities must consider the role they must play and, as the legal owners of this potential fuel, whether they will participate in direct investment and operation, or in collaboration with private enterprise. This chapter seeks to examine the challenges facing the introduction of WDFs and other reclamation processes.

It is fairly safe to assume that over the next 2–3 years several viable WDFs will have been successfully researched and marketed. The evidence from the United States, France, Switzerland, and our own DoE-sponsored projects in Tyne & Wear and South Yorkshire County Councils, points to success and although there is still some way to go to solve completely some worries on the corrosion and combustion characteristics of the fuel, all in all the prognosis is good. In my view, these fuels will fall into two broad groups of 'fine' and 'coarse' WDFs.

The coarse fuels will cover a range of the unsorted and unclassified generality of domestic and commercial wastes, usually in a pulverized form, to a screened and air-classified material stopping short of pelleting and briquetting. The coarse WDFs will be aimed at major coal-burning kilns and chain grate stokers and we already have evidence of success in this field with the work of Imperial Metal Industries, Blue Circle Industries Limited, the

British Steel Corporation and other organizations including the brave attempt of Babcock Product Engineering Limited in their ideas to convert the CEGB power station at Llynfi to burning WDF for power generation. Another major advance by the private sector is the Eastbourne WDF plant. Based on a process developed by Buhler Brothers of Switzerland this pelleting process was funded by private venture capital. The plant is now operated by East Sussex County Council who have formed a special operating company for this purpose.

The fine WDFs will tend to be aimed at smaller but more diverse energy users such as medium-sized industrial plants, whose space and process heating loads are provided by packaged boilers such as those manufactured by Parkinson Cowan GWB Limited. This company's Power-Master range of boiler is designed to burn a whole range of fuels from coal, oil, gas, propane,

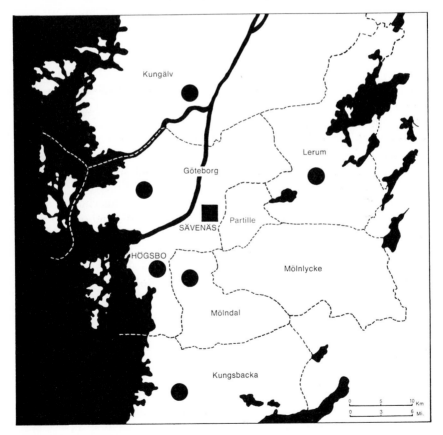

Figure 9　Map showing the location of satellite transfer stations and central incineration plant in Göteborg, Sweden. (Courtesy of the GRAAB Organization, Sweden)

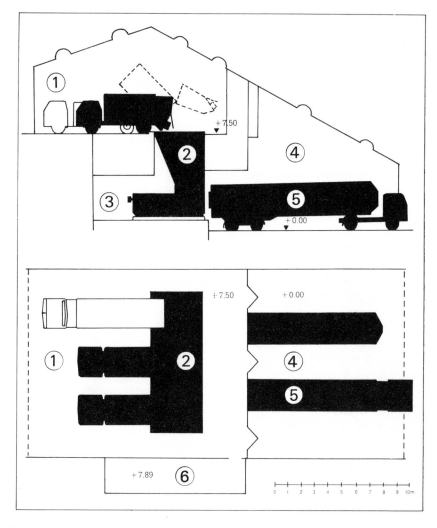

Figure 10   Cross-section through a 200 tonnes/day transfer station in Göteborg,
          Sweden. (Courtesy of the GRAAB Organization, Sweden)

including pelleted WDFs. This flexibility should achieve better utilization
factors than those expected from heat-recovery incineration plants tied to
single seasonal space heating loads. Parkinson Cowan and others have
demonstrated that in package boiler installations WDF can be considered as
fuel in its own right and not just in a supporting role to coal, as is the case in
the coarse fuel kilns and stokers.

I see both forms of WDF coexisting rather than competing as they aim at very different energy users. There is a substantial leap in capital and operating costs, as well as energy consumption, to make the transition from coarse to fine fuel, i.e. from pulverized waste to pelleted waste, but this will be compensated for by the ease of packaging, storage, handling, and diversity of uses to which the fuel can then be directed. The coarse fuels, with their economy and simplicity of process, have much in their favour, but their customers will be fewer and further between, and in general they will require concentrations of waste that can arise only from close proximity to densely populated urban areas. There is a case and a place for both variations of fuel and one should reject arguments that suggest an exclusive role for either of them.

THE FINANCIAL REALITIES

The substitution of waste reclamation processes in place of the established waste disposal techniques will be a formidable financial challenge. The inherent economy of sanitary landfill, even when this is coupled to the necessity to bulk transfer wastes to remote landfill sites, is such as to suggest that an authority must be faced with a disposal problem of moderate severity before such a substitution of process becomes feasible. The WDA must at least be faced with an inescapable decision to invest in some capital-intensive solution to its problems. The most common of such decisions will be the construction of a bulk transfer station conveying wastes to remote landfill sites by road, rail or river links. Both DoE-sponsored waste reclamation plants now under construction in Tyne & Wear and South Yorkshire County Councils will be seeking to establish not only the technical feasibility of the various separation processes they will employ, but to ensure that the ultimate cost of separation and reclamation of fibre, metals, and other products, and the discharge to landfill of only the untreatable residues, is more economic than the simple transfer of those wastes to landfill. If they can achieve this they will have been a resounding success. The same applies to the other private sector/public sector collaborating schemes such as those of the Imperial Metal Industries Limited and West Midlands County Council and the collaboration between the company Asthall Holdings Limited and the East Sussex County Council.

The established and proved treatment costs per tonne of these processes and the charges (if any) levied on the Waste Disposal Authorities will establish the band of disposal solutions against which they can compete. This, in turn, will define the rate at which reclamation processes can be substituted for other established disposal systems.

LOCAL AUTHORITY CAPITAL RESOURCES

Capital projects in waste disposal must be financed from limited 'locally determined' resources. These resources are scarce and, leaving aside technical arguments in favour of one solution or another, there is increasing pressure on WDAs to implement policies that tread lightly on 'locally determined capital'. This fact, plus the institution of corporate management multi-discipline assessment of priorities, is forcing a move away from the high-capital projects like incineration to lower-cost alternatives like transfer stations to landfill or similar co-operative ventures with the private sector. At all times the planning of these projects takes account of the need to economize on collection services by the construction and location of waste disposal plant at the 'centre of gravity' of the collection area.

The entry of the mineral extractors

The mineral extraction companies are now becoming more actively involved in waste disposal. They are changing their role from that of passive providers

Figure 11   A view of the GLC rail transfer station at Brentford. (Courtesy of the GLC)

Figure 12   Rail transfer of wastes to remote landfill sites. The GLC Brentford transfer
station. (Courtesy of the GLC)

of voids to be exploited by others to a more active stance of direct commercial exploitation using their own resources. These resources are considerable, the big mineral groups possess, 'in house', significant professional talent and well-established expertise in quarry management, the handling of plant and machinery, and proven marketing skills. More important they possess and have access to considerable financial resources necessary to develop the new landfills and treatment plants likely to emerge from the higher standards demanded by current legislation. Several of the biggest private waste disposal companies in this country are part of mineral extraction groups and in several well-known cases provide important waste disposal services to the WDAs. Redland Purle, London Brick, and Amey Roadstone are examples of these. The APCM as another case in its 1976 seminar spelt out very clearly its new attitude to the use of its mineral extraction voids. Direct exploitation and management were the key words, and annual reports from other mineral companies often make specific reference to waste disposal exploitation of worked-out sites. Restitution of mineral sites is a continuing theme of minerals subject plans in county structure studies, and a better appreciation of the part that can be played by the mineral companies as waste disposers must benefit the counties concerned.

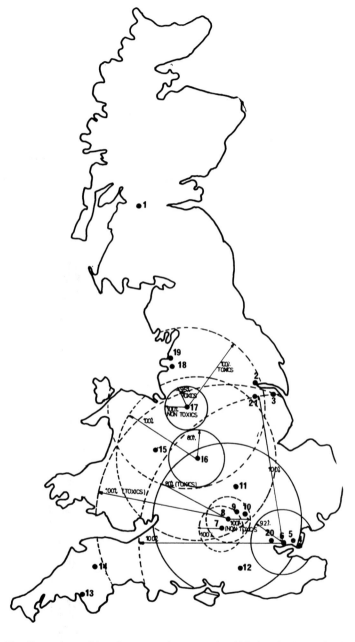

Figure 13   Operating orbits of some major waste landfill sites in the United Kingdom.
Solid lines: orbits for percentage of waste shown; dotted lines: orbits for 100% of
waste

Figure 14   Operating orbits of some major waste treatment plants in the United Kingdom. Solid lines: orbit for percentage of wastes shown; dotted lines: orbit for 100% of waste

Regional disposal decisions

The Local Government Act 1972 has been debated *ad nauseam*, particularly in respect of its split in collection and disposal duties, and the passage of the latter to the English County Councils. The English County WDAs have at least produced a greater professionalism in waste handling and allowed, at county level, better opportunities to plan disposal, use resources more sensibly, and regulate the deposit of industrial wastes. The private sector (at least the responsible parts of it) has welcomed these changes and NAWDC has played a full part in the deliberations and debates that have followed. So far so good, but the question is now posed (as this is a forum considering waste disposal and planning) whether the County WDAs are themselves big enough to decide reasonably the issues involved in treating and licensing landfills and treatment plants that serve several counties, sometimes regions, and occasionally the whole country. As illustration, Figures 13 and 14, prepared in 1977, show the operating orbits of some of the principal landfills and treatment plants in the UK. Almost exclusively in the hands of the private sector, these are truly regional facilities and one is forced to conjecture whether full note is taken of this in planning decisions. There are of course inter-county contacts, waste disposal officers meet at regular intervals, but is this enough?

Some parallels must be drawn to European practice where, particularly in West Germany, it embraces regional waste disposal planning and collaboration between the public and private sectors. Joint-venture companies are formed capitalized from both public and private funds. These operate as commercial entities paying their way, balancing income and expenditure, and raising their own funds. A good example of this is the Bavarian industrial waste disposal organization Gesellschaft zur Beseitigung von Sondermull Bayern (GSB). In like fashion the Ruhr is served by a regional environmental authority Siedelungsverband Ruhrkohlenbezirk (SVR).

In Scandinavia cities like Copenhagen and Malmo in Sweden operate refuse collection and disposal by public sector/private sector joint venture companies. These services extend to the operation of heat-raising refuse incineration plants and district heating schemes.

Our legislation allows and encourages public and private sector co-operation, and the future will hold good opportunities for jointly capitalized and operated sites and plants. The treatment of hazardous and difficult wastes is a particularly fruitful area for such co-operation, and one can visualize joint ventures between the WDA and the private sector receiving domestic, commercial, and industrial wastes. We may be a long way from the Regional Waste Disposal organizations now being developed in West Germany and Scandinavia—indeed we may not wish to travel that road—but some of their policies and facilities are worthy of study.

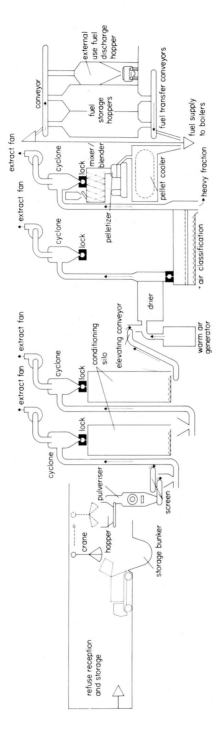

Figure 15 Flow line of the 'Societing' waste-derived fuel process developed in France. (Courtesy of Societing, France)

Figure 16   An external view of the Byker Waste Reclamation Plant, Newcastle upon Tyne. This photograph illustrates that modern treatment plants can blend into an industrial landscape. (Courtesy of Tyne & Wear County Council)

CONCLUSIONS

If operational economics are the main criterion, it is inescapable that the sanitary landfill of domestic, commercial, and industrial wastes will continue to be the predominant method of waste disposal in the UK. Even in those European countries like West Germany and Switzerland, where there are massive investments in incineration and composting plants, the operational position is that over 60% of domestic and commercial waste is still landfilled. Evidence obtained by the DoE shows that in a fairly active year surface mineral extraction in the UK creates about 250×10^6 m^3 of space. All the refuse produced in the UK could, with compaction, be accommodated in less than one-quarter of the depressions so created and still leave ample space for inert industrial wastes and construction industry debris. Leaving aside the accumulated reserves of despoiled land from earlier years of industrial activity, if only a fraction of this space is available for waste disposal, the reserves are such as to make one of the classic arguments for capital-intensive systems—shortage of space—difficult to support.

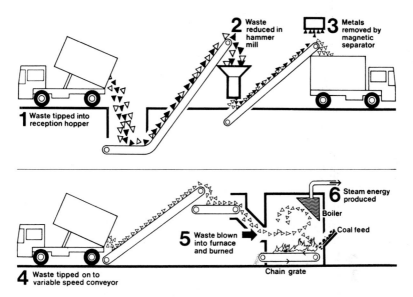

Figure 17 A simplified sketch of the Imperial Metal Industries pulverized refuse fuel plant that operated in Birmingham in the late 1970s. (Courtesy of Imperial Metal Industries)

The landfilling of domestic refuse, now dignified by the title 'sanitary landfill' can be a very different operation from the usual paper-littered, rat-infested eyesores that normally come to mind. It is becoming a science of its own. Knowledge of the geology and hydrogeology of landfill sites, the civil engineering and drainage works that are necessary, and the vastly improved mechanical engineering mobile plant are changing the face of this simple system. Plants are now under construction designed to press domestic refuse into high-density bales and thus allow a neat 'building block' pattern of landfill disposal. These systems, together with the pulverization of refuse prior to landfill, can perform useful land reclamation of derelict and despoiled land. Landfill solutions with their inherent economy have a powerful effect on the thinking of the WDAs, and materials and metal reclamation must fit into the realities of this situation.

There are many exciting innovations to be considered in the disposal of domestic and commercial wastes and some will be described in other chapters. Most if not all are still at embryonic stages of development and do not yet present viable and reliable day-to-day alternatives to sanitary landfill. As sensitive but realistic waste managers let us be aware of them and in due course give them a fair chance, but our responsible duty is to do that which is possible, safely and well, and this book tries to illustrate the realistic options.

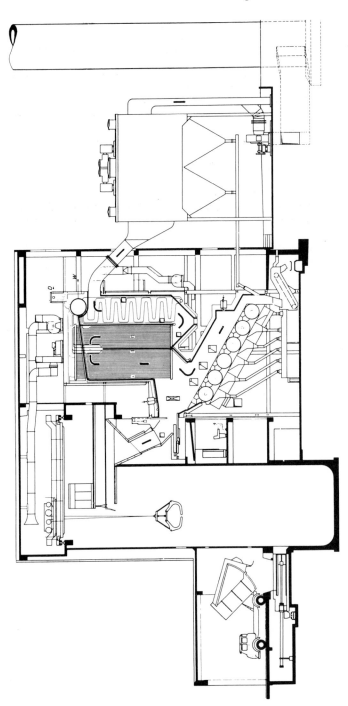

Figure 18   A cross-section through a VKW Dusseldorf system steam-raising incineration plant. (Courtesy of VKW, West Germany)

BIBLIOGRAPHY

The following papers and documents have been referred to in the preparation of this text.
1. Tunaley, C.I., 'Solid Waste Disposal—Problems Associated with Tipping and the Licensing of Sites'. *Chartered Municipal Engineer*, January 1979.
2. Society of County Treasurers Waste Disposal Statistics, 1979/80 estimates.
3. *The Estimation of Unweighed Domestic and Commercial Wastes*. Association of Waste Disposal Engineers, 1976.
4. Cope, C.B. *The Future Direction of Industrial Waste Disposal*. Public Works Congress, 1976.
5. Aspinwall, R. *Assessment of Landfill Sites—Point of view of a Private Consultant*. Public Works Congress, 1976.
6. Feates, F.S. *The Waste Disposal Problem*. Public Works Congress, 1976.
7. *Department of the Environment Working Party on Refuse Disposal*. HMSO.
8. United States Environmental Protection Agency, *Resources Recovery Catalogue of Processes*, 1973.
9. *Refuse Collection Statistics 1978/79 (England and Wales)*. Chartered Institute of Public Finance and Accountancy.

DISCLAIMER

The author wishes to make clear that the opinions expressed in this chapter are those of the author alone, and do not necessarily represent the views of Grand Metropolitan Limited or its subsidiary companies.

Practical Waste Management
Edited by J.R. Holmes
©1983 John Wiley & Sons Ltd

2

Planning for industry—future development control of waste management— A personal overview

R.G.P. HAWKINS, MA, MWM
Barrister, Fellow of the Royal Society of Arts

ABSTRACT

Public health, development control, and environmental law after the Industrial Revolution is summarized; the effect of the Town and Country Planning Act 1947 is outlined, together with its effect on waste management and engineering practices. The causes and effects of the Deposit of Poisonous Waste Act 1972 are related together with its complementary relationship with the Control of Pollution Act 1974 Part I; the relevant sections of the latter Act are explained.

The disciplines in the 1980s for land management and engineering in relation to refused materials are explained, together with some UK and EEC statistics. The interrelationship with planning and pollution control in the 1980s is recounted, together with the effect of the Local Government Act 1972 and the Town and Country Planning (Prescription of County Matters) Regulations 1980.

A national strategy is suggested and the use of environmental impact assessment is summarized. The conclusions of the House of Lords Select Committee on Science and Technology Report on Hazardous Waste Disposal are discussed, together with the Committee's lacunae, in particular that of detriment and benefit analysis and risk percipience.

The appendices afford checklists for the practical operator in the waste management and engineering fields and the complementary role of the Health and Safety at Work etc. Act 1974 is emphasized by actual safety policies for a leading waste disposal company. The relevant pollution control statutes in tabular and interrelated form are included.

HISTORICAL BACKCLOTH

I have never felt that the contemplation of the past, with the knowledge that it cannot come again, need be a source of sorrow.

DUFF COOPER, in *Old Men Forget*

Legislative concern over problems of refuse disposal and public health predates the Industrial Revolution. There were provisions in mediaeval statutes which called for the removal of refuse on pain of forfeits: see an Act of 1388, C.13.12 Richard 2; *plus ça change...*; *vide* the Control of Pollution Act 1974 (CoP) S.9(3) and S.97, sections implemented since 1974 which give default powers to the Secretary of State.

Before the nineteenth century, however, polluting deposits of waste or refuse more frequently led to private litigation than to public prosecution. The common law tort of nuisance gave a landowner the right to bring an action for damages although his chances of success had to be measured by a standard of nuisance or annoyance that was inevitably crude and subjective as well as being exercised against a declining feudal background.

The general community, as distinct from the private landowner, did not receive protection in modern times until the enactment of a statutory code of public law which began with the Public Health Act of 1848. This code is the first source of general public control over the activities of people depositing waste in modern times; a similar Act of 1875 gave enlarged powers to public authorities. The Public Health Act of 1936 (the statutory successor to the 1875 Act) enacts that any accumulation or deposit prejudicial to health or a nuisance is a statutory nuisance: see section 92(1)(c). Public Health Authorities have the power to serve abatement notices and to prosecute; this power will remain extant after the full implementation of the Control of Pollution Act.

The enactment of the public health code introduced a system of regulation but not of prevention. Local authorities were given no power to stop deposits being formed; they could only act after a nuisance had been created. Moreover the powers given under both public health and housing codes related only to individual sites which were in some way insanitary or sub-standard. There was no scope to take a comprehensive view of an area as a whole and regulate the pattern and formation of land uses therein. By the early twentieth century it was plain that local authorities needed more wide-ranging powers of planning. Town and Country Planning law began with the Act of 1909, giving very limited powers to authorities to make schemes so as to control development on land as yet undeveloped. For example, in 1929 a report commissioned by the Ministry of Health made wide-ranging and valuable recommendations for the more efficient operation and administration of all of London's refused materials; there is, however, no reference to overall national or regional planning in its final recommendations. Despite, however, the enactment of many town planning statutes, there was no complete control over land use until the Town and Country Planning Act of 1947 came into operation.

In general the statutory powers to control or prevent land uses judged to be environmentally harmful, were so restricted prior to 1947 that some local authorities sought other means to prevent a proliferation of refuse dumps (*sic*) within their boundaries. Such a practice had resulted in environmental disasters such as the South Hornchurch Dumps which after 35 years were some three-quarters of a mile in length, one-third of a mile in breadth, with further layers of refuse added even after the summit had reached 90 feet. Several county councils were given powers by local private Acts of Parliament to prohibit the deposit within their areas of waste and refuse emanating from outside their boundaries. Such powers can be found for example in the Surrey County Council Act 1931 section 94, the Essex County Council Act 1933 section 146, and the Herefordshire County Council Act 1935 section 26. These Acts all had the same object and gave much the same control. If anyone wished to deposit waste emanating from outside, but onto land in, a county, he had to seek permission from the authority. The permission, if given, could be subject to conditions regulating the operation of any disposal site. In most cases applicants for permission had a right of appeal either to the Minister or to an arbitration tribunal in case of refusal to grant, or decision to withdraw, permission. These Acts did not, however, stop a waste contractor from disposing of waste arising within the same county or district council area.

The criterion was not, however, that of geographical origin, but site suitability; refused materials are often environmentally helpful to improve, for example, river marshes and sea defences and such engineering has always been encouraged by central government; there is no geographical control over the movement of refined hydrocarbons, *ergo* there is no rationale on environmental, societal, or economic grounds to restrict the movement of discarded materials providing extant laws are obeyed. The USA Supreme Court has also ruled that it is unlawful for one state to restrict the trans-frontier movement of such materials.[1]

THE UK ENVIRONMENT AND WASTE DISPOSAL

Initial disciplines

Ten years ago and *a fortiori* previously, the disposal of materials which seemingly had no further use was a localized service throughout the UK, the responsibility of small organizations with few resources. The deposit of industrial or other wasted substances was either undertaken by the manufacturer himself, often simply dumping the wastes within his factory's curtilage, or by a small local quasi-specialist haulier who would be contracted to collect the materials for which no further use had been found and which were probably no more precisely defined than in a tea-stained crumpled docket 'chemical wastes', and left to entrepreneurial skills to arrive at a suitable grave. The contractor therefore provided what was essentially a transportation service and the majority had little technical understanding of the nature of the wastes or the environmental danger that could result from their disposal in such an uncontrolled fashion.

Many thousands of disposal sites were in use and the wastes received therein in the same disparate form as the consideration invested to obtain their entrance. The total gulf which existed between the owner of the industrial waste and the owner of domestic refuse, the collection and disposal of which was the responsibility of the former 1,400 local authorities, meant that there was a marked formal reluctance (but not necessarily reflected in the pragmatic attitudes displayed by some urban or rural district council tip foremen), to allow private sector industrial waste on public sector domestic sites. Sites were selected for their proximity and their convenience more than their suitability,[2] their interrelationship with the considerable national problem of the restoration of derelict land was wholly ignored and many of the sites were extremely small and quite incapable of being administered efficiently or in an environmentally acceptable manner in relation to their contamination potential.[3]

Early in 1972 an incident involving the concealment of cyanide wastes in the industrial Midlands of England rapidly became a national issue. Although no damage to persons, stock, etc. nor bruising to the environment had occurred—and the statistics show that there has been, albeit fortuitously, very little evidence of harm resulting from these historically crude practices—there were demands for tighter legislative control. The preamble to the Deposit of Poisonous Waste Act 1972 (DPWA) set out its intentions, *viz.* to penalize the depositing on land of poisonous, noxious, or polluting wastes so as to give rise to an environmental hazard, and to make offenders liable for any resultant damage; to require the giving of notices in connection with the removal or deposit of waste, and for connected purposes.

By S.1 it became an offence to deposit or abandon poisonous, noxious or polluting waste on land where it might give rise to an environmental hazard. The waste must be regarded as a hazard if it were present in sufficient concentration or quantity to threaten death or injury to persons or animals, or alternatively, if it were to threaten to contaminate a surface or underground water supply. Deposit of the waste in containers was not to be taken as excluding the risk of such a threat; the degree of risk was to be assessed by having regard to any measures which the owner or occupier might have taken to minimize the risk and from preventing children or other people from tampering with the waste. The phrase 'poisonous, noxious, or polluting' occurred in the Rivers (Prevention of Pollution) Act 1951 but it has never been the subject of close judicial analysis and the DPWA 1972 provided no further enlightenment; nor was there provided any definition of contamination, although the Regional Water Authorities (RWAs) would today agree that unacceptable contamination amounts to pollution.

A person charged with an offence under the DPWA could have broadly pleaded that either he acted under instructions from an employer or that he relied on information supplied by others, always providing that he had no reason to believe that the waste was of such a kind that it would be an offence to deposit it. Alternatively he might show that he had taken all such steps as were reasonably open to him to ensure that he had not committed an offence. Whilst the DPWA was repealed on 16 March 1981 as a result of the introduction of a new form of notification system under the Special Waste Regulations,[4] the offences and defences described above were re-enacted in S.3 and S.4 of the Control of Pollution Act 1974 (see below).

S.3 of the DPWA established a system of notification procedures to alert the authorities of vehicle and materials movements; thus it became an offence not to notify that land might be contaminated even to a level below causing an environmental hazard unless notices specified:

(a) the premises from which it was to be removed and the land on which it was to be deposited;
(b) the nature and chemical composition;
(c) the quantity to be removed or deposited, together with details of the number, size, and description of any containers;
(d) the name of the person who was to undertake the removal.

A sub-section excluded from this requirement was waste of a prescribed description or waste deposited in a prescribed way (see SI No. 1017 of 1972). The authorities to whom notification was to be made included the local government authority and the RWA (or River Purification Board in Scotland) both in the area from which the waste was to be removed and in the area in which it was to be deposited. A person charged with failing to give proper notice could plead that, with no reason to suppose otherwise, he had relied on information that the notices had been given by others. By S.4 it became an offence for the operators of commercial land to fail to give notice within 3 days that the waste had been deposited.

The effect on national and local government

The DPWA, despite its shortcomings, had a number of significant effects, viz.

(a) Under the notification procedures if (and a big if) local authorities and/or the RWA properly and assiduously collated the detailed data, then the scale and nature of the potentially hazardous waste problem in England could, to a large extent, be determined for the first time. The fact that the Dolcis Syndrome occurred (put everything into shoe boxes and forget it) has inspired the suggestion that now the notification procedure is changed because of the implementation of S.17[4] of the COP, there should be an offence of inadequate collation of information supplied. (Indeed, Regulation 17 states that the disposal authority for the area in which special waste is produced, or into which it is imported from abroad, shall so supervise the keeping of records under these regulations as to ensure the requirements are complied with.)

(b) The elevation of the administration of, and assessment of, the industrial waste disposal problem to a regional, and in some instances, to a national level.

(c) The direct involvement for the first time by the county and metropolitan councils in the surveying and planning of the problems of regional and industrial refused materials management.

The effects on the management of the waste materials' industry

(a) Increased professional standards and disciplines within the waste management industry.

(b) The closure of many unfenced, fly-blown, rodent-infested, litter scattered sites, which were simultaneously a children's playground, a totter's heaven, and also the local public dump, and which could eventually be replaced by fully prospected, engineered, and controlled licensed landfill sites and treatment plants.

(c) The development of applied techniques for the selection and evaluation of landfill sites suitable for the disposal of all contaminating materials.

(d) Technical developments and the commercial assessment of specialized regional plants for the treatment or destruction of all forms of the more intractable materials.

(e) Increased movement of contaminated wastes from the point of origin to the point of disposal, sometimes involving transportation over very considerable distances, which necessitated the enhanced safety standards in the design of equipment for their safe handling and transportation together with suitable training and integration with the emergency services.

The Control of Pollution Act 1974 summarized

Controlled waste (see S.30) which will arise within the catchment area of a WDA must be surveyed in order to prepare waste disposal plans under S.2. Each WDA must include in its plan the kinds and quantities of materials including soils which the WDA expects to dispose of itself and which the WDA expects to be disposed of by other persons together with the methods of such disposal in association with any reclamation possibilities (and see S.20) together with the estimated costs of the methods of disposal. This does not mean that each WDA must make its own arrangements for waste arising or being imported into its county;[5] its duty is to ensure that arrangements exist for such land or materials which are, or are likely to be, contaminated in its area. In order to ensure that its fiat runs through its own county, the DPWA offence of environmental hazard is repeated under S.3 together with a second offence under the same section of the depositing of any controlled waste or, indeed, the use of any plant or equipment to achieve the same object for the disposition of controlled waste unless the land has been licensed under S.5; the penalty for depositing waste which is poisonous, noxious, or polluting and its presence on land which is likely to give rise to an environmental hazard and it can reasonably be assumed to have been abandoned, is marginally more severe.

The requisites of a site licence are stated under S.6 and are further amplified in Waste Management Paper No. 4 and Circular No. 79/77. If the provisions of the Waste Management Papers, the COP 1974 and its subsidiary legislation are followed, then not only is it possible to contaminate land within the law but, quite properly, to receive the encouragement of the legislature and the executive in so doing.

Indeed, such would seem to be the worldwide practice. The Environmental Protection Agency (EPA) in the USA have published an account of the design, construction, and evaluation of landfills.[6] It is submitted that reasonable care would have been shown by a prudent site operator if he had reviewed the EPA summary of ideal conditions with his own site which should:

(a) be easily accessible in any kind of weather to all vehicles expected to use it;
(b) have safeguards against water pollution originating from the disposed solid waste;
(c) have safeguards against uncontrolled gas movement originating from the disposed solid waste;
(d) have an adequate quantity of earth cover material that is easily workable, compactible, free of large objects that would hinder compaction, and does not contain organic matter of sufficient quantity and distribution conducive to the harbourage and breeding of vectors (*sic*);
(e) conform with land use planning of the area.

In addition the proposed site:

(f) should be chosen with the highest possible regard for the sensitivities of the community residents;

(g) should be the most economic site available commensurate with the ultimate requirements for solid and other materials' deposit.

It should of course be large enough to accommodate the community wastes for a reasonable interval.

The standard of care required by recent UK legislation in the management of waste is essentially a *self-critical* process. Practical realization of this attitude of mind should be shown by the existence of:

(a) monitoring on a systematic basis;
(b) continuous technical audits;
(c) an efficient weather station;
(d) lysimeters for soil analysis and surface vegetation;
(e) a mathematical model for leachate production and quality forecasting;
(f) adequate cover material—*vide* the 1971 Sumner Report.

THE SIZE OF THE PROBLEM AND POSSIBLE OPTIONS

The UK

Every year about 3.5 million tonnes of materials notifiable under Statutory Instrument 1017 of the Deposit of Poisonous Waste Act 1972 are produced in the UK, chiefly by industry; these materials contain 300,000 tonnes of substances which, according to the EEC listings, are toxic or dangerous and merit special controls; 98% of the 3.5 million tonnes of this notifiable waste is managed by the private sector, through the routes shown in Table 1.

If one considers that the disciplines of landfill[7] are:

agriculture, chemical engineering, chemistry, civil engineering, climatology, cost and detriment analysis, ecology, electrical engineering, fill

Table 1

	Million tonnes
In-house or company-restricted sites	1.5
Private and public sector sites	1.2
Landfill	2.7
Incineration or chemical treatment	0.4
Sea disposal under MAFF licence	0.4
	3.5

Source: DoE submission to House of Lords Select Committee 1980.

sources and markets, horticulture, hydrogeology, landscaping, water engineering,[8] meteorology, metallurgy, microbiology, materials movement, reclamation, safe operating procedures, soil mechanics, transportation economics, hydrology;

and matters to be taken into account for marine and estuarial reclamation are:

tidal regime, current patterns, littoral drift, meteorology, wave climate, hydrology, hydrogeology, geology, soil mechanics, mining subsidence, drainage, reclamation, fill sources, sea defences, costs and detriment benefit, navigation, land ownership, riparian rights, wind speeds, dredging, emergency services, safe operating procedures, energy sources, effluent sewage outfalls, and cooling water systems;

the student of waste management in the early 1980s is therefore faced with a multidisciplinary subject embracing a wide range of skills, professional and scientific disciplines and to christen his operating base as a dump or a tip is an egregious misdescription which disserves those who strive for higher environmental standards. Not for the first time were the Californians ahead in titular imagineering.[6]

In contrast to Table 2, Table 3 sets out the best estimates in the EEC.

The EEC

The importance for a waste management engineer and his clients fully to comprehend development control procedures in the context of, particularly, landfill planning applications is again illustrated by Table 4. The recent Com(80)222 Progress made in connection with the EEC Environment Action

Table 2   Masses of solid waste arising in 1973 (million tonnes)

Coal mining	58
Mining other than coal	3
China clay quarrying	22
Other quarrying	27
Domestic and trade refuse	18
Industrial waste	23
Ash and clinker from power stations	12
Total	163

Sources: Royal Commission on Environmental Pollution Fourth Report 1974; Department of Environment Circular 53/1976 Part 1 (Waste on Land) Disposal Licences.

Table 3   EEC best estimates (million tonnes)

Extractive industries	300
Agricultural	1,000
Waste from manufacturing industries: metal, paper, rubber, plastics, textiles, and glass	80
Waste from manufacturing industries: chemical	45
Household refuse	90
Food waste	100
Total	1,615

Source: Paper to Joint Meeting EPCS–EEC/CEFIC March 1979, by J.T. Farquhar

Table 4   EEC chemical industry: Cost range of engineering waste by different methods (DM/tonne)

Controlled engineering to land	3– 30
Disposal to land in a site lined with plastic sheet	17– 75
Underground disposal by dropping into old wells or mines	30–300
Underground storage, e.g. in salt caverns	170–400
Land disposal after encapsulation either by mixing the waste with cement or other agent or by incarcerating whole drums in cement	17–150
Coastal sea dumping from ships or barges	6– 25
Deep-ocean dumping beyond the continental shelf	190–300
Simple incineration (without significant heat recovery)	50–180
Incineration with alkaline stack scrubbing	200–600
Incineration on board ship at sea	100–600
All types of chemical treatment and in particular:	
destruction of cyanide by hypochlorite	500–900
reduction of chromic acid	170–500
destruction of cyanide (catalytic)	350–400

Source: Author's researches in EEC countries 1979–80

Programme and assessment of the work done to implement it, together with the Second EEC Environmental Report of Autumn 1980, show once again that the Commission's work is primarily to provide a supranational environmental legislation for those countries that lack the plethora of UK legislation since 1945, together with its respected implementation. It is not so much that the environmental quality objective (EQO) approach is more suited to the UK's hydrogeology and estuarial hydrology *vis à vis* the uniform emission standard (UES) but that the Commission must understand that many an environmental and development control measure will derive its ultimate justification from a variety of factors which may be extraneous to the specific decision at hand, not least a precise correlation between the benefits arising out of the costs of the control measures, and not environmental control *per se*.[9]

The Commission should also realize that the aim of its environment action programme should be to avoid damage to ecological systems and not to eliminate barriers to trade (tweed is not indigenous to Tuscany and lemons will not ripen on Harris). Directives should only be raised where there is a definite need and will result in a significant effect; they should be based on sound scientific research and those concerning specific industrial sectors should be discouraged to avoid unnecessary duplication. There must be no repetition when, by March 1982, five separate actions have been brought by the Common Market Commission in the Court of Justice against Italy for failing to implement five EEC environmental Directives:

(a) Directive on the disposal of PCBs; compliance date 9 April 1978.
(b) Directive on the disposal of waste oils; compliance date 18 June 1979.
(c) Directive on bathing water quality; compliance date 10 December 1977.
(d) Directive on drinking water; compliance date 18 June 1977.
(e) Directive on waste; compliance date 18 July 1977.

RELEVANT DEVELOPMENT CONTROL LAWS

The interrelationship of planning and pollution control

The Town and Country Planning Act 1947 came into force on 1 July 1948. Thereafter planning permission was required for the development of any new waste disposal site or of any old site by extending its superficial area or height. This was the effect of Section 12(3)(b) of the 1947 Act now re-enacted in slightly more economical language as Section 23(3)(b) of the 1971 Act; likewise Sections 12 and 13 of the 1947 Act are now S.23 of the 1971 Act.

Certain limited types of waste disposal did, however, fall within the categories of permitted development set out by the General Development Orders. Currently Class VIII(2) of the 1977 Order permits an industrial undertaker to deposit waste from an industrial process on any land comprised in a site used for that purpose on 1 July 1948 *whether or not* the area or the height of the deposit is extended. Furthermore Class XIX(3) of the 1977 Order permits the deposit of waste by, *or by licence of*, a mineral undertaker in excavations made by that undertaker and already lawfully used for that purpose so long as the height does not exceed that of the surrounding land.

Permissions given to waste disposal operators could of course also be subject to conditions which would regulate operations on a disposal site.

Those conditions which related to the actual technical aspects of waste disposal were inadequate for the protection of the environment, the control of the pollution, or the general medium-term conservation of the local amenities. There are very few old soldiers[10] left in the public and private sectors who look back with misty and regretful nostalgia on past practices, but those who do could well be reminded that although man could tread the moon, in that very same year it was not on the UK earth considered necessary to include a hydrogeological profile of a potential waste disposal site in the papers submitted to a planning committee for their deliberation.[11]

The planning system has long since recognized that applications to change the use of land to that of a waste disposal tip tend to arouse controversy and residential opposition above the average. Such applications have been designated as one of the classes of development for which special requirements as to public advertisement of applications apply; see S.26 and Article 8 of the General Development Order 1977.

The object is to ensure that members of the public have the chance to make representations to the Authority before a decision is taken. Although only the exercise of more exigent and disciplined control in landfill site administration will provide the necessary practice and precept to prevent the present animus aroused by such a published planning application, only a disciple of Job would presently apply for permission to develop a dump, or it is submitted, a tip[6] (*see above*).

A local planning authority is also under a duty to consult with the appropriate regional water authority before deciding a waste disposal application; see Article 15(1)(f) of the GDO 1977.

From 1947 until 1974 the local planning authorities were the councils of the administrative counties and of the county boroughs. However, county councils were authorized to delegate functions relating to the control of development and the enforcement of planning control to county district councils and most of them did so. In practice, this meant that planning applications would be considered at district level in the counties, i.e. at the lower tier of the two-tier system. The county boroughs, covering the large town and city areas, were unitary authorities not subject to a two-tier structure.

The Local Government Act 1972 replaced the confusing and untidy pattern of local authority areas by a simplified two-tier system of county and district councils, whether metropolitan or non-metropolitan. County planning authorities are responsible for setting out general strategic policies in new-style development plans called structure plans just as they were responsible for the old-style development plans. As before, the lower-tier authorities handle the vast majority of planning applications. In Wales, the waste disposal, waste collection and waste planning activities are all at the second tier level; likewise in Scotland but with the added imprimatur needed from the particular Regional Planning Council. At the risk of incurring devolutionary chagrin but if only because of the necessity of a simple approach to such an inchoate subject, and also because site licensing was not introduced into Scotland until 1 January 1978, any observations on current law and practice should be viewed in the context of the English countryside.

The 1974 reorganization of local authority areas and functions did not lead to any marked change in the format of applications for permission to use sites for waste disposal purposes. However, the fact that planning applications were before the lower-tier district authority whilst the upper-tier authority was the waste disposal authority charged with a duty to make adequate arrangements for the disposal of all waste arising within their area, led to unhealthy tensions between county waste disposal officers and district planners. The district council was inevitably more responsive to the claims of amenity in considering a waste disposal application; the county's attitude was to a greater extent shaped by a consciousness of the need to provide facilities at reasonable cost, i.e. more landfill sites. In January 1981 this difference of approach has been eliminated by Regulations[12] providing that applications to deposit waste or refuse are now classified as a 'county matter'. (Previously only the deposit of mineral waste was automatically classified as a county matter: see Schedule 16, para. 32 of the Local Government Act 1972). The county council is now responsible for deciding the application. In the past there was a system of consultation between county and district over the question of finding suitable landfill sites, yet the county even with its more qualified and experienced officers had no power to override the district's decision.

A waste disposal application will now not only be considered at county and district level (the latter as a consultee) but it will also be submitted to a regional water authority whose veto on grounds of likely water pollution is in practice a decisive consideration. It is, however, regrettable that the Water Act of 1973 did not prepare or garb the RWAs with this important planning role of the long stop of last resort.

S.1 of the COP 1974 imposes a duty upon a WDA to ensure 'that the arrangements made by the authority and other persons for the disposal of waste are adequate for the purpose of disposing of all controlled waste which becomes situated in its area after this section comes into force and all controlled waste which is likely to become so situated'. Because of these Ss. 1 and 2 duties, i.e. to survey and make a plan, county WDAs must needs consult not only with their own district planning authorities about likely sites but with neighbouring county authorities. It is obvious that there will exist a shortage of strategically sited landfill sites; alternative methods of waste disposal where appropriate should be encouraged, but are substantially more costly. It is equally clear for both waste producer and consumer that not surprisingly there is as an uneven geographic distribution of the best natural sites for landfill as there is of high carbonaceous clay for brick-making. Counties which include centres of mineral extraction[13] are perhaps likely to have a greater supply of sites; it is possible that such supply could be used to help counties facing a situation of acute shortage or through accident or incidence of topography or commerce, have other facilities mutually to share.

'When it is not necessary to change, it is necessary not to change'

LORD FALKLAND

A national strategy

Let us therefore reflect on the fundamental requirements of a waste disposal strategy.

(a) The process which gives rise to the waste is accepted by the community, that is the product is reckoned to be worth any detriment arising from the waste.

(b) Since it is accepted that the waste is permitted, or indeed encouraged, to arise, the options for its disposal and minimization are identified. These options will include in-house re-use, recovery, and reclamation.

(c) For each option a model will be formulated expressing the system of operation necessitated by the option; the model will include environmental behaviour of the waste constituents.

(d) Since for each model the detriment to human health and to the environment will be estimated, then it may be seen that some of the options must be eliminated because the detriment is unacceptable, e.g. it may not conform with standards statutorily laid down or incorporated in codes of practice.

(e) There must then be a determination of the monetary cost of the remaining options and the elimination of the options which have both higher monetary cost and higher detriment than other options. Then:

(f) If only one option remains, the decision is clear-cut; that is the preferred option.

(g) If more than one option remains, they must be compared on a cost/ benefit basis. The detriment must be converted to a monetary cost and the resultant cost of each option estimated, i.e. cost of operations + cost of detriment and finally:

(h) The preferred option is that which has the smallest total cost.

It should be remarked that the necessary estimates depend on present knowledge; there can be no other basis. Although caution should be exercised in making judgment, over-caution is to be avoided in making comparisons of options; it is acceptable in an absolute risk assessment but not in a comparison.

Step (g) above inevitably involves a subjective judgment; one of the advantages of a careful step-by-step approach is that the subjective judgment is written down for critical appraisal by representatives of the community, i.e. paid officers and/or elected representatives. Although the approach is mathematical, the models and estimates may be crude. This is not a criticism of the methodology; it merely reflects the limitation of our knowledge and may point to research which can be done to improve it.

As a result of the absence of a national waste strategy understandably not having been imposed after the Industrial Revolution in the UK, it is sometimes argued that the present distribution of treatment and disposal facilities results in an over-concentration of refused materials at a limited number of sites. Whilst no evidence has been adduced to show that unacceptable contamination has taken place, or that there has been an intolerable convergence of heavy traffic via unsuitable roads, there is said to be unjustifiable mismatch between the areas which produce materials for treatment or for landfill as opposed to those areas which have the statutory responsibility for final deposit.

An effective national strategy would, however, require an extension of central powers, e.g. co-ordinating waste disposal plans, establishing data collection systems and monitoring arisings and disposals. Depending on the precise changes in the separation of functions under the strategy some local costs might be reduced while others might increase as a result of new requirements from the centre to ensure that the strategy was firmly based and effective. It is difficult to see how a significant erosion of local control and responsibility could be avoided. The establishment of any new body, particularly with a regional activity by the Department and its Regional Officers, would amount to a reorganization of one sector of local government.

There are reasonable grounds to suggest that local authorities are better placed to perform functions which might be transferred if there was a national strategy, since they:-

(a) are able and accustomed to monitor arisings and disposals in their areas;
(b) are better placed through experience, location, and current powers to balance local interests and concerns and undertake public consultation exercises;
(c) are responsible for dealing with the disposal of other controlled waste and it may be inappropriate to separate responsibility for potentially difficult waste from household discards, albeit classification by point of arising is misleading in the assessment of environmental effect;
(d) have built up a considerable degree of experience, particularly since the 1972 and 1974 Acts.

THE EIA AS A SOCIETAL AND CORPORATE CHECK LIST

UK and EEC view

In the context of waste management, environmental impact analysis is a process in which the impacts of a development proposal such as the reclamation of mineral extracted derelict land, on the natural environment and neighbours' enjoyment of landscape, health, comfort, convenience, and future employment are identified. The impacts deemed to be relevant to such a development decision are therefore analysed and evaluated by reference to existing operational practices.

The DoE's Report and Research II, *Environmental Impact Analysis*, at p. 67 found that there was a need to employ a system of environmental impact analysis for *some* kinds of development in England, but with the present flexible system of planning, development proposals could be dealt with as rigorously as the planning authority thinks fit. Projects can occur from time to time, possibly 25–50 in the course of the year, when the nature of the development, taken in conjunction with the existing environmental conditions, could be such that large-scale conflicts and environmental impacts will occur, which are difficult to comprehend and analyse, except by a systematic process in which the necessary specialist skills are employed, e.g. the Vale of Belvoir, Sullom Voe, the Walton to Grangemouth ethylene pipeline.

The EEC view, per contra, is that EIA[14] will:-

(a) provide the Commission with a key instrument for the implementation of the community environmental action programme;

(b) serve the centrist planning of the community, by helping to avoid the creation of economic investment conditions that vary from one region to another and threaten to distort competition (*sed quaere*);

(c) give public authorities the most exhaustive advance date possible, as well as the basis for designing alternative proposals;

(d) have a prophetic and economic value because it will 'help to save both operators' cost of protective measures which may turn out to be necessary, and to save society the social cost resulting from the wrong decisions which could do irreversible damage to the environment';

(e) be the best way of counteracting opposition resulting from 'obscure decision-making procedures which could lead to very expensive delays';

(f) act as a constant incentive to improve the methodology of environmental forecasting and assessment and assist in the optimization of urban areas and land use;

(g) help the community and member states to deal with the transfrontier impacts of development; and

(h) promote administrative co-ordination of environmental policy in the member states and eventually rationalize administrative procedures, cutting the time needed to plan and authorize development and activities.

Future preparation for planning applications and appeals

In the context of UK waste management, the extant planning procedures at present provide adequate machinery thoroughly to assess high and low technological applications; the cognoscenti will be aware that an individual impact assessment will be made by the following:

(a) county planning authority;
(b) county highway authority for density of, and approach by, future traffic flows;
(c) district council for public health, effluvia, noise, and smell;
(d) county and district recreational and parks for after-use;
(e) local councillors for psephological record;
(f) the Fourth Estate for national and local circulation;
(g) Regional Water Authority for hydrogeological profile;
(h) Institute for Geological Studies;
(i) Health and Safety Executive for standard operating procedures (WMPs as codes of practice have a quasi-force of law);
(j) nature conservancy and local ecological flora and fauna groups;
(k) central and local representatives of the manufacturing industries which will be providing temporarily discarded materials to resource the site.

Industrial development control ought not be justified *per se*, but only when it advances industry, the people who work in it, and their local government. The societal benefit of this control should arise from a decision-making process which adequately balances the considerations of the developers and the community (as expressed through the local authority)—and which is reached in the least possible time. It follows from the extensive planning and technical control systems already in operation in the UK (and some other EEC member states) that the most appropriate EEC action in the 1980s would be to issue a general mandate to all states to operate a planning system, with some guidance as to how and what methods should be considered; UK existing planning and environmental controls are detailed, additional legislation and administration unnecessary. It has always been extremely difficult to judge the economic effects, both in terms of cost and benefit, of environmental development and pollution control measures; for example it seems, to the EEC at least, a self-evident truth that there would be a net creation of jobs resulting from an environmental control policy; yet such a trend may shift resources from productive to service industries.

The House of Lords from August 1980 to July 1981 made as comprehensive a study[15] as possible in the essentially very limited time available in order to identify some of the problems of the identification, documentation, transport, consignment, monitoring, management, and engineering to landfill, airfill, and seafill of potentially difficult waste materials.

It would have been surprising within this imposed time-scale which was never justified, if this Committee of able and often healthy men, could have identified priorities,[16] and this they have not.

The Report also has one fundamental overriding flaw. For every decision in the 1980s to implement *further* environmental Great Britain legislation or codes of practice (the latter often having statutory force under the Health and Safety at Work etc Act 1974 (HASAWA)), the following analysis must be achieved:

(1) the probability of alleged detriment to the environment must be identified;
(2) all possible solutions researched;
(3) assessments must be made for reducing the risk of alternative strategies;
(4) an estimate must be made of the proposed benefits to accrue to both society and the environment, from the optimum strategy selected;
(5) the disbenefits and financial costs of achieving the best solution should be accurately estimated;
(6) the total disadvantages of change should be weighed against the proposed benefit—then, and only then, the appropriate decision can be made;
(7) however, an audit as formal as an EEC proposed Environmental Impact Assessment should be undertaken within 1 year as to whether the regulations, codes of practice, etc. are accomplishing their intent, and if any additional resources, particularly manpower, are needed and can be sanctioned and to what extent the environment has benefited to that date;
(8) a summary and report, together with the lessons learned, should be sent to the originating authority for formal acknowledgment.

Alas, throughout the whole of their Lordships's Report, the 34 conclusions and recommendations, their seven-fold strategies and seven appendices, the principle that the benefit of any regulations should be at least as great as the costs is neither distinctly enunciated or applied.

Objectors therefore to a development, and in particular one of waste management engineering (if the above procedure has been formally or informally followed) must not be afforded repetitive opportunities any longer to delay a proposed location of incremental societal wealth. The burden of proof has shifted in the 1970s largely as a result of a lack of disciplined homework subsumed by meretricious doomwatch hyperbole from those who demand evidence of industry to show that its operations do not unacceptably bruise the environment to those particular environmentalists (after all, we are all environmentalists), to show cause that industry and its services should be further restrained. This shift in onus can only be sustained by a vigilant waste management industry through a continued assumption of standards of care and control beyond that of strict statutory liability; they offend this canon at theirs and their comrades' peril whilst hazarding the goodwill of the industry which they have been invited to serve.

NOTES AND REFERENCES

1. *Journal of Solid Waste Management* (June 1980), **23**(6), 10 (see also Rhode Island judgment 1980 not reported)
2. Gray, Mather and Harrison, *Hydrogeological Guidelines.* Institute of Geological Sciences, 1974.
3. *Statistics on Derelict Land in England and Wales 1972 and 1973.* DoE.
4. The Control of Pollution (Special Waste) Regulations 1980.
5. Pitsea Landfill Inquiry Inspector's Findings and Secretary of State's letter of May 1979.
6. See The Resource Conservation and Recovery Act 1976. USA S.1004: 'The term "open dump" means a site for the disposal of solid waste which is *not* a sanitary landfill within the meaning of S.4004.'
7. *Co-operative Programme of Research on the Behaviour of Hazardous Wastes in Landfill Sites.* DoE, 1978; HMSO.
8. Circular 39/76: 13 April 1976: 'The Balance of Interests between Water Protection and Waste Disposal'.
9. See an interesting 1980 argument qualifying the British approach in Institute for European Environmental Policy, 'Memorandum submitted to Sub-Committee G of the House of Lords Select Committee on the European Communities on their general enquiry, based on Commission Communication COM (80) 222 to evaluate EEC environmental policy to date and its future direction'.
10. Hawkins, R.G.P. *Amendments to Control of Pollution Bill at the Report Stage May 1974.*
11. Hawkins, R.G.P. *Municipal Engineering,* 5 July 1974: 'Control of Pollution Bill and alarming imprecision'; 3rd para. and generally.
12. The Town and Country Planning (Prescription of County Matters) Regulations 1980.
13. *Planning Control over Mineral Workings*—Report of the Committee under Sir Roger Stevens: HMSO, 1976: and see also papers of the Land Reclamation Conference 1976 published by Thurrock Borough Council; the follow-up to this very successful conference will take place on 26–29 April 1983 at Thurrock Borough Council.
14. European Communities: European Parliament Working Documents 1981–1982: 2 February 1982, Document 1-569/81/rev. Report drawn up on behalf of the Committee on the Environment, Public Health and Consumer Protection on the proposal from the Commission of the European Communities to the Council for a directive concerning the assessment of the environmental effects of certain private and public projects.
15. *House of Lords Select Committee on Science and Technology Report on Hazardous Waste Disposal*: September 1981, Volumes I, II, and III. HMSO, £26.35.
16. Statement by Dr Mostafa K. Tolba, Executive Director, United Nations Environment Programme to the Environment Committee of the European Parliament, Brussels, 25 November 1981: 'The Stockholm Action Plan (following on from the 1972 Stockholm Conference) also contained within it some seeds of failure. It is clear, now, that, in the light of funds that were made available, it was overly-ambitious. Implementation might have gone ahead more quickly, if the Plan had conveyed a clearer sense of priorities.'

Appendix 1: The implementation of the Control of Pollution Act 1974: Some important dates in chronological order

1972

12 April	Circular 37/72	Review of Waste Disposal Facilities
19 July	Circular 70/72	Deposit of Poisonous Waste Act

1974

5 December	SI No. 2039	Control of Pollution Act 1974 (Commencement No. 1) Order 1974
20 December	SI No. 2169	(Commencement No. 2) Order 1974

1975

21 February	SI No. 230	(Commencement No. 3) Order Relates to Scotland
10 December	SI No. 2118	(Commencement No. 4) Order

1976

2 January	Circular 1/76	Guidance on Control of Pollution Act Part I
12 January	Circular 7/76	Guidance on (Commencement No. 4) Order
		Pollution of the atmosphere
27 February	Circular 2/76	Implementation of Part III—Noise
26 March	Circular 3/76	Introducing Waste Management Papers 1, 2, 5, and 6
13 April	Circular 39/76	The Balance of Interests: Water Protection and Waste Disposal

    (a) The present legislation summarised
    (b) Waste arisings and disposal
    (c) Technical consideration of relevant functions
    (d) DoE Landfill Research Programme

12 May	SI No. 732/1976	Made 12 May 1976
		Laid before Parliament 21 May 1976
	as amended by SI No.	Made 14 July 1977

	1185 of 1977	Coming into operation 14 June 1976 (a) Industrial waste definition (b) Excepted cases under S.3(1) (c) The method of Appeal to the Secretary of State (d) Particulars of the Register
12 May	SI No. 731/1976	The Control of Pollution Act 1974 (Commencement No. 5) Order 1976
21 May	Circular 55/76	The Control of Pollution Act 1974 Part I (Waste on Land) Disposal Licences
2 June	Circular 57/76	Guidance on (Commencement No. 5) Order
10 June	SI No. 956/1976	The Control of Pollution Act 1974 (Commencement No. 6) Order 1976
7 July	SI No. 1080/1976	The Control of Pollution Act 1974 (Commencement No. 7) Order 1976

1977

1 March	SI No. 336/1977	The Control of Pollution Act 1974 (Commencement No. 8) Order 1977 Relates to old Inner London Boroughs
11 March	SI No. 476/1977	The Control of Pollution Act 1974 (Commencement No. 9) Order 1977
17 August	Circular 79/77	Guidance on Licensing of Waste Disposal (Amendment) Regulations 1977
21 September	SI No. 1587/1977	The Control of Pollution Act 1974 (Commencement No. 10) Order 1977 Relates to Scotland
20 December	SI No. 2164/1977	The Control of Pollution Act 1974 (Commencement No. 11) Order 1977 Relates to S.2 in England and Wales

1978

5 April	Circular 29/78	(1) Introduction of Commencement Order No. 11 (2) EEC Toxic and Dangerous Waste Directive (3) Report on Hazardous Wastes in Landfill Sites

31 May	SI No. 816/1978	The Control of Pollution Act 1974 (Commencement No. 12) Order 1978 Relates to S.2 in Scotland
4 July	SI No. 954/1978	The Control of Pollution Act 1974 (Commencement No. 13) Order 1978
27 July	Circular 47/78	Guidance on the Control of Pollution Act 1974 (Commencement No. 13) Order Relates to S.13(3), (5), (6), (7), (8) in England and Wales

1980

| 1 July | SI No. 638 | Control of Pollution (Supply and Use of Injurious Substances) Regulations 1980 |

1981

17 February	SI No. 196	The Control of Pollution Act 1974 (Commencement No. 14) Order
20 February	Circular 4/81	Guidance on Special Waste Regulations
16 March	SI No. 1709	The Control of Pollution (Special Waste) Regulations 1980

1982

| 4 January | SI No. 252 | The Control of Pollution (Special Waste) Regulations (Northern Ireland) 1981 |

NB: The Refuse Disposal (Amenity) Act 1978 came into force on 23 April 1978. Under S.12 and Schedule 2 of this Act, para. 25 of Schedule 3 of the The Control of Pollution Act 1974 was repealed.

Appendix 2: Terms of service

1. Cleanaway undertakes

(1) to service the Waste in a proper and efficient manner according to Cleanaway's Quotation and Notifiable Waste Specification and subject to the following conditions contained in these Terms of Service.

(2) to comply with all special site and/or plant conditions and safe working procedures notified to, and acknowledged in writing by, Cleanaway before the date of the quotation in accordance with the obligations of the Customer under the Health and Safety at Work etc. Act 1974.

(3) to comply with all laws and regulations of any central or local governmental body or authority relating to the performance of Cleanaway's obligations and to the use of any equipment ("the Equipment") provided by Cleanaway in connection with servicing the Waste.

2. The Customer undertakes

(1) that the Waste is properly described in the Notifiable Waste Specification and will at all times correspond IN ALL RESPECTS with the description.

(2) that the constituents of the Waste shall be compatible and stable and no hazard will arise through the mixing of such constituents.

(3) that the Equipment will not be overloaded or improperly loaded and that save for the Waste, no explosive, dangerous, poisonous, noxious or polluting substance will be placed therein.

(4) to allow Cleanaway's authorised representative free and immediate access to the Customer's premises whenever requested for the purpose of inspecting the Waste and taking away samples provided that any such inspection or the subsequent sampling and analysis of the Waste shall in no way affect or lessen the Customer's undertakings upon which Cleanaway absolutely relies.

These undertakings are absolute and fundamental terms and/or conditions and shall continue unimpaired notwithstanding any inspection by Cleanaway of the Equipment or of any substance placed therein.

3. Changes in Waste description

(1) In the event of either party ascertaining that the description of the Waste has changed such party shall immediately notify the other with full details of the same whereupon the servicing shall be suspended until the Waste conforms with the description in the Notifiable Waste Specification without prejudice to Cleanaway's right to receive payment of any Equipment rental charges, any additional charges specified in the Quotation and/or otherwise

incurred or suffered by Cleanaway consequent upon the suspension of the servicing until the servicing is resumed.

(2) In the event of the Customer wishing change the description of the Waste or in the event of the Customer planning any change in production techniques or schedules likely to result in any such changes the Customer shall give Cleanaway full notice and details thereof to enable Cleanaway to consider whether the Waste as so changed is capable of being serviced under the Terms of this Contract.

(3) Before the service is resumed in respect of the Waste as so changed the parties shall agree in writing on a revised description of the Waste, revised charges and any other necessary amendments to the Quotation and Notifiable Waste Specification and this Contract shall thereupon be construed as if the Waste as so changed was the Waste described in the Notifiable Waste Specification and as if the revised Quotation and Notifiable Waste Specification were those referred to in these Terms of Service.

(4) Until the parties agree on the matters referred to in sub-clause (3) the provisions relating to Cleanaway's right to receive payment as provided in sub-clause (1) shall continue to apply.

4. Service in statutory notices

The Customer is responsible for ensuring that the provisions of Section 3 of the Deposit of Poisonous Waste Act 1972 or any re-enactment or amendment thereof or any order regulation or title made thereunder or any future statutory provision of similar effect are complied with. Three clear working days before the day of servicing the Customer shall ensure that there is served on Cleanaway or its nominated subsidiary a copy of any notice to be served by the Customer under the said Section 3, and where Cleanaway is not the operator of the site where the Waste is to be deposited or treated, on the disposal site operator notified to the Customer by Cleanaway.

5. Safety

(1) The Customer shall provide a suitable and a safe means of vehicular access for the servicing of the Waste.

(2) The Customer shall be responsible for giving proper notice in writing to Cleanaway's employees and agents of any special site conditions and safe working procedures in any way affecting the discharge of Cleanaway's obligations under this Contract and shall be responsible for the proper supervision of loading and/or collection of the Waste into the Equipment. Cleanaway reserves the right to refuse to service any order if it reasonably considers that the work required might place at risk any person, vehicle, equipment or property.

(3) The Customer shall be wholly responsible for the safety of all persons

(including the employees and agents of Cleanaway) within the curtilage of the Customer's premises.

(4) The Customer shall bear all risks involved in connection with the siting and use of the Equipment and failure to comply with any of the Customer's obligations in these Terms of Service.

The Customer shall indemnify Cleanaway against all proceedings and claims for any loss, damage, personal injury or loss of life arising from any of the Customer's undertakings and obligations under this Contract howsoever caused except where the Customer can affirmatively establish that the circumstances giving rise to the same were solely attributable to the negligence or wilful act or default of Cleanaway or any of its employees or agents provided that the liability of Cleanaway, its associated and subsidiary companies, shall be limited to the sum of £500,000 in respect of any one incident or series of incidents arising out of the same event.

6. The Passing of Title to the Waste

(1) Title to the Waste and general responsibility for its disposal shall pass to Cleanaway at such times as its collection vehicles leave the curtilage of the Customer's premises, provided there shall have been no breach of the Customer's undertakings.

(2) Cleanaway shall not be responsible for any property (including personal effects) deposited by the Customer or any other person in the Waste or the Equipment and shall not be bound to return the same nor be liable for any loss or damage thereto.

7. Insurance

The parties shall each maintain at all times during the period of this Contract at their own expense the insurance necessary to provide adequate cover in accordance with sound business practice and in respect of the indemnity referred to in clause (5) above, in connection with the operation of the Contract and each party shall if so required by the other, produce evidence of any applicable policy for inspection.

8. Servicing of Waste

(1) Cleanaway will endeavour to adhere to any servicing programme detailed in the Quotation unless delayed or prevented from doing so by circumstances beyond its reasonable control. Cleanaway shall not be liable to the Customer or to any third party for any direct or consequential loss caused by any delays in the performance of its obligations.

(2) The Quotation is given on the basis that the disposal point for the Waste will remain unchanged during the continuance of the Contract. In the event of

Cleanaway having to change the disposal point the charges specified in the Quotation shall be renegotiated.

(3) In the event of the Customer wishing to change the collection point for the Waste the Customer shall give Cleanaway full notice and details of the proposed collection point to enable Cleanaway to consider whether the Waste is capable of being serviced from the proposed collection point.

(4) Before the collection point is changed the parties shall agree in writing on the proposed collection point, revised charges and any other necessary amendments to the Quotation and/or Notifiable Waste Specification and this Contract shall thereupon be construed as if the proposed collection point was the collection point specified in the Notifiable Waste Specification and as if the revised Quotation and Notifiable Waste Specification were those referred to in these Terms of Service.

(5) Until the parties agree on the matters referred to in sub-clause (4) Cleanaway shall be entitled to service the Waste from the collection point detailed in the Notifiable Waste Specification.

9. Responsibility for Equipment

(1) Any Equipment supplied by Cleanaway shall at all times remain its property and it may replace the same with similar (but not less suitable) Equipment.

(2) The Customer shall not burn anything in the Equipment nor place any marking on nor sublet nor part with possession of any Equipment and shall be responsible to Cleanaway for any loss or damage to the Equipment (other than ordinary wear and tear) and for the cost of repairs and expenses resulting from the Customer's failure to take reasonable care of the same.

(3) Where the Equipment is placed (whether by Cleanaway on the Customer's instructions or otherwise) on a highway (whether public or private) or any public place, the Customer shall be absolutely responsible therefor and for the siting and lighting thereof and for obtaining all necessary permissions and licences (including those under the Highways Acts) and for ensuring observance of the terms and conditions thereof.

(4) The Customer shall ensure that all drums and other sealed containers are sound, suitable for the Waste and clearly marked with a legible and accurate description of the Waste placed in them, and where applicable with a legible warning that the Waste is inflammable or otherwise dangerous or hazardous. The Customer shall also ensure that no other marking is placed or allowed to remain thereon and that all statutory regulations relating to the marking thereof are complied with at all times.

10. Payment terms

(1) Cleanaway's terms of payment are full settlement within 30 days of the date of invoice. The Customer shall not be entitled to defer or withhold

payment to Cleanaway on the grounds of a claim or counterclaim. In the event of delay in payment beyond such period Cleanaway may at its discretion charge interest at the rate of 2% per calendar month from the date of invoice to the date of payment.

(2) No forbearance or indulgence by Cleanaway shown or granted to the Customer shall in any way affect or prejudice the rights of Cleanaway or be taken as a waiver of the terms of this, or any other, clause in these Terms of Service.

11. Early termination

(1) Should the Customer default with any payments due hereunder or (being a body corporate) have a receiver appointed or pass a resolution for winding up (other than for reconstruction or amalgamation purposes) or a Court order be made to that effect or (being an individual) have a receiving order made against him or enter into any composition or arrangement with his creditors, Cleanaway may thereupon at any time determine this Contract forthwith by written notice to the Customer.

(2) Either party shall have the right to determine this Contract forthwith if the other party shall commit a breach of its terms and such breach shall not have been remedied within thirty days of written notice specifying the default and requiring the same to be remedied.

(3) The foregoing rights of termination shall be without prejudice to the rights of the parties at such termination and upon any such termination Cleanaway may remove the Equipment and shall have the right to enter the Customer's premises to do so.

12. Assignability

Neither the Customer nor Cleanaway shall sublet or assign the benefit of the Contract except that (without in any way altering the rights or obligations of the parties) Cleanaway may employ such of its approved independent contractors or agents as it thinks fit to perform all or any part of its duties hereunder.

13. Application of terms

All servicing shall be subject to these terms which together with the Quotation and Notifiable Waste Specification supersede all other written and verbal agreements arrangements and representations made at any time between the parties and any conflicting terms of purchase or order sought to be imposed by the Customer.

No other terms, conditions or warranties express or implied shall be of any effect whatsoever unless in writing and signed by or on behalf of the Customer and Cleanaway in each case by a duly authorised representative.

FORM A	NOTIFIABLE WASTE ENQUIRY	Part 1 to be issued by	WS No.

A.O.D. Program **REDLAND PURLE LIMITED** | Compiled by: | Valid Unit

Removal Regional Water Authority

| | | Transport Depot |
| | Date sent | Season Period |

Removal Waste Disposal Authority | Part 1 Notice will be issued by Redland Purle for their customers unless specifically requested otherwise

Waste Producer Name and Address | Customer Name and Address (if different) Carrier if not RP

Contact and Tel. No

WASTE SPECIFICATION | Contact & Tel. No

1 General Desciption

2 Process Source
3 Principal Components

4 Physical Form
5 Qty & Service Frequency
6 Normal (max) concentrations for significant constituents

7 Are any of the following present (Yes/No) — Indicate level and nature

Cyanide/Sulphide .... Corrosives .... Flammables ....
Phenol/cresol/other biocide .... Strongly odorous material ....
Toxic metals .... Chlorinated hydrocarbons ....
Mercury/Antimony/Arsenic/Selenium ....
8 Client Safety/Handling advice (Attach standard data sheets if available)
9 Max Temperature
10 Max Density
11 Collection points & method of Site Storage
12 Is it necessary to enter a tank or confined space during waste collection? (Yes/No)
13 Is sample being sent with this Form .... separately .... to other location (state where) ....

FAILURE TO ANSWER ALL QUESTIONS WILL RESULT IN DELAYED ASSESSMENT AND APPROVAL

FOR LABORATORY/TECHNICAL CONTROL USE ONLY

CV BTU's/lb	Ash % W/W	Chlorine % W/W	Sulphur % W/W	Viscosity	Flash Point	pH	SG							

14 Provisional disposal site .... Confirmed/Alternative ....

15 Laboratory analyis. Date completed .... Lab. sig. .... T.C. sig. ....

DEPOT TO COMPLETE

Customer A/C No.   Depot Code

Depot Ref.   Customer's own ref.

| Date Received .... | Incinerator   Tech. Control

Cleanaway Limited

NOTIFIABLE WASTE SPECIFICATION

Review Date

Transport Depot

Earliest Permitted Date of Deposit

Season Ticket

EX WDA

Waste Producer Name & Address

Customer Name/Address (if diff). Carrier if not Cleanaway

Contact & Tel. No

WASTE SPECIFICATION for which the Waste Producer accepts responsibility
General Description

Contact & Tel. No.

Process Source
Principal Components

Physical Form
Quantity & Frequency
Permitted
concentrations for
significant
constituents/
chemical analyis

PLEASE READ YOUR UNDERTAKINGS ON THE ATTACHED TERMS OF SERVICE SHOWING THE ABOVE WS NUMBER

Max. Temperature
Max. Density
Collection point &
method Site Storage
Remarks

The Control of Polution
(Special Waste) Regulations 1980

It is out technical opinion that the
materials described in this
specification are NOT subject to the
provisions of the above Regulations

Handling and Safety Notes/Safe Working Procedures/Special Conditions SC No.

Acceptance of this specification
confirms your agreement

If you have any doubt you must
please contact your service depot.

DISPOSAL SITE

DISPOSAL AREA

TO WDA

V	TC	UN No	Hazchem

NOTE   This analysis in particular and all analyses in general are undertaken by Cleanaway Limited solely for
their internal control purposes and are on no account to be relied upon by the waste producer for his
own discriptions or operations

Cleanaway Limited

Date. Time and Document Number	Driver's Name		WS No.
	Signature		Review Date
Earliest Permitted Date of Deposit			Transport Depot
	Vehicle Regn. No.		Season Ticket
EX WDA			LRC Signature

'Waste Producer Name & Address

Customer Name Address (if diff.) Carrier if not Cleanaway

Contact & Tel. No.

WASTE SPECIFICATION for which the Waste Producer accepts responsibility
General Description

Contact & Tel. No.

Process Source

Principal Components

Physical Form

Quantity & Frequency

Permitted

concentrations for

significant

constituents

chemical analysis

AFTER APPROVAL HAS BEEN RECEIVED AND NOTED HEREON. THIS FORM OR AN IDENTICAL READABLE COPY MUST ACCOMPANY ALL LOADS IN RESPECT OF THE ABOVE WS NUMBER TO THE DISPOSAL SITE

TECHNICAL CONTROL SIGNATURE

Date prepared

Max. Temperature
Max. Density
Collection point &
method Site Storage
Remarks

YOU MUST HAVE READ AND ABIDE BY THE "DISPOSAL SITE INSTRUCTIONS TO DRIVER OPERATORS"

Handling and Safety Notes/Safe Working Procedures/Special Conditions SC No.

DISPOSAL SITE	DISPOSAL AREA	SITE USE				
TO WDA	LRC Approved Date	V	TC	UN No	Hazchem	

Redland

Health and Safety at Work

The health and safety of employees is an important factor which contributes to the well-being and efficiency of any works or site.

It is the mutual responsibility of management and employees at every level to ensure that the improvement of standards of hygiene and safety at the workplace is a primary objective of this Company. Safety representatives, safety committees and training, all have a special contribution to make in fulfilling this objective.

The Personnel and Training Officer (or Safety Officer, if appointed) has responsibility for advising management on all aspects of health and safety within the workplace and of any potential hazards arising from our operations which might affect our customers and the public at large.

Foreman and Managers have specific responsibilities for maintaining high standards of housekeeping, vehicle loading and unloading, safe access, safe tools, machinery guarding, lighting, ventilation and safe systems of work. All employees are encouraged to suggest methods of reducing hazards and eliminating accidents and to co-operate by using the clothing and personal equipment provided for their protection.

It is vital to the proper growth of the Company that full co-operation on health and safety is not only maintained but practised. Such an attitude of mind can only bring benefit to the organisation in the form of improved efficiency, reduced costs and a better environment.

Appendix 3: Health and Safety at Work etc. Act 1974, Section 2(3)

HEALTH & SAFETY POLICY STATEMENT

Redland Purle Limited recognizes a prime duty to protect the health and safety of its employees in the course of their employment. Similarly it recognizes a duty to protect other employed persons whilst on Company premises and also its duty towards the general public whose health and safety may be affected by its work activities.

The policy is to discharge this duty through a clear definition of duties and responsibilities of all employees, by the Company supplying relevant information and training, and through joint consultation.

Management and supervisory staff are required to provide and maintain safe conditions and systems of work. They are required to comply with statutory and Company regulations and codes of practice and acquaint employees with these regulations and codes.

P. J. Jansen (signed)

Claydons Lane
Rayleigh
Essex

July 1979

ORGANIZATION AND ARRANGEMENTS

I. *The Company Safety Manager* is responsible to the Company Technical Manager for:

   (i) Recommending a safety policy to the Company and the organization and arrangements for effecting that policy.

  (ii) Keeping the policy, organization, and arrangements under review and recommending changes where necessary.

 (iii) Assisting management to devise and put into operation rules and arrangements relating to health and safety in their areas.

 (iv) In particular areas or types of activity ensuring awareness of regulations and codes arising from the Health and Safety At Work Etc. Act or other statutory regulations and codes relating to that area or type of activity.

(v) Acquiring, holding, and making available where required information on the potential hazards of articles and substances used by the Company.

(vi) Establishing and maintaining links with the Health and Safety Executive, suppliers of materials, safety organizations, and other recognized bodies on health and safety matters. The Safety Manager is the Company's link with the Health and Safety Executive on matters relating to the health and safety of employees.

(vii) Providing direction to the Divisional Safety Supervisors.

(viii) Promoting externally the Company image of a safe and responsible organization.

(ix) Investigating accidents and dangerous occurrences where his specialist knowledge and experience may assist management to determine the cause.

(x) Keeping, analysing, and commenting upon accident records and notifying reportable accidents and occurrences to the Health and Safety Executive.

(xi) Advising on, and where appropriate organizing training on, safety matters in conjunction with the Company Personnel Manager. Arranging for the testing and periodic re-testing of skills relating to health and safety matters.

(xii) Advising on the provision of safety equipment and protective clothing.

(xiii) Through line management, assisting safety representatives to develop and exercise their statutory rights and privileges.

(xiv) Participating in safety committees on health and safety matters.

(xv) Issuing an annual report.

The functions of the Company Safety Manager in no way relieve line management of their safety responsibilities.

II. *The Company Personnel Manager* is responsible to the Finance Director for:

(i) Making arrangements for such training on health and safety matters as thought necessary by the Safety Manager and agreed by management, special attention being paid to new employees and juveniles.

III. *The Insurance Officer* is responsible to the Finance Director for:

(i) Ensuring that all liabilities known and predictable are covered by Insurance (including self insurance).

(ii) Reviewing insurance claim records to identify unsatisfactory performance and recommending corrective action.

    (iii) Making available to the Company Safety Manager information on the cost of insurance and the direct cost of work accidents and loss or damage to Company property where these risks are carried by the Company.

    (iv) Making arrangements for 'engineering insurance'.

IV. *The Redland Group Medical Consultant* is responsible to the Chief Executive for:

    (i) Supplying specialist medical advice on specific problems identified and defined by the Company Safety Manager.

V. *Redland Purle Retained Medical Officers* are responsible to the Finance Director for performing medical examinations on employees according to the current Occupational Health policy.

VI. *The Company Nurse* is responsible to the Company Personnel Manager for:

    (i) Performing medical examinations on employees in the southern region according to the current Occupational Health policy.

VII. *Divisional General Managers*, in order to implement the safety policy, will be required to demonstrate a personal and direct interest in health and safety matters. In particular they will *make provisions for*

    (i) Compliance with Statutory Regulations, Licences, approved and official codes of practice.

    (ii) Competent supervision being available.

    (iii) Emphasis to be placed on the safety and health element within induction and job training.

    (iv) Regular inspection, examination, and where necessary testing to ensure no hazardous deterioration has taken place.

    (v) Acquainting contract labour and visitors of the hazards associated with the Company's premises and work places and relevant safety rules.

    (vi) Enforcing safety rules.

    (vii) Continuous examination of working methods and places of work to minimize the risk.

    (viii) Provision of safe systems of work such as 'permit to work', 'gas-free certificate', 'hot working permit'.

    (ix) Consultation with safety representatives on health and safety aspects of work.

    (x) Fire precautionary measures.

(xi) Contingency arrangements in the event of emergency, e.g. firefighting, provisions of first aid.

(xii) The provision of protective equipment where necessary: the storage, issue, and maintenance and training in use of the equipment.

VIII. *Departmental Managers, Depot Managers, and Site Managers* are responsible to their Divisional Heads for:

(i) Familiarizing themselves with the Company's safety policy, organization, and arrangements and the safety rules and the codes applicable to their activities.

(ii) Ensuring that persons for whom they are responsible are trained in their work and are aware of any hazards or risks to health associated with their work or work place.

(iii) Ensuring that safety and health provisions and procedures are understood and adhered to by all persons in the supervisor's area of responsibility (including visitors and contractors) e.g. fire evacuation, safe work systems.

(iv) Visiting all areas of responsibility regularly.

(v) Maintaining a high standard of housekeeping.

Certain Managers will require to be designated to:

(i) Report accidents and dangerous occurrences and investigate their cause, involving safety representatives in the investigation.

(ii) Accompany representatives of the Health and Safety Executive or other statutory enforcement of bodies on their visits.

(iii) Chair Safety Committee meetings.

(iv) Bring to the attention of the Company Safety Manager and/or line management matters giving rise to concern.

IX. *Safety Supervisors (Transportation Division)* are responsible to their Unit Managers for:

(i) Liaising and agreeing with customers on safety aspects of waste removal operations from the customer's premises.

(ii) Supervising every operation judged to be sufficiently hazardous, especially 'one-off' operations.

(iii) Familiarizing themselves with all the liquid waste operations and performing sample inspections of these operations.

(iv) Carrying out safety inspections at Depots and reporting findings to the Unit Manager and Company Safety Manager.

(v) Performing other functions as may be determined at any time by the Company Safety Manager and the Management of the Transportation Division.

Safety Supervisors work under the professional direction of the Company Safety Manager.

X. *All Employed Persons* are required to:

  (i) Observe all safety rules at all times.
  (ii) Correctly use safety equipment and protective devices as necessary.
  (iii) Comply with instructions given on health and safety matters.
  (ix) Inform their immediate supervisor of all accidents or damage in their area whether persons are injured or not.

All employed persons are encouraged to participate in improving health and safety in the Company by making suggestions in these matters to their supervisor or safety representative.

XI. *Safety Representative SI.500 1977,* nominated by recognized trade unions, are responsible to their members for:

  (i) Investigating potential hazards and dangerous occurrences at the work place and examining the causes of accidents.
  (ii) Investigating complaints by any employee relating to that employee's health, safety, or welfare at work.
  (iii) Making representation to the employer on general matters affecting the health and safety or welfare at work of the employees he represents.
  (iv) Carrying out inspections in accordance with the regulations.
  (v) Representing the employees in consultation with Inspectors of the Health and Safety Executive.
  (vi) Receiving information from Inspectors.
  (vii) Attending meetings of Safety Committees where he attends in his capacity as a safety representative.

XII. *Safety Committees* may be formed at certain locations as recommended by the Safety Manager and agreed by local management. Such committees have a consultative function and in the event of failure to agree the dispute is to be taken through existing reconciliation machinery. Alternatively safety committees may be set up at the request of any two accredited safety representatives, as a statutory right. Safety committees in general should have the function of keeping under review the measures taken to ensure the health and safety at work of employees. Detailed advice and training will be given on planning and effectively executing the safety committees.

Redland Purle Safety Manual	SAFETY DATA	SD 1–01
		Sheet
		Issue 1
		Date

Subject HEALTH AND SAFETY POLICY STATEMENT

Written by D.G. John	Approved by	Issued by

Health & Safety Policy Statement

Ref. Health and Safety at Work etc. Act 1974 Section 2(3).

Redland Purle Limited recognizes a prime duty to protect the health and safety of its employees in the course of their employment. Similarly it recognizes a duty to protect other employed persons whilst on Company premises and also its duty towards the general public whose health and safety may be affected by its work activities.

The policy is to discharge this duty through a clear definition of duties and responsibilities of all employees, by the Company supplying relevant information and training, and through instruction, supervision, and joint consultation.

Management and supervisory staff are required to provide and maintain safe conditions and systems of work. They are required to comply with statutory and Company regulations and codes of practice and acquaint employees with these regulations and codes. The organization and arrangements for effecting the policy are detailed in the Company Safety Manual.

D. G. JOHN
Managing Director

Claydons Lane
Rayleigh
Essex

October 1980

CLEANAWAY

Head Office
Claydons Lane
Rayleigh, Essex SS6 7UW
Telephone: Rayleigh (0268) 775599
Telex: 99291

Health and Safety at Work etc. Act 1974

Health and Safety Policy Statement

Cleanaway Limited recognizes its moral and legal duty to protect the health and safety of its employees in the course of their employment. Similarly it recognizes a duty to protect other employed persons whilst on the premises of Cleanaway Limited and on the premises of other companies and also its duty towards the members of the public whose health and safety may be affected by its work activities.

In recognizing this responsibility, the Company's Health and Safety Policy is to:

(a)   provide and monitor safe conditions and systems of work;
(b)   clearly define the duties and responsibilities of all employees;
(c)   consult with employees and trade unions on aspects of safety;
(d)   provide safe working procedures;
(e)   continually monitor the safety procedures and performance.

Management and supervisory staff are required to provide and maintain safe conditions and systems of work. They are required to comply with statutory and Company regulations, codes of practice, and standard operating procedures contained in the Company safety manual and to acquaint employees with these regulations, codes, and procedures and monitor performance with respect to them.

This policy requires that all employees recognize the responsibilities under the Health and Safety at Work etc. Act 1974 to act responsibly and to do everything they can to prevent injury to themselves and other persons and to co-operate in the implementation of this policy.

C. Hoskisson
Managing Director   8 July 1981

Site Safety Organization

On the..........................site of Cleanaway limited, the person responsible for the arrangements necessary to implement the Company Health and Safety Policy is ....

The resources provided to assist him to do so include:

The Company Safety Manager	H.G. Pullen
The Company Occupational Health Nursing Officer	Mrs B. Anderson
The Assistant Safety Managers	D.J. Putt
	R. Rawson
	K. Smith

The above may be contacted by telephone: Rayleigh (0268) 775599

Safety Representative at this site ....

Copies of the policy and organization documents are available from your manager.

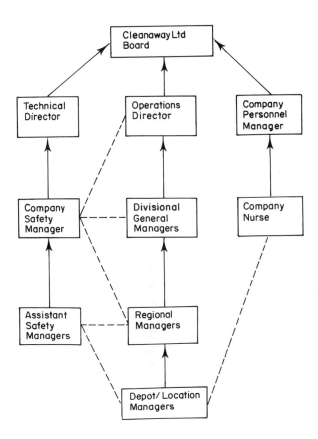

Appendix 4: Waste management papers published as at 1 March 1982

No. 1 Reclamation, Treatment and Disposal of Wastes.
No. 2 Waste Disposal Surveys.
No. 3 Guideline for the Preparation of a Waste Disposal Plan.
No. 4 The Licensing of Waste Disposal Sites.
No. 5 The Relationship between Waste Disposal Authorities and Private Industry.
No. 6 Polychlorinated Biphenyl (PCB) Wastes.
No. 7 Mineral Oil Wastes.
No. 8 Heat Treatment Cyanide Wastes.
No. 9 Halogenated Hydrocarbon Solvent Wastes from Cleaning Processes.
No.10 Local Authority Waste Disposal Statistics 1974/75.
No.11 Metal finishing Wastes.
No.12 Mercury bearing Wastes.
No.13 Tarry and distillation Wastes and other Chemical based Residues.
No.14 Solvent Wastes (excluding Halogenated Hydrocarbons).
No.15 Halogenated Organic Wastes.
No.16 Wood Preserving Wastes.
No.17 Wastes from Tanning, Leather Dressing and Fellmongering.
No.18 Asbestos Wastes.
No.19 Wastes from the Manufacture of Pharmaceuticals, Toiletries and Cosmetics.
No.20 Arsenic-bearing Wastes.
No.21 Pesticide Wastes.
No.22 Local Authority Waste Disposal Statistics 1974/75 to 1977/78.
No.23 Special Wastes: a technical memorandum providing guidance on their definition.

Appendix 5: Statutory enactments in the context of waste management and engineering

1892 ff.	Burgh Police (Scotland) Acts
1906	Alkali etc Works Regulation Act 1906 Local Acts
1928	Petroleum (Consolidation) Act
1933 ff.	Sundry Local Acts e.g. Berks, Bucks, Essex, Surrey
1936	Public Health Act
1937	Public Health (Drainage of Trade Premises) Act
1945 ff.	Water Act
1947 ff.	Town and Country Planning Acts
1951	Rivers (Prevention of Pollution) Acts
1951	Salmon and Freshwater Fisheries (Protection) (Scotland)
1956	Clean Air Act
1957	Occupiers' Liability Act
1958	Litter Act
1960	Clean Rivers (Estuaries and Tidal Waters) Act
1960	Radio Active Substances Act
1961	Rivers (Prevention of Pollution Act)
1961	Factories Act
1961	Public Health Act
1965	Rivers (Prevention of Pollution) (Scotland) Act
1967	Public Health (Recurring Nuisances) Act
1967	Civic Amenities Act
1967	Farm and Garden Chemicals Act
1968	Clean Air Act
1971	Town and Country Planning Act
1971	Oil in Navigable Waters Act
1971	Dangerous Litter Act
1972	Defective Premises Act
1972	Deposit of Poisonous Waste Act and SI No.1017
1972	Local Government Act
1972	Road Traffic Act
1973	Water Act
1974	Dumping at Sea Act
1974	Road Traffic Act
1974	Control of Pollution Act
1974	Health and Safety at Work etc. Act
1977	General Development Order
1978	Refuse Disposal Amenities Act
1979	Public Health (Recurring Nuisances) Act
1980	Local Government Planning and Land Act 1980

Appendix 6: Existing EEC Directives in the Field of Waste Management which should have been implemented by member states by 1980

1. Disposal of waste oil (78/439). Requires member states to ensure the safe collection and disposal of waste oil and its recycling.
2. Wastes (75/442). A framework directive, setting an objective on member states to dispose of waste without endangering human health or the environment.
3. Toxic and dangerous waste (75/319). Sets out arrangements for the proper disposal of toxic and dangerous wastes.
4. Disposal of PCBs (76/403). Sets out requirements to be met in disposing of PCBs.

Proposals for Directives in the Pollution Field include:

1. Water pollution by paper mills (paper pulp).
2. Air quality standards for lead.
3. Sulphur content of fuel oils.
4. Framework directive on harmful substances in the work place.
5. Discharges of mercury into the aquatic environment.
6. Discharges of drains.
7. Major accident hazards of certain industrial activities—known as post-Seveso proposal.
8. Environmental assessment of development projects.
9. Trans-frontier transportation of waste.
10. Beverage containers.
11 Sewage sludge.

Practical Waste Management
Edited by J.R. Holmes
©1983 John Wiley & Sons Ltd

3

The storage and collection of waste

A. E. HIGGINSON, MBE, FIWM, FCIT, MIENVSC, MRSH

Consultant, Solid Wastes Management and Environmental Services

ABSTRACT

An authoritative review of the historical growth and current state of the art in the storage and collection of solid wastes. This chapter looks at solid waste generation rates, composition, and future trends. Systems of storage, the range of bins, containers, and compactors are linked to a review of refuse collection vehicles and plant. Innovative pneumatic systems of collection, safety standards, waste reclamation, and operational performance levels are set down. The use of public civic amenity sites is given special emphasis.

PREAMBLE

Traditionally it is accepted that the storage and collection of waste is fundamental to public health, a sense of well-being, and for environmental protection. The responsibility for the provision of these services and for their efficient operation is a statutory duty placed upon District Councils and London Boroughs. It is a service closely linked with the daily life of the community, thereby being under close scrutiny by the public.

Although significant technical advances have been made in the provision of vehicles and equipment it is mainly dependent upon an adequate labour force operating with flexibility and mobility under all weather conditions.

There are a number of general principles to be observed in developing the service, the yield and composition of the waste to be collected, its density, and any likely variations in the constituents, as storage is planned. Team size, frequency of collection, vehicle selection, or whether separation at source of recyclable materials is economically viable and must be allowed for.

It is important that analyses of the waste are regularly collated so that the variations can be tracked to make sound decisions on whether compaction should be introduced in high-density dwellings to achieve economic payloads on vehicles or determine the type and size of the storage containers and equipment.

AMOUNT OF WASTE

The amount of waste produced per household per week has slowly declined from 11.7 kg/m^3 in 1970 to 11 kg/m^3 in 1980 but the density of occupants during the same period has also decreased. This tends to confirm that the output per capita has remained fairly static and is a view expressed in the British Standard Code of Practice[1] after studying current statistics available at the time. The figure quoted as the best estimate per output per dwelling is 12 ± 1.2 kg per week at a bulk density of 133 ± 13.3 kg/m^3. The bulk density which showed such rapid decrease during the 1960s has tended to slow considerably, as the comparison indicates, of 146.2 kg/m^3 in 1970 to 141.02 kg/m^3 in 1980. These figures have been obtained from actual analysis of domestic wastes carried out during the period quoted from national sources.

Although output and density appear to be almost static the constituents of domestic wastes have changed with a sharp increase in the plastic content and a corresponding decrease in paper and board products (Appendix 1).

To cater for the normal requirements of household waste the storage container has evolved over a long period of time, and although it has some limitations it contributes to an efficient collection system. To accommodate the amount of waste produced on a weekly basis the most common sites used are: 2½/3¼ ft^3 (0.071/0.091 m^3) metal or plastic dustbins with similar sizes for disposable sacks, paper or plastic; 1¼ yd^3 (0.96 m^3) containers. Although the disposable sack appeared to be supplanting the conventional domestic dustbin approximately only 35% of households have this type of storage container in spite of the positive recommendation of the DoE Working Party on Refuse Storage and Collection in 1967 that a disposable sack system was a more hygienic method to operate. About half the 600 million paper and plastic sacks used annually for refuse collection are issued by local authorities, the remainder are from commercial, industrial, and institutional premises.

LEGAL ASPECTS

The legal basis for waste collection has been regulated by the Public Health Act 1936 which recognized two types of refuse: 'house' and 'trade', but conflicting interpretation of the exact classification identified in the Act produced a confused national picture. Case law successfully argued that it was the nature of the waste which determined its category, resulting in local authorities collecting waste domestic in character without charge, from what were obviously trade premises.

Further complications were discerned from separate public health legislation for London dating from the 1890s embodied and consolidated in the Public Health (London) Act 1936 which contained definitions of house and trade refuse which differed from the rest of the country. In the metropolis case law had determined it was the character of the refuse, which resulted in its classification exempting hotels and restaurants and similar premises from 'trade' refuse charges. The only opportunity for local authorities to make a charge was for any additional collections required (S.74) above the normal service rendered.

The deficiencies and anomalies depicted in the Public Health Act 1936 and other unsatisfactory legislative aspects were reviewed and carefully assessed resulting in recommendations to be made by Government Working Parties to be incorporated in the Control of Pollution Act 1974.

This Act has re-defined categories of waste as: household waste arising from a domestic dwelling, residential home or from premises forming part of a university, school, part of a hospital, nursing home or other educational establishments. Except in prescribed cases the Act places a duty upon the collection authority to collect such waste without charge. Commercial waste consists of waste from premises used wholly or mainly for the purpose of trade, business, sport, recreation, or entertainment. Farm and quarry wastes are excluded from this particular category. An important departure from previous legislation is that on request a collection authority must make arrangements for collection and make a reasonable charge unless it is considered inappropriate. A clearer definition emerges on industrial waste which is described as waste from any factory within the meaning of the Factories Act 1961 or from a nationalized industry. A collection authority is not obliged but may make arrangements for the collection of industrial waste and must make a reasonable charge.

Previous ambiguities are removed because the new definitions provide that waste cannot belong to more than one class although judgment must be exercised in determining borderline cases between household and commercial and industrial wastes to prevent excessive or unreasonable demands for the commercial service. The regulations under S.12(3) proposes to prescribe certain types of household waste which need only be collected on request and for which a charge may be made.

One important change which has been irksome to many local authorities is the waste arising from hotels, restaurants, public houses, and boarding houses which formerly was contested as domestic and given the free collection but obviously existed as being profitable business enterprises. Now the waste is defined as commercial waste although domestic in character.

Some problems will remain after Ss.12–14 are implemented, particularly in regard to mixed hereditaments in which a part of the property is occupied as a private dwelling. In the DoE consultation document the Department claims

Figure 1   Advertisement showing range of dustbins

that most of the anomalies may be ironed out by reference to the general rate
coding for particular premises. If the General Rate Act 1967 is studied a
mixed hereditament is described as one in which the greater proportion of its
rateable value is attributed to that part used as a private dwelling, and it is
suggested this method of assessment is followed.

The CBI in tabling charges made by different local authorities for wastes collected from commercial and industrial premises point out the great disparity in levying charges for the service given and suggests a more equitable method of charging especially when mixed hereditaments are considered.

Charging is directly linked with the classification of wastes but the CBI refers to the rating system and the substantial contribution business premises contribute to the General Rate Fund and when Ss.12–14 are implemented they contend their financial contribution should be fully evaluated in the assessment of charges for the collection of commercial waste. They have asked that in considering charging policies, carefully defined criteria should be used and be nationally applicable. This does not conflict with peculiarly local circumstances which can be admitted when applying a national charging policy. A moot point arises in the interpretation by which a collection authority may waive charges for the collection of commercial waste where it considers a charge inappropriate.

Obviously the CPO Act 1974 has endeavoured to smooth out anomalies by clarifying the definition of wastes and the classifications of premises from where the waste arises but many problems appear unresolved and the future of trade refuse is still conjectural.

STORAGE OF WASTE

To provide adequate storage of waste at or in domestic dwellings is to assess the amount of waste generated and frequency of collection given by the local authority. The statistics collated during recent years have indicated that the general mix of refuse produced per capita is remaining reasonably static with density only marginally decreasing. In effect this means that the most common bin in use domestically conforms to BS 792 1973 mild steel dustbins manufactured in four sizes, 0.03, 0.06, 0.07, and 0.09 m^3, and has remained unchanged for some time. The choice of size depends upon household density, social habits, and frequency of collection which has meant the medium-size bin being the most popular in general use.

Criticism of noise by clanging metal bins and lids during the collection service caused an introduction to be made of rubber lids and foot rings as specified in BS 3735 1976 to be fitted to bins. A major alternative to metal bins is the lightweight BS 4998 moulded plastic dustbins, manufactured from polyethelene, plasticized PVC, or polymerized rubber compounds which produces a tough solid made bin in four nominal sizes 0.06, 0.08, 0.10, and 0.12 m^3. The important characteristic developed in the construction of these bins is that they prevent deformation of shape, avoid impact damage, and can be specifically designed to meet customer preferences.

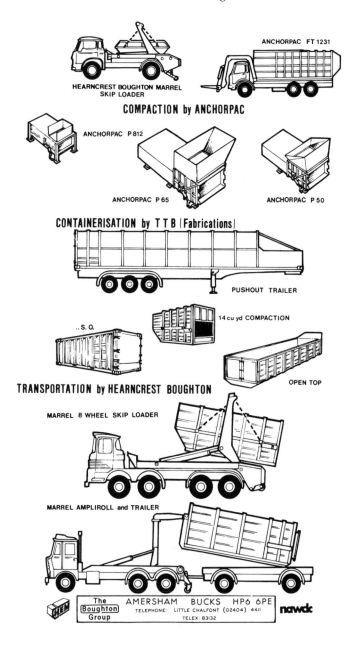

Figure 2   Advertisement showing the range of waste-handling systems available.
(Courtesy of The Boughton Group)

Figure 3   An example of a modern hydraulic waste packing unit designed for use in industrial premises

Some domestic dwellings are so arranged that the means of collecting waste is from a communal storage container sited some distance away, and for these conditions a metal bin 0.014 m^3 capacity conforming to BS 1577 mild steel refuse containers was designed to be used in the kitchen for daily use. Latterly the metal bin has been superseded by the pedal-operated plastic bin which is not only lighter and easier to keep clean, but has a variety of pleasing colours giving an attractive appearance in an important area of the household.

Arising from the recommendations of the Working Party Report on Refuse Storage and Collection 1967 was the use of disposable plastic or paper sack system as an alternative to the ordinary dustbin as being a more hygienic method of storage and collection. An expendable sack system evolved providing for either a paper or plastic sack being attached to a metal frame or used as a bin liner inside the type of bin in use. This system is clean, hygienic, noiseless, and labour-saving as a new container is provided each week thereby removing the criticism of smells and offensive odours from the traditional bin often dirty with decaying food scraps adhering to the bottom or sides. The method has a further advantage, as during holiday periods additional sacks can be issued to give extra storage and provide a much greater flexibility in maintaining a regular collection service.

Sack-holders and guards can be wall-mounted or free-standing, the former being cheaper to install and comprising a galvanized mild steel lid, clamping arms, and a sealing ring to secure the top of the sack.

In the free-standing unit the guard should be manufactured from cold-reduced mild steel, welded and galvanized after fabrication.

There appears to be no satisfactory standard yet introduced for paper sacks except they should be made from two-ply wet strength kraft 70–90 litre capacity to fit standard-sized bins.

Plastic sacks are made from both high- and low-density polyethelene of different gauges, practical tests on various types of micron thicknesses for failure rates has led to improvements in manufacture of high-density sacks or

Figure 4 Storing office wastes. An example of an office waste compactor. Simple and easy to use it can contain a wide variety of office wastes. There are two methods of entry. A chute door at arm height for normal throwaway items and waste-paper basket contents and a large-capacity ground-level bin to receive heavier or awkward loads. This system is marketed by Columbus Dixon Inc. of the USA

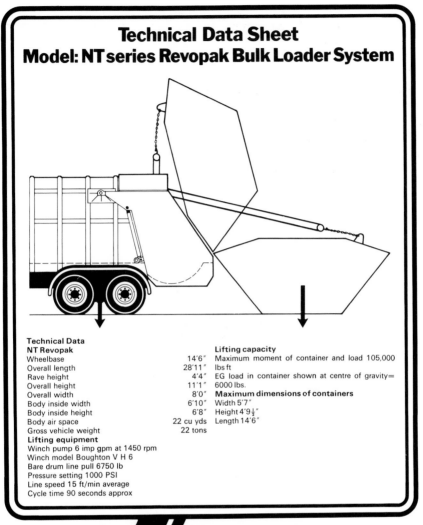

Technical Data Sheet
Model: NT series Revopak Bulk Loader System

Technical Data
NT Revopak

Wheelbase	14'6"
Overall length	28'11"
Rave height	4'4"
Overall height	11'1"
Overall width	8'0"
Body inside width	6'10"
Body inside height	6'8"
Body air space	22 cu yds
Gross vehicle weight	22 tons

Lifting equipment
Winch pump 6 imp gpm at 1450 rpm
Winch model Boughton V H 6
Bare drum line pull 6750 lb
Pressure setting 1000 PSI
Line speed 15 ft/min average
Cycle time 90 seconds approx

Lifting capacity
Maximum moment of container and load 105,000 lbs ft
EG load in container shown at centre of gravity= 6000 lbs.
Maximum dimensions of containers
Width 5'7"
Height 4'9½"
Length 14'6"

Shelvoke and Drewry Limited, Municipal Vehicle Division
Icknield Way, Letchworth, Herts SG6 1EN, England
Telephone: Letchworth 6555 Telex: 825556

A Butterfield - Harvey Company   Specifications subject to alterations without notice

Figure 5   An example of a technical data sheet for a modern refuse vehicle designed to handle bulk containers

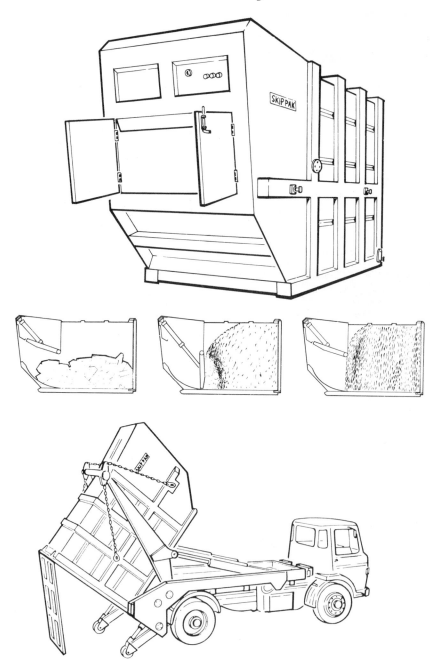

Figure 6   Example of an industrial waste-handling system. (Courtesy of Powell Duffryn Engineering Ltd)

Figure 7   Bulk 1.6 m³ refuse bins being emptied into a modern refuse collection
vehicle. (Courtesy of Perth & Kinross District Council)

liners resulting in lower gauge being suitable for general use. The advantage is
that plastic sacks can be made in various colours of different thicknesses and
sizes with printed instructions so that storage of refuse, removal of salvage
materials or trade waste can be easily organized by colour selection. This has
led to simplification in removing commercial or trade waste in areas where a
few premises can be integrated with domestic collection and avoid special
facilities having to be provided.

The disposable sack system may cost more in financial terms although there
are substantial economies made in labour and use of vehicles, but the cost
may be justified for amenity reasons such as improved methods of storage,
better hygiene, and flexibility in collection with quietness, an added advantage.

As fuel costs mount and expenditure in new vehicles increases substantially, the cost disparity is gradually lessening and provided a realistic work
performance is obtained the near future may see the sack-collection system
predominating.

The storage of waste at houses and bungalows presents no problems by
providing individual containers for each dwelling with suitable access for

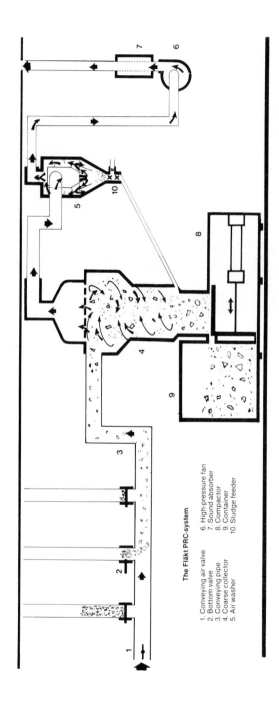

The Fläkt PRC-system

1. Conveying air valve   6. High-pressure fan
2. Bottom valve   7. Sound absorber
3. Conveying pipe   8. Compactor
4. Coarse collector   9. Container
5. Air washer   10. Sludge feeder

Figure 8   Pneumatic refuse conveying. The chutes are emptied fully automatically in accordance with a preset time programme comprising one or several emptying cycles per day. An emptying cycle is initiated by the conveying air valve (1) being opened and the high-pressure fans (6) being started. The bottom valve (2) closest to the receiving station is then opened and the refuse drops into the conveying pipe (3) by gravity, but assisted by the suction of the conveying air. The conveying air transports the refuse to the coarse collector (4)

Figure 9　Examples of modern easy-handling refuse containers widely used in Europe. (Courtesy of SSI Schafer)

collectors but where dwellings are in separate but multiple-occupation units, problems arise on the general principles of the suitable facilities to be provided.

Whatever facilities are contemplated they must be adequate in size and space, have maximum convenience for the tenants and collectors, by hygienic, safe from fire risk, and soundly insulated. The composition of domestic waste contains a substantial proportion of potentially valuable constituents that could be recycled and provision should be made for the storage and selection of suitable components within the area of refuse storage.

It is well to follow the Code of Practice BS 5906 in the classification of systems for the handling of domestic refuse, although the classification can be extended to commercial wastes.

Simple storage and direct collection:

(1) individual waste storage containers;
(2) communal waste storage containers;
(3) communal waste storage containers with chutes;
(4) bulk waste storage containers.

Reference has already been made to domestic dustbins and disposable sacks for individual waste storage containers but a considerable number of dwellings are in residential blocks of varying numbers of storeys. Where chutes are not installed and the provision of communal containers is preferred to individual bins, the former should be housed in storage chambers with doors and provided with a platform to give access to the top of the container. Ventilation is essential and all waste storage chambers, their location, dimensions, construction, lighting, cleansing, road access, and approaches to the buildings carefully documented in BS 5906 (formerly CP 306 Part 1), the details of which should be scrupulously followed in any developments contemplated.

In all dwellings with four or more storeys it is recommended that communal chutes be installed conforming to British Standards not only for construction purposes but the containers to be used, BS 1136 or BS 3495 applying. The waste storage containers for this type of development are combined with chutes based on the notion that the occupier should not be expected to carry refuse more than 30 m distant. This requires chutes to be spaced at not more than 60 m intervals, BS 1703 specifying the requirements for the chutes, which should be preferably not less than 450 mm in diameter.

With the changing circumstances of social habits and degrees of affluence a considerable proportion of bulky waste requires removal. One of the existing problems of high-rise dwellings and storage allocated for waste in the chambers is the scarcity of space for these unwanted articles. Where such communal chambers are installed separate enclosed accommodation at ground level is recommended for these bulky articles or even salvageable materials, a minimum space of 10 m^2 or 0.3 m^3 per person should be allowed with a headroom of not less than 2.3 m.

Many chutes serve the Paladin type 1 m^3 container which evolved from the ordinary bin and side-loading container, but this has serious disadvantages, especially difficulty of cleaning and the light weight carried of modern refuse, particularly if any bulky items such as empty cartons are placed inside without prior crushing. The Paladin container, owing to its configuration, is not capable of allowing compaction of the refuse within the container because of practical difficulties of compression and finally discharge. This led to considerable investigations being carried out by the Building Research Station in 1970. Measured densities of wastes from four-storeyed blocks of flats, chute fed, were found to have densities of only 80 kg/m^3 against the national

Figure 10   A modern refuse collection vehicle working in the City of Coventry.
(Courtesy of Coventy City Council)

average domestic density of 146.2 kg^3. These revelations hastened the
application of on-site volume reduction to secure more economic payloads in
storage containers. Among the first schemes, where space permitted, the
Deva machine was installed where the discharge chute repeatedly filled the
receptacle before compression and by the use of a rotary turntable achieved
about 4 : 1 reduction in volume. As the receptacles were sacks and had to be
manually handled a limitation on the weight to be carried was important from
a health and safety angle. Unless the collection vehicle was fitted with
auxiliary lifting equipment the manual employees would be unwilling to
collect heavy sacks weighing 22–34 kg. Apart from the dangers of cumulative
strains to manual employees if lifting the sacks were attempted, damage could
be done to the vehicle compression mechanisms or to the hopper or
bodywork by handling dense unyielding sacks.

The suitability of properly designed auxiliary lifting equipment operating
from almost ground level and fitted to refuse vehicles is vital if compaction of
waste is to grow. For this reason alone early consultation by architects or

other developers of domestic dwellings with the local authority's cleansing service is essential before any volume reduction equipment is fitted in any type of building, be it domestic flats or commercial or industrial premises.

Because of the limitations of the Paladin type containers a new range of bulk containers were introduced for the storage of waste materials. This stemmed from industrial sources using the open skip type of container for the storage and removal of large quantities of wastes. The application to local authority operations was principally for markets, precincts, and civic amenity sites but heavy capital expenditure for the transport servicing unit made it more suitable for specialist private waste contractors.

Hygienic storage for domestic waste is fundamental but the open type container, not being totally enclosed, is only suitable for a limited range of operations and not for domestic storage. As many local authorities give regular service to commercial premises, hotels, and restaurants where storage is a problem the compaction of refuse has resulted in considerable improvement. Not only is less storage required but a better collection service often results. Refuse storage in most new buildings will be subject to the constraints of on-site volume reduction using suitable compaction equipment. Architects must consider this essential need in all design considerations.

VOLUME REDUCTION

On-site volume reduction

The low weight/density ratio of modern waste has demanded that large office blocks, shopping areas, commercial premises, markets, and high-rise housing units all require some form of on-site volume reduction to achieve economic collection costs. Multiple stores and offices with a high paper/packaging content, frequently up to 80% in composition, require compaction of these light bulky materials to economize on valuable floor space. Similarly hotels, restaurants, and supermarkets need compaction facilities. This has led to compaction of waste over a wide field of applications. Basically the principle of compaction is a platen pushing waste materials into a container. It can be driven hydraulically, pneumatically, or by a screw mechanism working in a horizontal or vertical manner using disposable or re-usable containers.

Many types of compactors are available suitable to fit most types of premises; they are usually classified as small, stationary, and portable. A good example of a small compactor is the 'Orwak' unit, a vertically operated machine in which the ram compresses the refuse within a steel cylinder into plastic or paper sacks. If the refuse from a hotel or restaurant is very wet, where 33% of the waste is composed of food waste, it can be discharged into steel bins with fitted lids. Headroom is important as about 3 m height is required but as a 4 : 1 reduction in volume can be achieved it is an important installation. The refuse may weigh 27–32 kg after compaction and needs to be wheeled by a trolley and mechanically hoisted into a collection vehicle.

Stationary compactors compress refuse into bulk containers suitable for use in commercial and industrial premises where large quantities of wastes regularly arise and space is limited. A whole range of containers from 7 to 11 m^3 are available but the container chosen must be matched to the servicing vehicle used. For premises where output of refuse is lower a smaller container of approximately 2 m^3 capacity can be used but the container must be wheeled to a vehicle fitted with container loading which gives a flexible system as the vehicle can serve many premises without discharging its bulk load.

Recent developments have seen an extension of portable compactors comprising a power unit, compactor, and container all integrated as one unit. They are easily transported and being self-contained moved to different sites with a minimum of dislocation and can be handled by conventional skip hoisting vehicles.

On-site treatment of waste

The low weight/density ratio of domestic waste makes on-site treatment an economic possibility for volume reduction apart from creating a better method of handling crude wastes. Four methods of on-site treatment are usually listed: incineration, shredding and crushing, baling, and compaction.

As compaction is so closely allied to storage and use of containers it has already been dealt with under volume reduction.

Incineration is a volume reduction method which has the advantage of achieveing 10 : 1 ratio of the original input volume. Individual gas incincerators have been used with success but capital expenditure, running costs, and the service needed for the residues and incombustible materials has made the system unacceptable with so many other options available; the exception being hospital use for clinical and pathological wastes where cross-infection is always a possibility unless wastes are dealt with promptly and effectively. Even the well-designed, operated, and carefully maintained incinerators of larger capacity have found difficulties in meeting the air emission standards required by local authorities under the Clean Air Acts. Coupled with high capital and maintenance charges incineration for domestic refuse in large-scale housing projects is not favoured at the present time.

Shredding and crushing

Shredding of domestic wastes is usually a central facility confined to waste treatment and disposal processes at transfer stations or incorporated at specially designed reclamation plants concerned with waste-derived fuels or fibre recovery. But in big hotels, commercial premises, and restaurants,

Practical waste management

Figure 11   Part of a range of lightweight waste-handling systems based on the 'Ant' three-wheeled vehicle. (Courtesy of BTB Engineering Ltd)

shredders are installed to reduce the large volumes of food cans, glass bottles, plastic containers, office papers, and confidential wastes. Even offcuts, scrap metal, or faulty production products can be shredded either to improve reclamation options or reduce transport costs of moving light bulky wastes. Under these circumstances no further compaction is necessary as considerable volumetric reduction is achieved by the shearing action lessening voids, as the packing action of shredded materials tends to reduce volume by the weight of its own static head giving a good density/weight ratio.

Baling

Balers are able to compress a whole range of materials such as cartons, loose papers, containers, and offcuts, especially where large daily waste products have to be dealt with. By applying higher pressures than normal compactors better density is obtained and when the bales are strapped they become firm and easily manageable.

Space saving in all types of buildings is a very important costing exercise and balers, either chute- or hand-fed, can crush and bale a whole range of bulky wastes and occupy surprisingly small floor areas for servicing. Quick cycle times reduce large volumes speedily and in baled form wastes can be conveniently stacked to await collection. Fire risks are considerably reduced, hygienic storage improved, and ease of operation by unskilled employees makes baling of wastes a great advantage to stores, supermarkets, or large office complexes where the composition of the wastes may contain up to 80% of paper products making volume reduction an essential requirement.

Where paper or cardboard reclamation is required the waste streams from premises can be separated into the recoverable items for baling and other wastes consigned directly to compactors or skips.

PIPELINE COLLECTION METHODS

The pneumatic system of refuse collection. both large and small bore, is the most promising future hygienic development designed to handle all domestic wastes except the large bulky items. Based on the Swedish principle Centralsug, the City of Westminster was the first local authority to install the pneumatic system which involved a large number of medium-rise housing blocks and some terraces of town houses totalling 1,576 dwellings.

The system comprises 56 standard 450 mm diameter vertical refuse chutes fed by conventional hoppers, each chute terminating in a pneumatically operated plate valve above which the refuse is stored pending collection. These chutes are linked to 36 purpose-built chute valve stations (16 single- and 20 double-chute installations). Pipes are 517 mm bore and approximately 15 mm wall thickness, the total length of pipework being 1.8 km and

Figure 12   A modern vertical-shaft refuse pulverizer in Sheffield, South Yorkshire.
(Courtesy of Newell Dunford Engineering Ltd)

maximum conveying distances 600 m. The whole system is automatically
controlled from a central plant room in which, at predetermined intervals (at
least twice a day) or when activated by sensors, the 250 b.h.p. electric motors
are accelerated until the correct vacuum is achieved. When the required
speed is obtained the plate valve is opened by pneumatic action and the waste
falls from the chute and is sucked into the transporter pipes and conveyed to a
40 m^3 bulk silo sufficient for 1 day's storage for the whole estate.

Originally refuse was discharged from the silo by a reciprocating hydraulic
ram on to a conveyor belt and fed to an incinerator of modern design.

Figure 13   Pulverized domestic and commercial waste fed via hydraulic compactors into 20 m³ bulk containers en route to a remote landfill site. (Courtesy of South Yorkshire County Council)

Problems of feeding and blockage from the silo plus poor incinerator performance were carefully assessed, resulting in a decision to replace the incinerator with an on-site compactor with suitable vehicular access for collection purposes.

The present capital costs for the pipeline installation of pneumatic conveyance of wastes is a deterrent for local authorities in times of financial restrictions but the automated system shows that the cost balance in the City of Westminster scheme is in favour of piped refuse against the conventional methods of collection by a substantial margin. The relatively low cost

Figure 14   A 'Paladin' type 1.0 m^3 bulk bin commonly used in commercial premises, schools, and flats being handled by modern refuse vehicle

alternatives of on-site compactors are a serious competitor for those premises producing large volumes of light refuse. Although business premises appear to be suitable for the installation of pipeline systems it is unlikely that many schemes will develop owing to the initial capital requirement and the effectiveness of alternative compaction methods.

WATERBORNE SYSTEMS

A clear distinction is necessary between the individual sink grinder and the Garchey system, although both use the kitchen sink outlet for the removal of wastes.

A sink grinder is a useful piece of equipment to deal with the putrescible content of waste but as it is only required to dispose of about 15% volumetrically of the total domestic output it scarcely lessens the collection service to premises. Also sink grinders increase the consumption of water and place an additional load on the drainage and sewage disposal plant. Care is needed in fitting the unit in the kitchen as considerable vibration is caused unless proper gaskets and point-mounted suspension is used to allow independent movement of the machine. Owing to their limitations in handling other components of domestic waste they will make little future difference to the collection service.

In the Garchey system refuse falls into collection units placed under each sink and is flushed down special pipes to a central collection chamber independent of the drainage system. The system is only suitable for large housing developments whereas sink grinders can be fitted to individual premises and macerated waste flushed down the main drainage network.

As the Garchey system can only deal with about 50% of the domestic waste output a secondary collection service is needed, making it an expensive method to install.

In catering establishments the prompt removal of food waste is essential if good hygienic standards are to be maintained; food waste disposers are therefore frequently installed. These machines are heavy duty, specially designed to handle all kinds of wastes, grinding them into almost liquefied form and discharging direct into the drainage system.

Advice is sometimes required for larger premises in the delicate problem of disposing of sanitary towels. Completely automatic systems are available whereby a motor is started as the flap is opened, water flows in as the flap is closed, the towels are shredded to small particles and flushed into the drainage system.

WORK STUDY

British Standard 3138 1969 defines work study as 'A management service based on those techniques, particularly method study and work measurement, which are used in the examination of human work in all its contexts and which leads to the systematic investigation of all the resources and factors which affect the efficiency and economy of the situation being reviewed in order to effect improvement'.

Method study as applied to refuse collection is the systematic recording and critical examination of all the factors related to proposed ways of doing the work involved with a view to applying easier and more cost-effective means including the use of labour, vehicles, and equipment. It is necessary to demonstrate first how the service is to be organized, quality standards to be imposed, the economic considerations, human factors, vehicle selection, job evaluation, productivity, and financial rewards.

The basic steps in critical examination will involve observing the present practices, breaking down distinctive elements for analysis and measurement. Most method study investigations have objectives in reorganization and the

Dimensions and Weights

Item	Bulkmaster II 70 (Crew Cab)
Wheelbase	4439mm (14'6¾")
Overall Length	7899 mm (25'11")
Overall Width	2426mm (7'11½")
Overall Height	3455mm(11'4")
Turning circle (approx.)	17.6 m (58')
Rear overhang (closed)	2133mm (7'0')
Rear overhang (open)	3200mm (10'6")
Overall height (open)	4724mm (15'6")
Tyre size (standard)	10.00 × 20 (16 ply)
Unladen Registration weight (Estimated)	8890kg (8tons 14cwt)
Front Axle (plating weight)	5583 kg (5 tons 10 cwt)
Rear Axle (plating weight)	10160 kg (10 tons)
Gross (plating weight)	15743 kg (15 tons 10 cwt)
Height of loading rail	1041mm (3'5")
Body Space	13.2m³ (17¼ cu.yds.)
Payload	6096kg/6604kg (6/6½ tons)

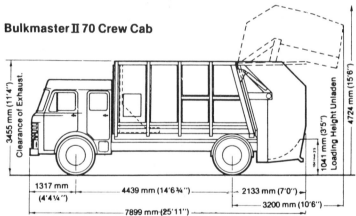

Bulkmaster II 70 Crew Cab

Figure 15   A typical technical check sheet for a modern rear-end-loading refuse vehicle. (Courtesy of Hestair Dennis Ltd)

introduction of a financial incentive scheme for higher productivity. Where failures have occurred in schemes adopted they have been due to faulty recording of the facts, as refuse collection requires extensive, laborious analysis of existing activities and fact-finding on location spread over a reasonable length of time.

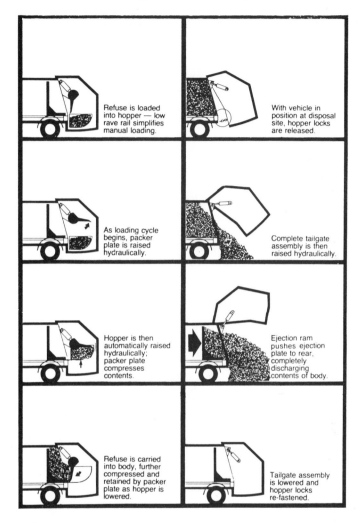

Figure 16   Sketches illustrating the operating sequences of a refuse collection vehicle. (Courtesy of Hestair Dennis Ltd)

The all-important documentation of the established work load needs accurate details of all properties to be serviced, manner in which the 'rounds' are organized with well-defined and disciplined routes, and quantities of trade and commercial wastes if collected with domestic premises.

Allied to assessing work loads is detail regarding the actual carrying capacities of vehicles used, and distance to disposal points. In preparing future strategies forecast trends of waste output are necessary, anticipated life of existing landfill sites, and present and future use of depots either for garaging, servicing, or combined with workshops.

In planning organization and methods the impact of the Health and Safety at Work Act 1974 requires full attention to be given to safety of personnel and the public during the whole operation, and any undesirable practices rigidly cut out.

When decisions on the organization and methods to be employed have been detailed, work study follows, setting standard times for jobs, planning work schedules, assessing manning levels, measuring performance, and providing management with information for any necessary actions subsequently.

The techniques available are well known, involving breaking down each job into distinctive elements for observation, measurement, and analysis. Rate of working, observed timing, allowances, and output are defined elements adjusted to a standard rating and performance. A specification of all the conditions such as establishment of work values, basis of incentive bonus payments, completion of work schedules, safety, quality of work performed, deductions for unsatisfactory or uncompleted work, trial period, and above all revision of values and periodic review.

Criticism of work study results from many schemes being out of date and in need of extensive review. Conditions change, rounds differ, traffic routes are modified, modern vehicles contribute to less fatigue, and rationalization is essential for a fair assessment to the public and employee alike. The results of work study need reassessment owing to increased inflexibility of labour attitudes to changing conditions, the problems of transferring employees to other work, and the decline of standards engendered by 'task and finish' involving serious underuse of expensive vehicles, plant, and equipment.

MANNING LEVELS

Although conditions vary from district to district the best team size appears to be four loaders and a driver per vehicle. The use of work study techniques should produce variations in team size dependent upon the types of premises being serviced. It may be that a smaller number is more economic where short 'carries' are established, and perhaps a team of five may be better employed in certain residential areas. Careful study of all the factors involved are necessary in determining the best use of labour and extensive vehicles, and a flexible approach is the right stance to take.

Allied to team size the determination of frequency of collection is necessary and by assessing the climatic conditions in the UK a 7-day interval appears satisfactory for most residential areas. High-density housing with limited storage accommodation requires its own frequency interval commensurate with hygienic considerations. The frequency of 7-day intervals is consistent

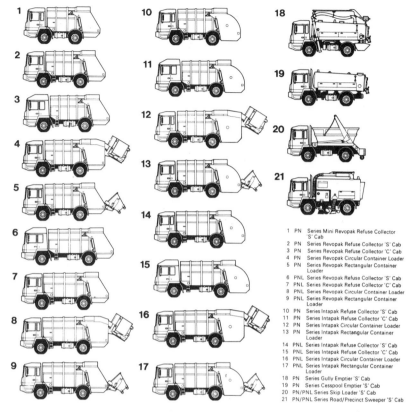

Figure 17   Typical advertisement showing a wide range of waste-handling vehicles used in municipal cleansing services. (Courtesy of Revopak Ltd)

with the domestic output of waste per week, the storage required, hygienic standards, and the methods of collection.

HEALTH AND SAFETY

The purpose of the Health and Safety at Work Act 1974 provides a legislative framework to promote a safety awareness and prevent risks, not only to employees but the general public, which may arise from work activities.

In refuse collection every vehicle is classed as a place of work and must conform to the statutory conditions laid down. Domestic premises are also classified as places of work for the purpose of the Act so that refuse collectors must at all times take care, not only for the safety of himself but of other persons who may be affected by his acts or omissions. The Act lays duties and responsibilities on the employers to provide such information, instruction, training, and supervision as is necessary to ensure, as far as is practical, the health and safety at work of all employees.

Written safety policies must be prepared and kept up to date; they must be communicated to familiarize all staff regarding safety rules, procedures, and activities. This includes a Code of Practice or firm instructions on how refuse is to be collected and deposited in vehicles. (See Appendix 2 for typical Safety Standards).

CIVIC AMENITIES

Under the Civic Amenities Act 1967 a duty was placed upon local authorities to provide places for the disposal of any matter whatsoever, except refuse falling to be disposed of in the course of a business, whether organic or inorganic. This means liquid or solid waste and this may be deposited free of charge. With regard to business waste the local authority may require the waste to be deposited at a civic amenity site on such terms as it feels reasonable.

It is obvious that the prime intention was to create a much-needed service to persons resident in the area of the authority and to other persons upon such charges (if any) as the authority thinks a fit amount for depositing waste. The site was to be reasonably accessible to residents and to be available at convenient times including parts of the weekend.

In order to cope with various types of wastes, powers were given to provide plant and apparatus for treatment or disposal of the refuse being deposited. These powers were not meant to curtail the collection of bulky household wastes by district councils and London Boroughs, although a steady decline was observed in bulky waste collections since the Act was introduced with a concomitant growth of refuse at civic amenity sites. The increase of bulky waste at civic amenity sites could be interpreted as being a conscious reduction in the collection service which was the major factor in the physical changes experienced.

Opportunities to use civic amenity sites as treatment centres for wastes have been tried with varying results. Oil tanks for the receipt of waste oil from DIY motorists as part of a recovery programme has met with only partial success, likewise cullet, glassware, and metals.

The control of large numbers of cars, vehicles, and people using the site has indicated the need for the service but requires careful assessment in terms of location, visual amenity, ease of unloading, etc.

The situation is intensified by the heavy use made of the facilities at weekends, resulting in new site layouts to include safety aspects and assist in traffic movement. The problems are accentuated by the need for extra containers to cover peak flows of refuse, also to guard against traders trying to

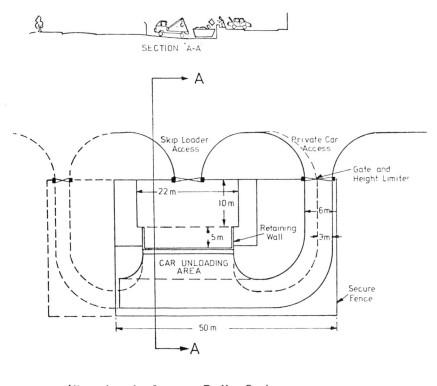

SECTION 'A-A'

————Alternative for One-way Traffic System

Figure 18   Plans of a typical local authority household waste site for private car and van-delivered wastes. (Courtesy of the DoE)

use the sites free of charge. To appreciate intensive use, at a GLC site at the peak hour on a Saturday morning, 110 vehicles per hour were recorded and on Sunday morning, 205 vehicles per hour. The population density plays a major part in usage, also the condition of housing accommodation, i.e. number of inhabitants per dwelling and distance between location of civic amenity sites and the services offered by the local authority in the collection of bulky household wastes. The importance of civic amenity sites is highlighted by the fact that about 10% of the total domestic waste output is attributed to bulky household or civic amenity wastes.

The economic climate is making an impact on collection authorities in the free service given for any waste produced by a householder, but experience has shown there is a need for both the provision of civic amenity sites and special collections for bulky and other unwanted wastes, which requires co-operation between districts and WDAs in rationalization.

VEHICLE REQUIREMENTS

Fleet size is determined by the number and types of premises, output of refuse, and frequency from the district to be serviced. The premises will need to be segregated into domestic hereditaments with an accurate bin count, commercial and industrial premises enumerating bulk containers, skips, and whether compactors are in use. An essential prerequisite is an analysis of wastes indicating a true volume and density. It is only with such basic data that details of vehicle requirements can begin to be collected.

The cost and complexity of modern refuse collection vehicles demands that the final selection meets the necessary criteria in operational efficiency availability, minimum fuel consumption consistent with the payload carried, utmost safety to loaders, noise control, and reduced smoke emissions.

Vehicle manufacturers must incorporate in their products technical principles to meet existing legislative requirements, any extra refinements being added for customer preference to meet operating conditions. The principles usually embrace maximum compression ratios, ease of loading, large-capacity rear hopper, body appearance, capacity commensurate with chassis compatibility, manoeuvrability, strength of materials, and adequate discharge facilities. Warning systems are fitted for safety aspects, cab sizes depend on client's requirements but must fall within the design criteria, visibility for drivers, clean loading, easy control mechanisms for all operational needs. These are the main criteria for selection but after-sales services and costs of spares to maintain fleets at the highest level of efficiency without carrying excessive spare parts plus ready assistance from company service engineers all need to be added to the list of requirements for final selection.

The extent of vehicle demonstrations to examine vehicle propensities under actual local working conditions should be determined at an early stage; this means careful statistical recording of all relevant items. After demonstration, and after a guarded selection has been made, enquiry from colleagues operating similar vehicles with guidance on unit costing, payloads, workshop reaction on maintenance or replacement of any vital parts, down-time, any faulty components, will all help to make the final vehicle selection the correct choice.

MAINTENANCE

An item of considerable importance in a council's budget is maintenance of vehicles. There are a number of items that are unavoidable, and in no way relate to operational efficiency, although they appear as expensive charges; these are licences, insurance, depreciation, drivers wages, garage accommodation, and they account for about 75% of the total expenditure. The remaining 25% covers tyres, repairs, and maintenance but the ratios shown

TYPICAL "CIVIC AMENITY" SITE

SHOWING SOME DESIRABLE FEATURES
Where site is at an existing enclosed and fenced
local authority depot, additional fencing may
not be necessary.

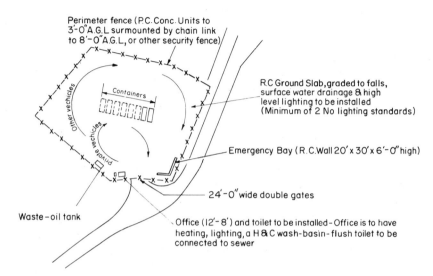

Perimeter fence (P.C.Conc. Units to
3'-0"A.G.L surmounted by chain link
to 8'-0"A.G.L, or other security fence)

R.C Ground Slab, graded to falls,
surface water drainage & high
level lighting to be installed
(Minimum of 2 No lighting standards)

Containers

Other vechicles

Private vechicles

Emergency Bay (R.C.Wall 20' x 30' x 6'-0" high)

24'-0" wide double gates

Waste – oil tank

Office (12'- 8') and toilet to be installed - Office is to have
heating, lighting, a H & C wash-basin- flush toilet to be
connected to sewer

Figure 19   A government-recommended layout for a household waste site. (Courtesy
of the DoE)

can be distorted unless the utmost economy is exercised in maintenance to ensure maximum availability and minimum cost.

The low hours worked by collectors on refuse collection vehicles means a considerable proportion of a sophisticated vehicle's time is idle, and therefore the utmost protection should be given to avoid general deterioration. This means providing covered garage accommodation observing basic principles of construction, such as a clear floor space, and if heating is not included some form of thermal insulation, good falls for drainage, conveniently spaced airline sockets for maintaining correct tyre pressures, and if electric vehicles are part of the fleet, facilities for daily battery charging.

To maintain maximum vehicle efficiency planned maintenance is essential; this enables compliance with legislation and avoids expensive breakdowns. Cleanliness is a daily requirement as constant appearance in public places brings vehicles into consistent focus of residents so that routine maintenance such as vehicle washing, oil changes and topping-up procedures, chassis greasing (especially those components subject to hard abrasive wear), and

Figure 20 A poster encouraging the salvage of household waste elements. (Courtesy of the Keep Britain Tidy Group)

periodic inspections coupled with necessary adjustments, keep the vehicles operating in a neat and efficient manner.

Routine maintenance can be carried out by internal workshop staff or by arrangements with private contractors. If the latter course is followed then a strict specification needs to be prepared in accordance with the vehicle manufacturers' requirements and maintenance manuals.

The degree of involvement by a full maintenance staff or the driver carrying out routine cleaning, greasing, battery topping-up, and correct tyre pressures is a matter of organization and methods applied to the council's fleet generally. There are sensible arguments to be heard for both systems. With larger fleets it may be necessary to employ workshop staff on a shift basis. This can be important in reducing down-time if a vehicle develops a fault late

Figure 21   A glass bottle reclamation scheme in a British City. (Courtesy of the Glass Manufacturers Federation Ltd)

in the working day as attention can be given and the vehicle can then be available at normal starting time the following day. In all routine maintenance skilled supervision is necessary to oversee that the vehicles are maintained to standards. This means that the periodic inspections must be thorough with a systematic check on the general conditions. Correct working of all instruments, clutch, steering, and brakes must receive particular attention.

However thoroughly vehicles are maintained, occasions arise when a vehicle is off the road and the use of a spare becomes imperative. A spare vehicle should always be ready to replace a vehicle breakdown but the quantity of spares which should be kept is very important. It has been thought prudent that one standby for ten in service is a reasonable precaution, but this assumes standards of vehicle maintenance are good.

Dependent upon the size of the fleet will be the workshop area allocation with adequate space provided for ease of manoeuvring awkward-sized and shaped vehicles. Likewise plant and equipment requirements are needed to cover vehicle repairs, tyre maintenance, body repairs, painting, and electrical bays if size demands, and electronic fault-finding equipment to quickly discover erratic running or failure of components.

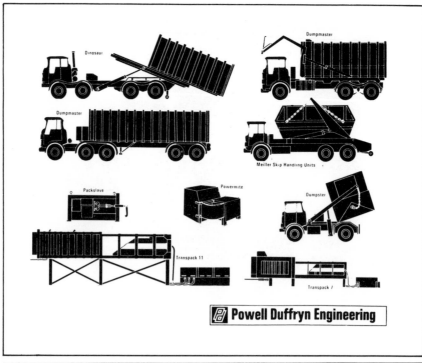

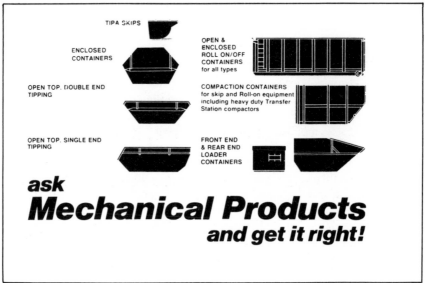

Figure 22   Advertisements illustrating a wide range of waste container and handling systems

Figure 23   The traditional method of refuse collection and emptying—usually on a once-per-week collection

Figure 24   A more modern 'dustless' automatic emptying of wheeled and lidded refuse containers—a system widely used in continental Europe. (Courtesy of SSI Schafer)

Figure 25　A large housing installation in Denmark using pneumatic refuse collection systems. (Courtesy of Brunn & Sørensen AS)

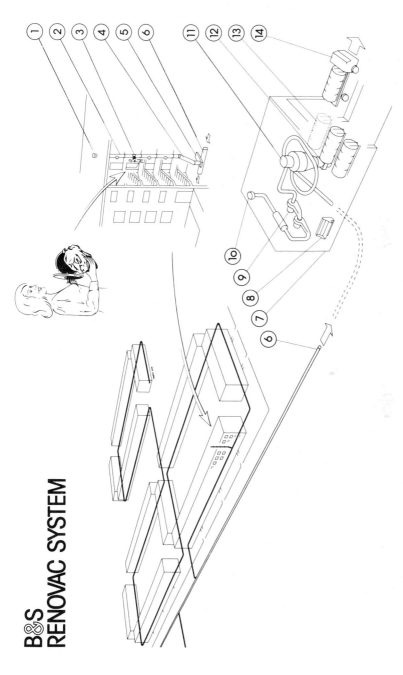

Figure 26 A sketch illustrating pipework systems for pneumatic refuse collection. Explanation of numbers: 1. Air inlet. 2. Refuse chute. 3. Inlet door. 4. Store. 5. Chute valve. 6. Transport tubes. 7. Suction central. 8. Control panel. 9. Blowers. 10. Air outlet. 11. Cyclone. 12. Compactor. 13. Container. 14. Removal for disposal. (Courtesy of Bruun & Sørensen AS)

ACROWPACTOR

on site waste compaction for high rise buildings

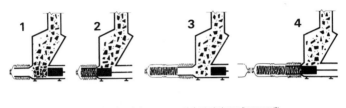

Operation is automatic. When the volume of waste in the compaction chamber builds up to a level sufficient to break an infra-red control beam for more than 4 seconds, Acrowpactor switches on automatically (1). It will remain in operation until the waste falls below the control beam. Compacted waste is forced through a discharge tube into a 5 metre long plastic sleeve (2), and, as the sleeve is filled it

When the sleeve is full it is divided into separate bags by pulling open the pleats

progressively extends along a roller conveyor (3). On completion of the filling operation the Acrowpactor automatically switches off and signals for attention. The sleeve can then be easily sub-divided into 10 individual bags of waste by the attendant (4). With the fitting of a new sleeve Acrowpactor is again ready for operation.

The end of each bag is then tied off and the bags cut apart.

Figure 27(a)

A transport stores plays an important part in vehicle maintenance and careful investigation is needed so that minimum stocks are carried. The layout with portable bins and suitable racks, so that spares can be quickly handled and issued, requires thought; it should also be decided whether special tools should be kept for use and issued only by careful arrangement. Tyres need a cool, clean area, away from strong sunlight, dampness and contact with oil; similarly special conditions are required for the storage of diesel or gas-oil.

To round off the brief survey of maintenance, documentation is necessary for costing and statistical records, such as vehicle hours worked, mileage, weight carried, fuel and oil issues, number of gullies cleaned, mileage of road swept, brush wear, and changes. Drivers should be properly trained to fill in daily log sheets, record defects, fuel consumption, hours worked, rest periods, loads carried, etc., to complete the whole cycle of operation and maintenance details. With workshop co-operation the life of a vehicle, its work performance, costs, availability, and so forth can be properly documented for the guidance of all concerned.

RESOURCE RECOVERY

It appears to be good sense to recycle scarce and non-renewable resources, but for a national strategy to be evolved a major political and economic decision will need to be formulated. If domestic and commercial wastes are

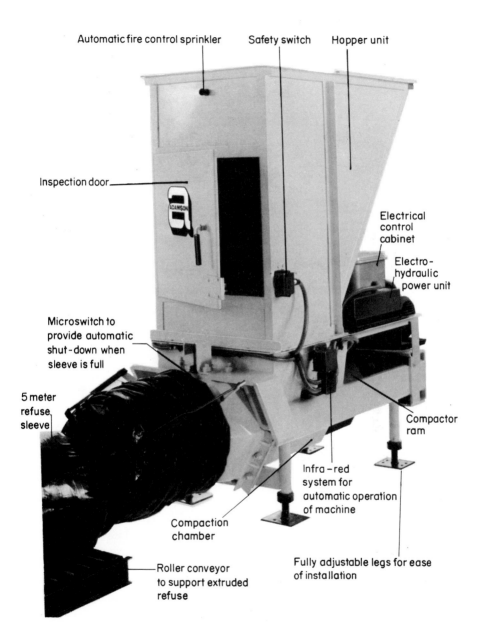

Automatic fire control sprinkler   Safety switch   Hopper unit

Inspection door

Electrical control cabinet

Electro-hydraulic power unit

Microswitch to provide automatic shut-down when sleeve is full

5 meter refuse sleeve

Compactor ram

Infra-red system for automatic operation of machine

Compaction chamber

Roller conveyor to support extruded refuse

Fully adjustable legs for ease of installation

Figure 27(b)

considered integrally some 23 million tonnes per annum are available from which usable components can be extracted for re-use in industry. The steeply rising costs of refuse collection and disposal emphasize the potentialities of recovery to reduce the burden of expenditure on these vital and continuing services.

In plain economic terms any cuts in imports of raw materials by self-sufficiency is a contribution to a healthy balance of payments and may be a significant factor in energy savings. Unfortunately most of the useful constituents of refuse, when recovered and marketed, are subject to competition from international trading and must be competitive in price and consistent in quality.

The refuse collection service is best geared to embark on resource recovery as they collect from domestic and commercial premises on a weekly basis and can easily transport salvaged items to a central recovery plant for processing.

At the present time industry is passing through a great recession, making it difficult to find markets even for a reduced tonnage of secondary materials.

The conversion of waste to fibre recovery or waste-derived fuels are economically feasible but only if given the correct national support and financial incentives.

Already waste oil is being recovered at civic amenity sites and should intensify as fuel costs rise significantly. Glass recovery is increasing, and bottle banks situated in convenient car parks and public places encourage householders to place their glass bottles and jars in the specially designed containers. The levels in which cullet is being utilized are increasing with considerable growth planned at glass treatment plants in various parts of the country. In well-run schemes a profit to the local authority is quite possible, apart from savings not only on disposal costs but in energy usage.

The recovery of food cans for tinplate and mild steel is unlikely from the normal refuse collection service although 850,000 tonnes per annum is found in the domestic waste stream. It can best be achieved centrally at transfer stations, recovery plants or landfill sites by magnetic extraction.

The greatest opportunity for recovery from domestic premises is waste paper collected by means of a trailer attached to the collecting vehicle. Much opposition from employees in the use of trailers has resulted in a number of authorities conducting an entirely separate collection service of waste paper, although costs are higher. Management is failing in its duties to the public if the best system of waste paper collection is by trailer attachment to collection vehicles and is prevented by unwarranted opposition. The use of trailers can assist in recovery of textiles from domestic premises and even collection of non-ferrous is a useful adjunct. If resource recovery in the future is to continue, district councils must consider attaching trailers to refuse vehicles, operate suitable bonus incentives, and encourage full co-operation from the public and employees concerned. When economic circumstances permit separate salvage schemes can be considered. They could be self-supporting and become an additional source of local employment.

REFERENCE

1. BS 5906 1980. Code of Practice for storage and on-site treatment of solid waste from buildings.

Standards applying to waste storage and collection

BS 792 Mild steel dustbins
BS 1136 Mild steel refuse storage containers
BS 1577 Mild steel refuse or food waste containers
BS 1703 Specification for refuse chutes and hoppers
BS 4998 Moulded plastic dustbins
BS 5832 Specification for compacted waste containers for lift-off vehicles

Appendix 1 National analyses of household wastes (Sources: Institute of Solid Wastes Management and Department of the Environment)

Classification		1935	1963	1967	1968	1969	1970	1972	1973 Jan.	1973 April	1973 July	1973 Oct.	1973 Average	1974	1975	1976	1977 Jan.	1977 April	1977 July	1977 Oct.	1977 Average	1978	1979
Screenings (2 cm) or dust and cinders	%	56.9	38.8	31.0	21.9	17.2	14.9	19.9	17.7	22.8	14.0	19.8	18.7	19.8	18	18	21	15	10	11	14	11	12
	lb	21.4	12.0	8.8	6.4	4.8	4.4	5.1	4.5	6.1	3.4	5.1	4.9										
	kg	9.7	5.5	4.0	2.9	2.2	2.0	2.3	2.0	2.8	1.6	2.3	2.2	2.1	2.1	1.8	2.1	1.6	0.9	1.2	1.4	1.2	1.37
Vegetable and putrescible	%	13.7	14.1	15.5	17.6	19.5	24.5	19.5	17.8	17.1	19.5	18.2	18.1	21.3	20	19	21	22	28	27	25	29	24
	lb	5.1	4.4	4.4	5.1	5.5	7.3	5.0	4.5	4.5	4.8	4.7	4.6										
	kg	2.3	2.0	2.0	2.3	2.5	3.3	2.3	2.1	2.1	2.2	2.1	2.1	2.3	2.4	2.0	2.1	2.3	2.7	2.8	2.5	3.2	2.59
Paper and board	%	14.3	23.0	29.4	36.9	37.9	36.8	30.5	31.0	32.3	34.2	32.8	32.7	26.8	30	24	26	26	26	26	26	27	29
	lb	5.4	7.1	8.4	10.8	10.7	11.0	7.9	7.9	8.6	8.4	8.5	8.4										
	kg	2.5	3.2	3.8	4.9	4.9	5.0	3.6	3.6	3.9	3.8	3.8	3.8	2.8	3.4	2.4	2.7	2.7	2.6	2.6	2.7	3.0	3.21
Metals	%	4.0	8.0	8.0	8.9	9.7	9.2	8.7	10.6	7.8	8.1	9.1	8.8	8.5	8	8	8	9	9	8	9	7	8
	lb	1.5	2.5	2.3	2.6	2.7	2.8	2.3	2.7	2.0	2.0	2.3	2.2										
	kg	0.7	1.1	1.0	1.2	1.2	1.3	1.0	1.2	0.9	0.9	1.1	1.0	0.9	0.8	0.9	0.9	0.9	0.9	0.9	0.8	0.88	—
Textiles and man-made fibres	%	1.9	2.6	2.1	2.4	2.3	2.6	3.0	3.2	3.1	3.2	3.0	3.1	3.5	3	4	3	3	3	3	3	4	4
	lb	0.7	0.8	0.6	0.7	0.6	0.8	0.8	0.8	0.8	0.8	0.8	0.9										
	kg	0.3	0.4	0.3	0.3	0.3	0.3	0.4	0.4	0.4	0.4	0.4	0.4	0.4	0.3	0.4	0.3	0.4	0.3	0.4	0.3	0.4	0.47
Glass	%	3.4	8.6	8.1	9.1	10.5	9.0	10.4	10.8	9.7	10.7	10.7	10.5	9.5	9	9	10	11	12	10	11	9	10
	lb	1.2	2.7	2.3	2.7	2.9	2.7	2.7	2.7	2.6	2.7	2.8	2.6										
	kg	0.5	1.2	1.1	1.2	1.3	1.2	1.2	1.2	1.2	1.2	1.3	1.2	1.0	1.1	0.9	1.1	1.2	1.2	1.0	1.1	1.0	1.13
Plastics	%			1.2	1.1	1.4	1.4	1.9	1.8	2.0	2.3	2.0	2.0	2.9	4	5	4	5	6	6	5	5	7
	lb			0.3	0.3	0.4	0.4	0.5	0.5	0.5	0.6	0.5	0.4										
	kg			0.2	0.2	0.2	0.2	0.2	0.2	0.2	0.3	0.2	0.2	0.3	0.5	0.5	0.4	0.5	0.6	0.6	0.5	0.6	0.76
Unclassified	%	5.8	4.9	4.7	2.1	1.5	1.6	6.1	7.1	5.2	8.0	4.4	6.1	6.9	8	14	7	9	5	8	7	6	6
	lb	2.2	1.5	1.3	0.6	0.4	0.5	1.6	1.8	1.4	2.0	1.1	1.5										
	kg	1.0	0.7	0.6	0.3	0.2	0.2	0.7	0.8	0.6	0.9	0.5	0.7	0.7	0.9	1.4	0.7	0.9	0.5	0.8	0.7	0.7	0.62
Total/premises/week	lb	37.5	31.0	28.4	29.2	28.0	29.9	25.9	25.4	26.5	24.7	25.8	25.5										
	kg	17.0	14.1	13.0	13.3	12.8	13.5	11.7	11.5	12.1	11.3	11.7	11.6	10.7	11.6	10.2	10.3	10.4	9.6	10.2	10.1	10.9	11.03
Density	cwt/cu wt	4.37	3.00	2.42	2.37	2.15	2.2	2.31	2.46	2.4	2.2	2.3	2.28										
	kg/m³	290	200	160	157	143	146.2	153.4	164	159.6	147.2	153.1	151.5	161	164	152	113	137	129	124	126	141.2	141.02

Note: in 1974, and subsequent years, analyses were carried out in October, except where stated otherwise.

Appendix 2: Safety Standards (City of Bradford Metropolitan Council; Work Specification for Refuse Collection, April 1974)

1. No refuse collector shall climb walls, fences etc., or walk across private gardens to effect short cuts.
2. Refuse collectors to wear gloves to avoid personal injury.
3. No refuse collector shall place dustbins awaiting collection so as to obstruct pedestrians whether on public or private thoroughfare.
4. Refuse collectors shall not ride on any part of vehicle except in crew cabs.
5. Refuse collectors shall take all reasonable steps to avoid lifting dustbins, refuse etc., in such a quantity as might impede their walk and cause personal injury.
6. No refuse collector shall compress refuse in the rear hopper of a compression vehicle with his hands or feet.
7. Drivers of refuse collection vehicles in their journeys to disposal points should observe all speed limits, statutory transport requirements, and rules of the road.
8. Drivers of refuse collection vehicles disposing of refuse at disposal points should keep to the speed limits laid down in traversing the sites.
9. Any other safety measures such as from time to time may be necessary depending on changing circumstances shall be observed by all members of the team.

Appendix 3:   An outline of refuse collection performance in England and Wales—1978/79

Selected extracts by courtesy of the Chartered Institute of Public Finance and Accountancy *Refuse Collection Statistics, 1978/79.*

Table 5   Costs of collection of waste, 1980–81

Class of Authority	1980/81 Gross cost per Tonne	1980/81 Net cost per Tonne	1979/80 Net cost per Tonne	1980/81 Net cost per Domestic Hereditament
London	35	31	28	28
Metropolitan	29	27	24	22
Non-Metropolitan				
—England	26	24	20	19
—Wales	17	17	16	20
All	27	25	22	21

Table 6   Waste collected per Head of Population

Class of Authority	Kg per Head of Population 1980/81	Kg per Head of Population 1979/80
London	345	336
Metropolitan	309	307
Non-Metropolitan		
—England	305	313
—Wales	471	437
All	321	322

Table 7   Waste collected per Vehicle and Operative

Class of Authority	Tonnes collected per Vehicle	Tonnes collected per Operative
London	1,330	415
Metropolitan	1,400	419
Non-Metropolitan		
—England	1,465	473
—Wales	2,006	746
All	1,460	463

Table 8   Other ratios

Class of Authority	Operatives per vehicle	Ratio of Vehicles & Vehicle Mtnce staff	Ratio of operatives & Tech & Admin.
London	3.2	4.4	13.2
Metropolitan	3.3	5.5	14.1
Non-Metropolitan			
—England	3.1	6.4	15.3
—Wales	2.7	6.1	12.0
All	3.2	5.8	14.4

Appendix 4.1: Sample typical analysis of refuse collection service

1. *Base data*
   Population   220,000
   Domestic premises   80,000

2. *Collection rounds*
   Refuse collection rounds 14
   Bulk container collection rounds 6
   Special collection crews 3
   Spare labour provision overall 10%

3. *Crew sizes*
   Domestic refuse crew Driver + 4 to Driver + 2
   Town centre refuse crew Driver + 2
   Special refuse collection Driver + 1
   Bulk container crew Driver + 1

4. *Performance*
   Domestic refuse collection crew 7000 bins/week
   Domestic refuse collection crew 9000 sacks/week
   Each household 1–1.2 bins/week
   Each household 1.5 sacks/week
   Bulk container crew 150 containers/day
   Special collection crew 20 jobs/day
   Skip vehicle performance 12 skips/day
   Rear-end loaders 2–3 trips per day

Note: These figures have been considerably improved by recent (1982) commercial refuse collection contracts. See Appendices 4.2 and 4.3.

Appendix 4.2: Some modern refuse collection performance norms

System definition	System container	Property performance collection times			Crew composition		Assumed effective collecting time 6.5 h/day (min)	Mean properties vehicle/day	Mean properties loader/day	Mean population served/vehicle (2.25 p/house) (1.2 sacks/house)	Mean payload per vehicle per trip (tonnes)	Mean payload per vehicle per day (2 trips) (tonnes)
		Terraces (min)	Detached (min)	Mean (min)	Driver	Loaders						
Front door	Sack	0.43	0.575	0.5	1	3	390	2340	780	26,325	6.5	13.0
Back door	Sack	0.64	0.86	0.75	1	3	390	1560	520	17,550	6.5	13.0
Front door	Bin	0.64	0.86	0.75	1	3	390	1560	520	17,550	6.5	13.0
Back door	Bin	0.80	1.15	1.0	1	3	390	1170	390	13,163	6.5	13.0

Note: Some recent performance norms suggest three full loads per day for a refuse vehicle equalling about 19.5 tonnes in special cases. 2.5 full loads per day is a more realistic norm.

Appendix 4.3: Some modern bulk bin collection rounds performance norms

Area	Bulk bins handled per day			Average bins per vehicle load	Average contents per bulk bin (kg)	Average payload per trip (tonnes)	Specific loading time per bin (min)		
	High	Medium	Low				High	Medium	Low
Metropolitan	140	130	120	50	100	6.50	2.78	3.00	3.25
County town	120	110	100	50	100	5.50	3.25	3.55	3.90

Practical Waste Management
Edited by J.R. Holmes
©1983 John Wiley & Sons Ltd

4

Street cleansing and litter control

A. E. Higginson, MBE, FIWM, FCIT, MIEnvSc, MRSH
Consultant, Solid Wastes Management and Environmental Services

ABSTRACT

Street cleansing and litter control, one of the key yardsticks by
which the public judge the efficiency of cleansing and sanitation
officers, is set down in this contribution to the book. The nature of
street debris, the legal background in the United Kingdom, and the
variety of services to be performed are described. Gully emptying,
street sweeping and washing, the type of mechanical and manual
tools and equipment used, and the organization of the service are
described. The control of litter and the contributions that can be
made by voluntary organizations are suggested.

STREET CLEANSING

Background to the services

Street cleansing services differ in standards of frequency in town, village,
urban, or rural surroundings dependent upon the amount of dirt, debris or
litter deposited or the intensity of traffic upon the highways serving the
respective communities.

Over a long passage of time sweeping and watering of streets has been
undertaken by local authorities as a public health service for the suppression
of nuisances that might arise from the deposit of any kind of rubbish upon any
street, highway, or passage. By-laws for this purpose were made under
various acts relating to the cleansing of footways and pavements, and the
prevention of nuisances arising from 'snow, filth, dust, ashes, and rubbish'.
Even later legislation such as the Public Health Act 1936 (S.76) specifically
referred to local authorities to provide receptacles for refuse in streets and
public places. This included litter bins, sand, and salt containers for which
planning consent was not required provided any such installation did not

131

cause an obstruction or become a nuisance. S.77 refers to sweeping and watering of streets, and states that the local authority may undertake cleansing and watering but the Minister can require cleansing although authorities are not obliged to carry out watering. They can cleanse private roads without prejudice to ownership but cleansing does not include gritting, maintenance, or gully cleansing work. S.343 of the Public Health Act 1936 defines a street to include all highways, road, lane, footpath, square, court, alley, or passage whether a thoroughfare or not.

Modern legislation (Highways Acts) makes specific references to offences affecting safety, obstruction, or nuisance arising from snow, soil, refuse, or anything which is a nuisance upon the highways and indicates the manner in which the offences are dealt with.

The Control of Pollution Act 1974 S.22(1) states: 'It shall be the duty of each highway authority to undertake the cleansing of the highways for which it is the highway authority so far as the cleansing of the highways is necessary for the maintenance of the highways or the safety of traffic on them'. (2) It shall be the duty of each local authority to undertake cleansing of the highways in its area so far as the cleansing of them appears to the authority to be necessary in the interests of public health or the amenities of the area: but that duty shall not include a duty to undertake cleaning of any special road which is a trunk road or any other cleaning falling to be done by a highway authority in pursuance of the preceding subsection.

It would be considered proper in the context of legislative aspects of street cleansing to mention litter and include the effect of S.24 COP Act when implemented:

(1) It shall be the duty of the council of each county in England and Wales and the local authorities of which the areas are included in the county, and, where the county includes land in a National Park, the Park authority, to consult from time to time together, and with such voluntary bodies as the council and the authorities consider appropriate and as agree to participate in the consultations, about the steps which the council and each of the authorities and bodies is to take for the purpose of abating litter in the county: and it shall be the duty of the county council—

    (a) to prepare and from time to time revise a statement of the steps which the council and each of the authorities and bodies agree to take for that purpose; and

    (b) to take such steps as in its opinion will give adequate publicity in the county to the statement; and

    (c) to keep a copy of the statement available at its principal office for inspection by the public free of charge at all reasonable hours.

Although S.24 applies specifically to England and Wales similar legislation in Scotland prescribes the duties for the council of each region and district councils regarding litter. It is appropriate to mention litter because it has been described as any rubbish which is in the wrong place.

The highway authority, normally the County Council, has the duty for providing safety measures for traffic upon the highway, including public footpaths, but argument often results between highway authorities and district councils as to frequency of cleansing for safety, public health, and amenity.

Looking at the legislation on street cleansing during the last century the emphasis on public health aspects remains unaffected, the sharp difference being safety and visual amenity that are now focused in public attitudes, but litter has become a subject in its own right.

Characteristics of street debris

Street debris has changed in character over the centuries due to social habits, amount of traffic and type, road surface developments, and levels of service given. When public cleansing services were not universally provided the street became the chief depository for all kinds of filth and rubbish. Improvements to refuse storage and frequency of collection lessened the burden of rubbish dumped upon the streets. Better road surfaces alleviated the abrasion of cobbled and waterbound macadamized thoroughfares and ousting of steel-rimmed wheels by the introduction of pneumatic tyres dramatically altered the sources of street debris. Dust, which is still a principal constituent of street sweeping, has declined in quantity as road surfaces with smoother surface finishes have improved, although other sources of debris such as sweet wrappings, packaging, paper, and containers of all descriptions have increased.

The decline of horse-drawn vehicles is nostalgically publicized but early in the present century the horse was the main contributor to street dirt and offensive smells, as each day it deposited ½ gallon (2 litres) of urine and up to 22 lb of dung upon the streets. Today the dog, with its excrement, befouls footways and verges to make the street objectionable and unhygienic.

Much of the rubbish found upon the streets today is avoidable even though legislation is available to punish blatant offenders. An analysis of modern street debris comprises dust from a variety of causes, some airborne and others from mechanical abrasion of road surfaces, vehicles, and tyres; fallen leaves, twigs, and branches from trees and shrubs; unwanted wrappings and food from 'take-away' shops eaten in streets and discarded with remnants of food; drink containers, metal, glass, or plastic; fruit peel, skins, and ice cream cartons. This is not a complete list as other contributory factors arise from badly loaded vehicles shedding all kinds of debris, particularly demolition wastes, sand, soil drawn out from sites on vehicles' tyres forming mud and dust upon road surfaces, and litter which embraces all kinds of waste to include large bulky items of varied descriptions.

For such a wide variety of discarded rubbish the street cleansing service has to be adapted to improve the appearance of town, village, or countryside; to make it safe for travelling and remove any hazard to public health or danger to the environment.

Street cleansing methods

Basically three methods are used in operating street cleansing services, they divide into manual sweeping (a) beat system, (b) gang or team system, and (c) mechanical sweeping. Combinations of methods are sometimes used, as for example in busy shopping areas a team will sweep debris from footways into the path of a mechanical sweeper operating in the roadway in order to quickly vacate the area before traffic builds up or parking forbids proper cleansing.

By studying each method the advantages and disadvantages can be detected so that in selected areas all systems may be applied to meet the diverse problems or peculiarities of the district.

(a) Beat system

In this system an area is allocated and the street orderly is provided with the necessary tools comprising shovel, channel path, or platform broom and street orderly truck or barrow. As there is a limited amount of street sweeping a man (or woman) can undertake satisfactorily, he is based at a sub-depot to which he can return to deposit sweepings during the day, unless serviced by a parent pick-up vehicle at determined times. The daily route will be planned by his supervisory staff and the main advantage in this system is the orderly knows his area, can take pride in its cleanliness, and give a regular identifiable service to the local residents. He can quickly remove broken glass which is a danger on the highway, or if a vehicle has shed debris he can effect immediate clearance. Weeds at the back of the path can be controlled and in wet weather he can unblock covered gully grids if leaves are a problem. The best areas to employ the beat system are high pedestrian traffic zones, shopping centres, seaside promenades, or where heavy litter occurs. The main disadvantage in the system is the limited area to sweep; street orderly barrows lack capacity, they are difficult to push if full, there is lost time in waiting for a pick-up service vehicle, and they are unsuitable for hilly districts.

(b) Team sweeping

The introduction of electric vehicles into the street cleansing service has tended to reduce the number of street orderlies employed. This is because greater output per man is achieved as the time lost by walking back to the orderly truck is eliminated. There are pedestrian electric vehicles (p.e.v.) and the ride-on types which require careful appraisal for ideal team size. The p.e.v.'s performance is governed by the walking speed of the operator in charge of the truck. The normal practice is for a team of three: one orderly for each side of the road sweeping path and channel, and the operator picking up the sweepings by criss-crossing of the road. Studies into productivity indicate that the maximum effective team size is three; even so the area of sweeping compares to that of four or five street orderlies working the beat system. Also instead of a pick-up vehicle being required to collect sweepings the p.e.v. can carry the sweepings for a full day's team effort to the depot or disposal point.

The advantage of the ride-on electric truck is that larger team sizes can be employed, up to five being a common figure. As the driver/operator picks up the sweepings and drives to where the heaps are formed work can be organized to maximize the truck's capacity. The four orderlies can be divided into two teams operating in different streets sweeping footpaths and channels and forming sweepings into convenient heaps, leaving the driver to pick up the sweepings with no lost time. The whole team can ride out in the morning at reasonable speed at the start of the day's work and commence sweeping promptly, making the system flexible, very mobile, and efficient.

(c) Mechanical and suction sweeping

The mechanical sweeper was developed from the rotary brooms which swept the road debris into the channel from its angled sweep, the debris being cleared by a street orderly. Later developments incorporated a collecting mechanism which swept the refuse into the channel and mechanically lifted the debris into the body of the vehicle.

Owing to the rapid increase in road traffic it was found that all vehicles tended to move any rubbish on the highway into the channel, making the wide sweep of the mechanical sweeper collector unnecessary. This fundamental change heralded the present range of road suction sweepers able to use high-pressure water sprays to control dust conditions as vigorous channel brushing takes place and high-velocity air to suck road detritus into the body of the machine. The versatility of these machines is demonstrated when large deposits of chippings are lodged in channels or excessive accumulations of autumn leaves are quickly collected and disposed of, relieving manual employees of extremely arduous work. A wandering hose mounted on a turntable with a working circular radius of about 3.35 m can be fitted and used

JOHNSTON 400 SERIE

The Johnston 400 Series unit is a significant advance in road suction sweeper design. It combines our world-wide operating experience gained from the production of over 5000 of the highly successful 200 Series, with the latest developments in engineering technology.

Every aspect of the road suction sweeper has been re-examined in detail and the 400 Series incorporates many totally new features. These ensure optimum performance and reliability with minimal maintainance needs for global operation.

The cab mounted Master unit gives finger-tip control of all sweeping functions and the auxiliary engine start system.

General view showing body in discharge position, channel brush/nozzle assembly, auxiliary engine compartment and wandering hose equipment.

Figure 1 Technical aspects of a modern street sweeping vehicle. (Courtesy of Johnston Brothers Ltd)

Auxiliary Engine/Fan System cowling gives complete protection. h level ducts control cooling for operating up to 52°C ambient iperature. Sound insulation ensures minimum noise levels. ivy duty air cleaner.

Channel Brush/Suction Nozzle system incorporates variable speed steel tine channel brush. Suction nozzle is rubber lined. All sweeping functions are cab controlled.

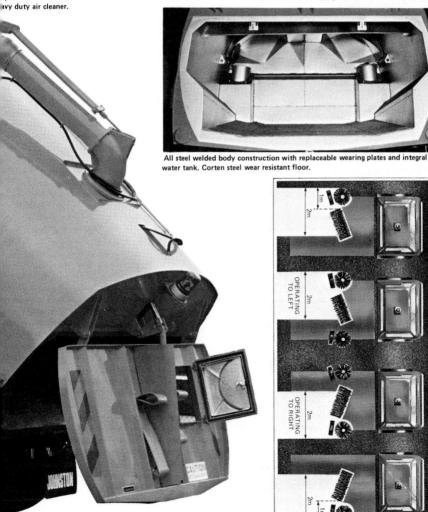

All steel welded body construction with replaceable wearing plates and integral water tank. Corten steel wear resistant floor.

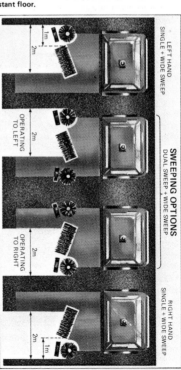

LEFT HAND
SINGLE + WIDE SWEEP

OPERATING
TO LEFT

SWEEPING OPTIONS
DUAL SWEEP + WIDE SWEEP

OPERATING
TO RIGHT

RIGHT HAND
SINGLE + WIDE SWEEP

Figure 2   Modern street sweeping vehicles

although such work requires the services of another man to operate the equipment in addition to the driver.

The large suction sweeper operates best on unobstructed trunk and principal routes but is handicapped where parked vehicles or narrow streets dominate owing to limitations of manoeuvrability. Thus developments into smaller types of machines have progressed rapidly in recent years. Three-wheeled versions have been introduced; two rear driving wheels and one front steering wheel to give improved sweeping on small-radius turns. Fitting three hydraulically driven brushes allows a choice of sweeping combinations to meet different road conditions.

Designs of modern shopping complexes and busy road intersections have introduced underpasses, pedestrian subways, and shopping precincts for which smaller compact yet robust machines could operate. To mechanically operate a sweeper under these exacting conditions has needed specially designed machines to work in confined areas, able to manoeuvre close to obstructions, clear debris around street furniture, and to mount kerbs where required without damaging equipment.

As a further development the pavement sweeper was introduced to mechanize footpath cleansing by pedestrian-controlled suction sweepers. These are effective on various surfaces and have a gross vehicle weight

Figure 3 Detailed view of rotating road brushes and suction assembly on street sweeping vehicle. (Courtesy of Johnston Brothers Ltd)

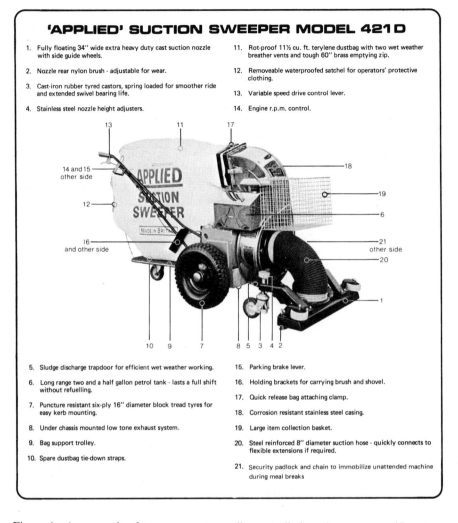

'APPLIED' SUCTION SWEEPER MODEL 421D

1. Fully floating 34" wide extra heavy duty cast suction nozzle with side guide wheels.

2. Nozzle rear nylon brush - adjustable for wear.

3. Cast-iron rubber tyred castors, spring loaded for smoother ride and extended swivel bearing life.

4. Stainless steel nozzle height adjusters.

11. Rot-proof 11½ cu. ft. terylene dustbag with two wet weather breather vents and tough 60" brass emptying zip.

12. Removeable waterproofed satchel for operators' protective clothing.

13. Variable speed drive control lever.

14. Engine r.p.m. control.

5. Sludge discharge trapdoor for efficient wet weather working.

6. Long range two and a half gallon petrol tank - lasts a full shift without refuelling.

7. Puncture resistant six-ply 16" diameter block tread tyres for easy kerb mounting.

8. Under chassis mounted low tone exhaust system.

9. Bag support trolley.

10. Spare dustbag tie-down straps.

15. Parking brake lever.

16. Holding brackets for carrying brush and shovel.

17. Quick release bag attaching clamp.

18. Corrosion resistant stainless steel casing.

19. Large item collection basket.

20. Steel reinforced 8" diameter suction hose - quickly connects to flexible extensions if required.

21. Security padlock and chain to immobilize unattended machine during meal breaks

Figure 4   An example of a pavement manually controlled suction sweeper. (Courtesy of Applied Cleaning Economics Ltd)

for accumulations of litter or leaves which the vehicle is unable to reach, unlikely to cause problems under the Vehicles (Conditions of Use on Footpaths) Amendment 1966.

In street cleansing organization the cleaning of footpaths presents the most difficult problems. Until 1961 local authorities (outside London) had no legal powers to use mechanically propelled machines on footways. A major step forward to mechanization was in the Public Health Act 1961 (S.49) which authorized mechanical footway machines for cleansing and certain other purposes. The Ministry of Transport could impose regulations prescribing conditions under which the rights conferred could be exercised. Regulations were introduced in 1963 and made provisions relating to construction, maximum weight, maximum speed, and the hours to be worked. These regulations were further amended in 1966.

As no developments specifically for local authority work had been carried out prior to legal agreement the machines offered were those engaged in private industry cleaning factory floors, workshop areas, and commercial buildings. Exacting conditions caused by irregular surfaces of footways, uneven building lines, street furniture, kerbs, crossings, and pedestrian rails all presented problems of operation with the industrial-type sweepers as further modifications were required for municipal work.

Perhaps the limitations still felt on the use of pavement sweepers is the action that can be taken under common law if damage or injury occurs through negligence in the use of mechanical equipment plus the problems of operating where pedestrians predominate.

Gully cleansing

Grit and dust which builds up from a variety of causes is constantly moved into the channels as fast-moving modern traffic clears it from the centre of the roads. Rain which falls is drained away from all paved areas to gravitate via street gullies to pass into surface water or combined sewers.

The gullies are made of salt-glazed ware, iron, brickwork, or latterly pre-cast concrete units but it is preferable they should be manufactured to a BSS. They serve two main functions:

(a) To prevent oversized solid matter located in the channels from passing into the main sewer where blockage may occur. This is done by placing heavy-duty grids over the opening, trapping large items, such as leaves, small wrappings, or cigarette cartons but allowing the rain water to enter the drainage system unimpeded;

(b) By being so designed that a water seal prevents any offensive gas from the sewer escaping to the atmosphere.

Although these precautions are taken, occasionally the gully becomes blocked with detritus and it is necessary to clear the obstructions for the gully to function properly, therefore a rodding eye is fitted to effect clearance without digging up the road.

Regular cleansing is essential for gullies to perform their role as part of the street drainage system as rapid run-off of heavy rain is important for traffic safety and to prevent flooding of premises or to avoid splashing of muddy water on to passing pedestrians. As part of regular cleaning the design of the

Figure 5   A modern pavement suction sweeper with carrying tray for bulky litter.
(Courtesy of Applied Cleaning Economics Ltd)

gully must conform to prime features, smooth sides and finishes, concave bottom, adequate capacity for detritus without blocking the outlet, well-fitting heavy duty cast iron grids, and ease of operation.

The dust, grit, and other solid matter accumulates at the base of the gully and must be removed at determined intervals. Based on experience a 3-monthly interval appears satisfactory, though many gullies may have to receive more regular attention, e.g. those at bottom of hills or inclines, which tend to silt up quicker; gullies near to construction sites where mud, sand, or soil is being washed in; or gullies which are spaced so far apart they carry a heavy discharge.

The organization of gully cleansing is conducted by employing very efficient vacuum tank machines which can suck up the detritus including beer cans and pieces of brick, and other heavy masonry, usually by single action with a powerful exhauster compressor. The advantage of modern gully machines is that they can incorporate a hydraulic grid lifter. This reduces the fatigue of the operator brought on by the weight, and usually tight fittings of the frame.

The number of gullies cleansed per day will vary owing to the position and difference in size, but assuming standard patterns a figure of 80–100 gullies per day for a machine operated by a driver and mate is reasonable. It is sometimes necessary to clean the odd cesspools from unsewered parts of the district, and a gully machine can be adapted to perform this task efficiently by modifications to the vehicle including tool racks to carry hose lengths up to 100 ft (30 m) or more to reach and cleanse the cesspool. The versatility of a gully machine with suitable modifications enables street watering, channel flushing, or market cleansing to be carried out by fitting with interconnecting valves and special jets. These special jets include ducks'-feet sprays for washing the road surface, fan jets for footpaths, and channel jets for channel cleansing. High-pressure sewer jetting is carried out by modern gully emptiers, saving the use of a separate vehicle normally used for this purpose.

Organization of street cleansing

The organization of street cleansing varies from district to district dependent upon the priorities given to frequency and degree of mechanization. The advantages of mechanical or suction sweeping, which originally reduced the number of street orderlies, now poses problems of operating with traffic density by increased number of vehicles; parking of vehicles for excessive periods, and restrictions imposed by shopping precincts, dual carriageways, and one-way streets. All these have tended to alter conditions. A combination of manual and mechanical sweeping with round-the-clock operations has been adopted in many places to deal with the increasing problems in the denser-trafficked urban areas.

Street cleansing in rural areas is simpler; long stretches of trunk and main traffic routes can be mechanically swept without interference, giving a lower unit cost per mile cleansed because of less debris, unrestricted sweeping speeds, and practically no parking problems. Mobile teams can cleanse lay-bys and concentrate on villages or urban areas, the street orderly being allocated to specific shopping districts of main centres of population.

The major towns have found in recent years that street washing formerly carried out during early mornings cannot now be satisfactorily performed. Work is severely hampered by parked vehicles resting firmly by the wheels in the channels. Many suburban streets suffer the same fate, making street washing or channel damping impossible; likewise the employment of mechanical or suction sweeping. Although attempts have been made to legislate for vehicles to be removed prior to street cleansing it appears that no authority has successfully implemented such legislaton. The Control of Pollution Act 1974 S.22 instructs highway authorities to undertake cleansing of their highways to keep them safe for traffic and S.24 to keep them clean in the interests of public health or the amenities of the area; S.23 refers to the prohibition of parking for street cleansing purposes. The regulations to be issued in this connection will be studied by highway authorities with great

Figure 6   A selection of street waste and litter plastic containers exported to a Middle
East city. (Courtesy of Glasdon Ltd)

interest, and the public's reaction to an incursion into their loved relationship with their motor car and its street resting place will also be of interest. It is obvious that street cleansing organization suffers because of an insurmountable problem; that of the parking of vehicles and the blocking of channels and gullies.

The speed and often insecurity of loaded vehicles has provided another problem of sweeping upon motorways as they leave a trail of dangerous debris in their wake. County Councils as agents have referred to the excessive quantities of debris swept up on motorways and the miscellaneous items collected. Novel arrangements for picking up dangerous items have necessitated on occasions police protection against the hazardous task involved. Such illegal behaviour of insecure loading of vehicles has resulted in high expenditure for preventive measures in the interests of road safety and protection of human life or injury. Shattered windscreens from stones flung up from motorway surfaces are frequent causes of accidents, requiring constant patrol to inspect and remove offending items.

Basically the organization of street cleansing will depend upon the type of town, urban area, or rural district and means selecting the best combination to meet safety and environmental needs. In dense city or shopping areas the street orderly with barrow and tools may be the most effective method of collecting debris and litter. The frequency of cleansing will depend upon the community's attitude to clean streets and an abhorrence of dirt and untidyness. The judicious use of mechanical and suction sweeping for main thoroughfares, one-way streets and trunk roads working under the best traffic conditions, be it night or day, is fundamental in obtaining the maximum use of expensive equipment. Careful planning of the routes on a daily basis is essential in cost-effectiveness. Team sweeping with varying team sizes and using carefully selected mechanical or electric vehicles dependent upon prevailing conditions will form a good basis for cleansing suburban areas. Teams can be quickly switched for special purposes such as leaf fall in autumn or to the vicinity of football grounds where litter is dropped in abundance.

The type of organization will depend upon many factors and the funds allocated to the service, but the range of equipment available should give any community a basic cleansing service able to develop a civic pride in providing clean streets.

Night or shift working

In securing the maximum use of expensive equipment night or shift operation for street cleansing is essential in some city centres or main thoroughfares. By this approach more intensive and less restrictive methods for street cleansing can be made, using suction or mechanical sweepers and gully emptiers to their maximum potential. Access to the channels to give uninterrupted sweeping or ease of location for gullies is fundamental in employing expensive machines at enhanced rates for drivers and mates during shift or night work. The presence of one parked vehicle as mechanical sweepers operate reduces the standard of hygiene and visual attractiveness, as a long swathe indicates that a sweeper collector has had to manoeuvre past a stationary vehicle. As most of the trapped dirt and litter is confined to the channels every effort for mechanical sweeping should be undertaken during the period when parking is not a problem.

One-way streets are best swept at night or during the late or early shift working owing to the dislocation to traffic of a slow sweeper operating at speeds varying from 5 to 12 mph. Dual sweep equipment is particularly suitable for one-way streets and dual carriageways, allowing the work to continue along the line of traffic movement.

Shopping precincts which carry heavy pedestrian traffic during the day should be swept at night or during the shift work, so that the operator can sweep uninterruptedly right up to the frontages, furniture, or other obstructions without danger of inconvenience to the public. Gully emptying should be confined to non-residential areas as the noise created by vacuum emptying is a serious source of disturbance.

Figure 7   Litter containers in public parks

Major city centres and streets can only be washed at night and equipment mounted on gully emptying machines with suitably designed and placed jets controlled from the driver's cab can channel-flush or street-wash without impeding traffic. The opportunity to double-shift expensive vehicles is both economic and desirable but the employment of night operations is likely to be confined to the larger municipalities. There are, however, some large cities which might benefit from detailed feasibility studies into night or shift operations to overcome the many problems of traffic density and parking of vehicles experienced in their areas with day work only.

Although only mechanical equipment has been mentioned for street washing the high standards of cleansing performed by hand washing from hose reels attached to suitable hydrants is relevant for night work.

Night or shift operations have a beneficial effect if salting and gritting of frosty roads is a responsibility of the department, as men and vehicles are ready to be switched to this service. Gritting and salting when traffic is clear of roads being treated is obvious, a good spread as a continuous operation, with the materials being laid in the right place, has many advantages in cost and safety.

Street cleansing equipment

The opportunity to select the right equipment covers a wide choice. For many decades manufacturers have investigated particular problems with operation-

Figure 8   Litter containers in garage forecourt—an accent on safety and appearance

al managers and introduced new designs or improved versions of existing appliances, vehicles, and equipment. They even design especially to meet particular local needs. Although a great deal of sophistication may have appeared to develop of recent years it is because of more involved technical requirements. This is manifestly so in gully emptying equipment as dry loads of detritus are required on landfill sites. Therefore compressing the contents to achieve this state means strengthening the rear door and clamping arms and using a powerful pusher plate, thereby altering the design concept to avoid stress and to give a dry load. It also means that vehicles can continue gully emptying for a full day without discharging, enabling maximum payloads and high performance.

The scope for purchasing vacuum sweepers for safe and reliable working for a variety of purposes is wide, as many features can be embodied to suppress dust, remove tenacious dirt, and provide access by a wandering hose to difficult places of access.

Practical waste management

Figure 9   A manual street sweeping 'orderly' trolley widely used in Britain. (Courtesy of Glasdon Ltd)

It is impossible to provide a schedule for this purpose within the confines of a chapter but the Institute of Solid Wastes Managemet provides an exhibition preview and a handbook in association with its Annual Conference covering the largest display of specialist vehicles, appliances, equipment, and protective clothing held for all services connected with waste management. Descriptions and salient features are briefly detailed in the handbook, giving interested parties an opportunity to assess any appliance, vehicle, or piece of equipment in which they may have a particular interest. It covers all the principal manufacturers and suppliers so that the needs of any potential user are well catered for.

LITTER

Although litter appears to be a modern description of 'rubbish in the wrong place' it is highlighted today because legislation introduced to control it specifically refers to litter in the title; e.g. Litter Acts 1958 and 1971. Local authority by-laws also refer to litter in describing an offence committed if a person throws down, drops, or deposits and leaves anything, etc. But the Litter Acts have not satisfactorily curbed the defacement of public places by unlawful depositing of rubbish, owing to the difficulty of enforcement of the

Figure 10 Twin post-mounted litter bins on public footpath

appropriate legislation. The weakness of the Litter Acts is proof of depositing, and as most offenders stealthily, sometimes under cover of darkness, dump unwanted articles without witnesses there is lack of evidence for prosecution. There are many practical difficulties, even if witnesses are present when the offence was committed, for those members of the public to become involved to enable a prosecution case to proceed.

The role of KBTG

As in other developed countries experiencing similar problems litter abatement is being tackled vigorously in the UK by an organization known as the Keep Britain Tidy Group, who act as a national umbrella to promote, among other activities, advertising; public relations; research; education; and providing information and assistance with promotional material to local authorities, schools, and voluntary groups. They also hold seminars, meetings, and workshops to discuss the behavioural approach to litter; educational programmes; involvement of industry and commerce; studies into sources of litter; outlining the role of voluntary organizations; and expanding on public relations and communications. Its purpose is to enhance the amenities of town and countryside in the UK, particularly by promoting the prevention and control of litter and environmental improvement schemes.

Through its important work and organizational skills the KBTG has become the national agency concerned with litter abatement and control, and as such recognized by the Government as the central body on this difficult aspect of social, environmental, and public service.

Figure 11 Post-mounted 25 litre capacity litter bin

Figure 12 Manual street sweeping—a simple but effective system (Courtesy of Glasdon Ltd)

Definition of sources of litter

There is a close relationship in describing the characteristics of street dirt and rubbish which defaces our environment and in defining litter. By international agreement litter is basically 'rubbish in the wrong place', but it includes materials of all kinds varying in size from sweet papers to bulky items such as unwanted furniture. It also includes any accumulation of garden waste, cuttings, grass clippings, twigs, or leaves deposited in a careless or untidy manner on vacant ground or on highway verges. The list continues with broad categorizing of the sources of litter as compiled by KBTG:

mishandled household waste, improperly containerized;
commercial waste;
construction and demolition sites;
sloppy loading and unloading of vehicles;
uncovered vehicles;
pedestrians;
motorists.

Examination of sources found that in the average American city the litter emanating from pedestrians and motorists sources accounted for only 20% of the total, the remainder being attributed to the remaining five categories. One source not mentioned specifically, but which causes widespread concern, is litter thrown overboard from ocean-going vessels and ending up on beaches and shorelines. Plastic bottles, containers, and metal cans found on beaches have been identified by survey teams as not originating principally from beach users but from passing ships. The effect of litter on beaches has a serious environmental impact but where glass is concerned there is a danger of cuts to unprotected feet, especially of young children.

Control of litter

The necessity of control of litter involves many factors: economic, environmental damage, loss of aesthetic amenity values, loss of community pride in surroundings, adverse effect upon wildlife, and accidents to children or motorists.

Millions of pounds are spent annually by local authorities through highway authorities or cleansing departments in removing litter (often dangerous) from roadside verges, derelict sites, streets, parks, beaches, or promenades at very high costs. This is a gross waste of financial and labour resources where their deployment to other urgent social needs is frustrated by unnecessary dumping of unwanted trash.

The public are aware that clean and attractive surroundings, be they in town or country, are quickly damaged by the dropping of litter by the dirty

Figure 13   Public relations and litter prevention: the *Cutty Sark*, Greenwich. (Courtesy of the Keep Britain Tidy Group)

and unsocial behaviour of a few. Visual amenity is immediately impaired with consequent loss of amenity values. The pride felt by people in the neat orderly streets, shopping areas, or roadside verges is quickly reduced to disgust at rubbish scattered indiscriminately to foul previously clean surroundings.

Damage to wildlife occurs through broken glass, jagged metal, unwanted canisters or containers often containing toxic substances deposited upon verges or in roadside hedges. Accidents to children frequently occur through discarding glassware on beaches, and motorists often sustain physical injury apart from expensive damage upon highways as objects such as wood, brick, or bits of hardcore are thown up by fast-travelling vehicles and shatter windscreens. These are the penalties endured by innocent members of the public from the careless, anti-social behaviour of litter scatterers. The control of litter is basically a change in public attitude and a new awareness that a costly, illegal, wasteful, and thoroughly dirty habit must be eradicated.

Equipment and education

To control disfigurement of streets and public places from the effects of unwanted litter requires properly designed and easily used equipment. Before litter receptacles are placed in strategic locations a careful appraisal of prospective sites is necessary to determine their accessibility, capacity, frequency of emptying, appearance, likelihood of vandalism, and usage.

The type of bin or container selected should be of sound workmanship, pleasing in colour and design, and be unmistakably for the reception of litter. It may be that the location of litter bins is subject to restraints or that their placement falls to the responsibility of another Committee of the Council and this feature should be the subject of initial enquiry although Public Health Act 1961 (S.51) empowers all local authorities to provide litter receptacles but County Councils may provide only in rural districts.

The number of bins, size, and shape to be fixed in any road or shopping area will be determined by whether the containers make a positive reduction to littering in the particular area selected. Expenditure can be caused in putting litter bins in what appears casually to be the right place, only find the litter is found some distance away. This is particularly noticeable in the location of take-away food establishments where the unwanted cartons are found away from the shop as people eat the contents walking in different directions. To determine accurately the sources of litter, and where initially dropped, is essential before locations are finally agreed.

As part of the street cleansing work schedules regular emptying of receptacles is important; overflowing bins cause disgust and complaint and militate against the whole concept of litter control. Whether receptacles are placed on lamp columns, on walls, or are free-standing depends upon density of pedestrian traffic, width of footpath, or likelihood of damage to persons colliding with bins in passing. These are problems of site location.

To cut expenditure a programme of sponsorship by business enterprises to provide litter bins should be investigated, or whether advertising should be allowed to offset the cost of placement and servicing. Whatever form of receptacles are provided it is prudent to affix the KBTG Tidyman logo on them, so that the symbol should immediately be recognized as part of the campaign for litter control.

There are other means of controlling litter which are especially aimed at motorists; for example the litter bag kept in the car for cigarette cartons, sweet wrappings, ice-cream containers, and food residues. They are usually issued by business concerns or civic groups and bear some advertising or printed matter upon them. A more frequent issue is desirable instead of one-off, providing evidence of good usage is made; a difficult problem to assess.

As the motorist is frequently cited as being responsible for careless littering some thought should be given to capacious bins at lay-bys, motels, and car

Figure 14   The practical aspects of keeping streets clean. A selection of street orderly barrows. Twin-bin 150 litre units or 4 × 75 litre bin units. Designed for use in shopping precincts, markets, or transport termini, for bulk collection of refuse. Lidded for hygiene purposes. The frame and bins are manufactured in mild steel and polythene respectively, and interstack for minimum freight costs. (Courtesy of Glasdon Ltd)

parks where the motorist can discharge unwanted trash with ease. Regular clearance from these sites is essential, otherwise overflowing bins defeat the efforts of effective litter control.

To abate litter requires more than a supply of neatly designed and frequently emptied receptacles; it means a continuous, orderly educational programme as outlined so manifestly by KBTG. The whole success of litter control rests upon changing the social behaviour of people by making everybody aware that littering of streets and public places is very expensive in money and labour resources; it is unnecessary; and it can be abated by adopting sensible ways of handling all kinds of rubbish be it in the home, street, or countryside.

A comprehensive programme is needed, and the local authority must play a predominant part; its street cleansing service must be adequate to deal with any additional bins or containers placed at strategic sites. To support the local authority the creation and maintenance of relevant information on litter legislation, the cost of removing litter from streets, and other educational aspects is essential. Only by a sustained approach of education, persuasion, or enforcement can the anti-social practice of littering be controlled as a dirty, ugly, potentially dangerous, and illegal habit. The KBTG is consistently investigating every aspect of litter and its control; help from them is readily available, and their new system of getting the facts, involving people, planning, etc., can be the desirable goal to be reached.

MAINTENANCE OF EQUIPMENT

The Public Health Act 1961 not only empowers local authorities to provide litter receptacles, but it specifically requires regular emptying so that the receptacle or its contents do not become a source of complaint or nuisance.

Figure 15   A small suction street sweeper based on the 'Ant' three-wheeled vehicle system. (Courtesy of BTB Engineering Ltd)

Figure 16   A selection of hard-wearing concrete litter containers

Figure 17   A paper sack system of waste storage and collection. (Courtesy of Reed
Medway Waste Handling Systems)

Practical waste management

Figure 18

Figure 19

Figures 18 & 19   An example of a small electric drive side-loading refuse vehicle used for street sweeping and litter prevention services. (Courtesy of Brent London Borough Council)

Over the years competitions for the design of street furniture, including litter bins and containers, has evolved in providing receptacles pleasing in appearance, robust, of sufficient capacity, and giving minimal interference to pedestrians by obstruction. Manufacturers have used their design techniques in utilizing materials that are easily cleaned, able to withstand rough usage, which blend in with the surroundings due to the range of colours, and which are easily emptied.

The maintenance of the equipment requires regular servicing, not only for emptying the contents but checking any damage, looseness of fastenings, replacement of sacks, repainting if necessary, and (where bins have become fouled) washing to remove dirt and odours.

If litter bins have been placed out by contractors regular inspection for damage or defacement is necessary and any bins damaged must be removed and replaced as required. These bins carry advertisements and are maintained by contractors, but a loss of visual amenity occurs by the advertising matter carried. Litter bins are part of street furniture design and British Standards are applied recommending selection and siting of bins whether they are free-standing, pole-mounted, or wall-fixed. An essential requirement is that the receptacle must be easily removed for emptying or be fitted with a loose inner bin, sack, or liner. Litter bins are part of the overall campaign for abatement. Efficient maintenance of equipment is essential to keep litter off the streets. An orderly programme of rapid repairs, repainting, and servicing is the key requirement for prevention.

Legal aspects

The problems of littering are historical, and legal means to prevent it are found in local by-laws and ancient enactments. Modern legislation is found in the Litter Acts of 1958 and 1971, Civic Amenities Act 1967, and in the Refuse Disposal (Amenities) Act 1978 (parts refer to litter and dumping); the Highways Act 1959, Public Health Acts 1936 and 1961, Caravan Sites Act 1968, and Control of Pollution Act 1974.

The whole purpose of the legislation is based upon reasoning that abatement is secured if legal sanctions are provided, but this view is at variance with the problems of obtaining successful prosecutions. There are all kinds of loop-holes from which an offender can escape, e.g. items of dumped rubbish are traced to a particular premises but this is not possible to prove in a court of law. The offender must be seen dumping the rubbish and any witnesses must testify to this effect. The reluctance of witnesses to testify in court is well known and makes enforcement a difficult problem. Enforcement of the law is essential, requiring consultation and co-operation with the police forces; public opinion must be secured, and local authorities need to examine and overhaul where necessary their by-laws on litter and its abatement. As some sources of litter arise from the Council's own operations in refuse collection and disposal a Code of Practice should be implemented defining the requirements for good housekeeping, e.g. lids to remain on bins awaiting collections in streets, firmness of satisfactory containers being used, prevention of waste being put outside premises in a loose condition to await collection, etc. Local authorities can only achieve success in legal enforcement of existing legislation if their own procedures are backed up by appropriate by-laws and handled by experienced legal practitioners.

THE ROLE OF VOLUNTARY ORGANIZATIONS

Community groups and voluntary organizations are well suited to assist in controlling litter as usually membership is concerned with the beauty of the countryside, beaches, and shorelines and the cleanliness of towns and villages. Their role in group action projects is essential and demonstrating how things can be done such as the clearing up of a derelict site and developing it for a recreational facility, cleaning up litter after outdoor events where wind blown debris can cause disfigurement over a wide area, selecting a rural area plagued as a dumping spot for clearance. These are positive steps in beautification and lead to extension of the aims and objectives in Keeping Britain Tidy. There are many ways in which voluntary groups can act in setting up community associations to foster new attitudes to litter prevention and to organize special projects all aimed towards a cleaner and healthier environment.

Appendix 1.1: A guide to Local Authority street sweeping performance

Methods	Manual sweeping performance. Road length per man-day* Mechanical km per hour	Where best employed	Other factors Affecting performance
Manual			
Beat system or (a) Street orderly	0.75 to 1.0 km per man-day	Shopping centres; principal streets; high pedestrian traffic	Whether supervised on beat sweepings, distance from depot
(b) ditto	1.61 km per man-day	Suburban roads	ditto
Team sweeping Pedestrian electric vehicle; 1 driver + 2 sweepers	1.80 km per team; 5.40 km per team	Suburban roads; mainly residential	Output affected by distances from depot or transfer station
Team sweeping Ride-on electric vehicles; 1 driver + 4 sweepers	1.93 km per man-day; 9.65 km per team	Ditto	Advantage: can ride to/from depot at road speed
Mechanical			
Pedestrian-operated footpath vacuum sweeper	3.22 km per hour	Principal streets; suburban roads, footpaths; shopping precincts	Output limited to walking speed of operator; distance from depot; clearance of sweepings
Mechanical sweeper collector, road and footpath	4.83 km per hour	Ditto	Weight limitation on footpaths; travelling time to/from depot; disposal of sweepings; servicing
Mechanical sweeper collector	6.44 km per hour	Principal road: suburban roads	Travelling time to/from depot; arrangements for servicing; discharge of sweepings
Road suction sweepers	6.44 km per hour	Principal roads; suburban roads; dual carriageways	Ditto

*Manual sweeping affected by type of surface on roads and footways, amount of litter, pedestrian, and vehicle traffic parking. Mechanical sweeping for surface area cleansed depends on sweeping width of machine. Average output per hour is determined by arrangements for servicing machine and effective sweeping time.

Appendix 1.2:   Some commercial street sweeping performance norms

System	Hourly performance	Daily performance	Typical crew sizes		Effective sweeping strength	Unit hourly performance per sweeper	Unit daily performance per sweeper
			Drivers	Sweepers			
Mechanical road sweeper	7.0 km per vehicle/h	45.5 km per vehicle/day	1	0	N/A	N/A	N/A
Cabac vehicle and crew	1.48 km per vehicle/h	9.65 km per vehicle/day	1	4	4	0.375 km per sweeper/h	2.41 km per sweeper/day
Pedestrian controlled vehicle	0.91 km per unit/h	5.91 km per unit/day	1	2	2.5	0.36 km per sweeper/h	2.36 km per sweeper/day
Manual sweeper barrow	0.25 km/h	1.625 km/day	0	1	1	0.25 km/h	1.625 km/day

Note: In 1982 the growth of commercial waste collection and cleansing services in the UK witnessed a considerable upward trend in accepted performance norms.

Practical Waste Management
Edited by J.R. Holmes
©1983 John Wiley & Sons Ltd

5

Planning landfill operations— A management view

P. N. O. CRICK

Director of Landfill Development, Redland Purle Limited

ABSTRACT

Coming from one of the biggest private waste disposal companies in the United Kingdom, this chapter takes a fresh view of the management factors involved in bringing into operation a major private-sector landfill site. Site search and selection, marketing factors, preparation and engineering, the consent of the statutory authorities, commercial evaluation, restoration, and after-use are all considered.

The planning and selection of landfill sites has gone through many phases during the recent past, from the days when the erection of chestnut paling was an extravagant engineering feat at the high end of the technology scale, via the golden years of searching for holes in light aircraft by international jet-setters, through the doom and gloom era where every site was bound to prove a disaster to somebody, to the age of comparative enlightenment in which we now find ourselves. Today the implications of what we do at a landfill are understood, and are reflected not only in our plans but also in the price which we charge our customers.

There are six phases to the planning of landfill operations, all of which are interrelated:

(1) site search and selection;
(2) plan site preparation and engineering;
(3) plan operational methods;
(4) plan site closure and restoration;
(5) consider requirements to obtain consent of the statutory authorities;
(6) evaluation.

This chapter will discuss each of these aspects in turn, giving an overview of the subject for the manager, without delving deeply into technical matters which are the province of specialists.

SITE SEARCH AND SELECTION

When setting out on a programme of landfill acquisition it is important to set out geographical priorities. Given that a landfill is capable of being operated safely and economically, it is its location, in terms of proximity to the market it serves, which determines its value. It is very easy, and entertaining to those of us who enjoy it, to tear about the country looking at holes in the ground, but much time can be wasted if a proper disciplined approach is not used. In this respect the objectives of the private and public sectors are different and therefore different criteria will need to be applied; these will be discussed more fully in the section devoted to Evaluation.

Although approximate geographic areas need to be determined, landfill search will always be essentially opportunistic. Holes in the ground exist where men dig them, not where waste disposal operators would like them to be, and even above-ground or cut-and-fill operations can only be located where they are acceptable to landowners and the Planning Authorities. Furthermore, there are several factors which cause many, indeed most, excavations to be unsuitable for landfill. These are discussed in more detail in the sections which follow, but the more notable ones, which can often rule out a site at a glance, are:

(a) small void space, rendering site uneconomic;
(b) unsuitable geology leading to surface water or ground water contamination;
(c) unsuitable water balance;
(d) proximity to housing leading to unacceptable environmental impact;
(e) poor road access;
(f) difficulty of site operations, e.g. due to topography.

The nature of the wastes to be deposited will have a major effect on site selection, since this will determine leachate quality. Much of the thought which goes into the evaluation and design of landfill is directed towards leachate control. There are basically two alternatives: the leachate can remain on the site or it can leave. For it to remain on the site the underlying strata must be impervious (usually clay), and care must be taken to ensure that the site does not fill to overflowing by ensuring that waste is deposited at a sufficient rate and that it is then sealed with an impervious capping. If the leachate leaves the site it may do so artificially by physical means, or naturally by soaking into the underlying strata. For this to happen without contaminating groundwater supplies, the geology must be granular (sand or gravel) with an unsaturated zone, the depth of which varies according to circumstances. Alternatively the site can be located over an unimportant or already contaminated aquifer, in which case the geology is less important.

The geological and hydrogeological considerations involved in individual situations are complex and it is not intended to elaborate further in this chapter. Some contracting companies leave the Water Authorities to pronounce on these matters, but it should be pointed out that the employment of specialists by contractors can enable problems of the future to be foreseen at the planning stage, whilst the standards of evaluation applied by Water Authorities, certainly in the past few years, have been distinctly inconsistent.

PLANNING SITE PREPARATION AND ENGINEERING

The first considerations in determining the engineering requirements of a site are determined by the hydrogeologist. He will carry out a water-balance calculation to determine the rate of ingress and egress of water in the form of rain, surface, and groundwater. He will point out that it is more difficult to protect surface water when operating a clay pit in Lancashire than a gravel pit in Essex, due to the difference in rainfall. In the former case he will probably suggest the construction of drainage ditches to prevent surface water entering the site. In the latter he may require some form of lining to prevent groundwater contamination if the base of the pit is close to, or below, the water table. Proximity to water abstraction points will also influence his decision, because of the dilution effect.

In the event that leachate needs to be removed from the site it may be pumped or taken by tanker to a sewage treatment works, or it may be treated by an on-site plant and discharged to water courses.

Having resolved the major environmental issue of water contamination, the other environmental topics need to be considered: visual and audible impact, gas control, and road access. Proximity to housing will probably be the major determining factor in the need for screening by planting or earthmoving, but the surrounding topography will also have a part to play. Gas control tends to be a lesser problem in the UK than in warmer climates, but deep sites with high rates of input can generate significant quantities of methane, and precautions need to be taken to prevent this becoming a hazard. Road access needs to be considered, both in the immediate vicinity of the site entrance, where sight lines must be assured by means of a suitable splay, and also in the general environs, with consideration being given to prohibiting traffic passing through small villages and narrow lanes.

In addition to road access from the public highway it is important to ensure that on-site roads are adequate to permit access to the tipping face in all weathers. The actual requirements will depend on the anticipated volume of traffic and the nature of the soil.

If liquid waste is to be received at the site particular attention needs to be paid to the method of reception and deposit. Trenches are to be favoured over lagoons, but with a large input volume it is most suitable to have a concrete reception sump, which can be most easily fenced and surfaced with a pumping system to allow deposit at a constantly changing location on the site.

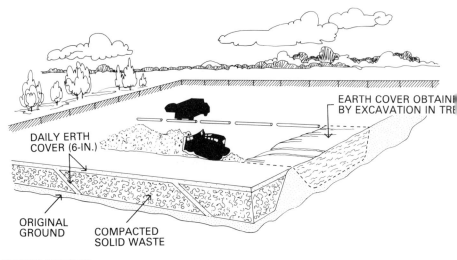

EARTH COVER OBTAIN[
BY EXCAVATION IN TRE

DAILY ERTH
COVER (6-IN.)

ORIGINAL
GROUND

COMPACTED
SOLID WASTE

TRENCH METHOD

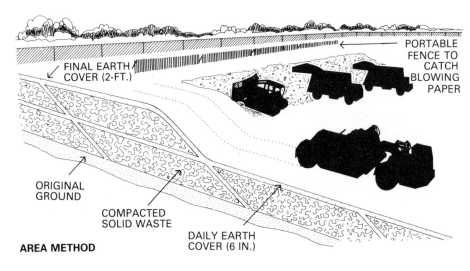

PORTABLE
FENCE TO
CATCH
BLOWING
PAPER

FINAL EARTH
COVER (2-FT.)

ORIGINAL
GROUND

COMPACTED
SOLID WASTE

DAILY EARTH
COVER (6 IN.)

AREA METHOD

Figure 1   Technical drawing and description of modern landfill methods. (Courtesy of the Caterpillar Company Ltd)

On a large site, especially when hazardous materials are involved, a radio communication system may be the most effective method of control. Consideration may also be given to double-handling facilities to avoid road vehicles travelling long distances across the site. Alternatively a 'short tipping' area

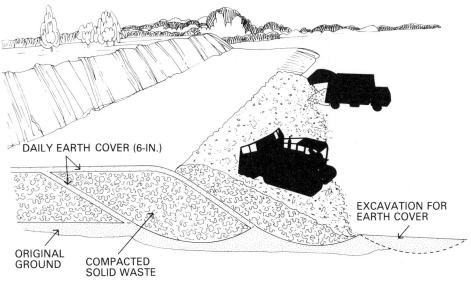

DAILY EARTH COVER (6-IN.)

EXCAVATION FOR
EARTH COVER

ORIGINAL
GROUND

COMPACTED
SOLID WASTE

RAMP METHOD

can be used in bad weather. Double-handling will inevitably be a feature if waste is transported other than by road, e.g. by rail or river, and this will probably involve the installation of craneage.

In order to ensure that leachate management goes according to plan, it is advisable to install water-monitoring points, frequently using concrete pipes laid as the level of the site rises with infilling. The level and quality of the leachate can then be monitored. As a precaution it is also worth carrying out background monitoring before commencing operations to establish water quality in surrounding water courses and the underlying aquifer, in order to determine the effect, if any, of the subsequent landfill operation.

Finally, the site needs to be adequately fenced, and services (electricity, water, telephones) and staff facilities must be provided. Security is an important aspect, and the location of the gatehouse needs careful consideration to ensure visibility of both the site entrance and tipping face.

PLANNING SITE OPERATIONAL METHODS

The actual plan for depositing the waste will be largely determined by the topography and hydrogeology, and the after-use of the land when it is restored. It is important to plan the sequence in which the areas will be filled, in order to foresee what problems, if any, will arise. Leachate considerations may indicate that a cell method will be most appropriate, leaving the remainder of the site uncontaminated, but this is not always necessary. If hazardous wastes and/or liquids are to be deposited, special techniques will be appropriate, and leachate management becomes more critical. Safety precautions, particularly relating to fire control and first-aid facilities, also become a matter for more serious attention. This will lead to a need for emergency procedures, which will require to be rehearsed at frequent intervals.

Having decided on the method of working, the staffing and plant requirements can be determined. When considering plant, particular attention should be paid to the need for initial compaction. The life of a small site can be maximized by using steel-wheeled compaction rather than bulldozers, and by on-face working. On sites with longer lives the natural decomposition of the refuse over a number of years may achieve the same result, with lower operating costs.

The availability of cover material is a major consideration for the operator. If there is not sufficient arising in the locality as a result of civil engineering works, etc., then it may be necessary to extract material from the site before covering the floor with waste.

Finally, the environmental issues of noise, dust, mud on roads, wind scatter, smell, vermin, and insects need to be considered and appropriate procedures adopted for their control.

PLANNING SITE CLOSURE AND RESTORATION

Nowadays, planning consents usually determine the essential end-use of restored sites. Frequently, however, they do not do so in any detail. This provides an opportunity for a little imagination in landscaping and planting to leave behind something of which the operator can be proud. There is nothing like a well-restored site to impress planning officers and committees when new infilling consents are being sought.

The quality of the final restoration will often depend upon good planning before commencing operations, ensuring adequate supplies of final cover and topsoil, and operating in such a manner that residual leachate problems do not arise. In this respect it is particularly important that impervious sites are adequately capped with clay to prevent further ingress of surface water, and that the capping is well compacted and not ruptured either by vehicles travelling on it, or by weathering.

CONSENTS OF STATUTORY AUTHORITIES

The ultimate consent required is that of the site licence, issued by the WDA, but planning permission and the approval of the Water Authority are precursors to this. It is advisable to commence informal discussions with officers of all three authorities early in the planning process, as their views will inevitably be valuable in determining requirements in all the stages already discussed.

In the case of the planning authority, which at present (1980) may be District or County, but should ultimately become County in all cases—to the great benefit of those seeking to provide waste disposal facilities—the requirements will relate mainly to the site's environmental impact and its after-use. This may affect the method of operating, and may influence capital and operating costs to some degree.

The Water Authority will need to be satisfied on surface and groundwater pollution protection. This may well require the sinking of boreholes as part of the investigative process—a matter of no mean expense. Where the investigation is complete, the engineering necessary to protect water resources will become evident, and this may well be the most significant single element of expense in both site preparation and operation.

EVALUATION

The ultimate evaluation of a landfill site must be on financial grounds. The public and private sectors will have different criteria, different objectives; nevertheless, the final decision must be a financial one. Although this will be tempered with other criteria, mainly environmental ones, these are largely dictated by the statutory authorities, and this chapter does not set out to discuss moral issues.

In the public sector the objective is to minimize the cost of refuse disposal to the ratepayer. The elements of cost are the same as in the private sector—transportation and disposal. Sites must be located in such a way as to minimize the sum of these two costs. The situation is simpler than that of the commercial operator, however, because the volume of arisings of waste is known more accurately, and it is not necessary to take account of the uncertainty of competitive activity.

In the commercial world, competition adds an additional level of complexity. Basically a contractor can find himself in three types of disposal situation relative to a market (which will in general be any centre of population)—positive, negative, and neutral. He will be in a positive disposal position if he has exclusive access to a site which is located nearer to the market than any which his competitors may use. In such a situation his competitors will be in a negative disposal position. If the nearest site were open to all-comers (e.g. a council-operated site admitting private contractors) then the disposal situation would be neutral. Clearly the contractor's first priority must be to eliminate negative disposal positions in markets in which he wishes to operate. A geographic representation of the concept is shown in the diagram. The exact boundaries of the different areas will be indistinct, and may vary

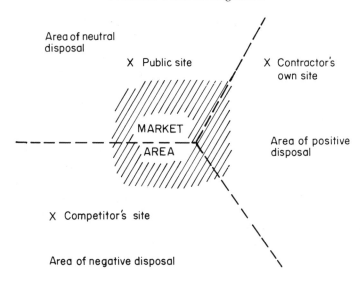

Figure 2   A management sketch illustrating marketing decisions in a landfill investment

slightly based on the relative costs of disposal at the three sites, which will typically be different.

When it comes to justifying an investment in acquiring a new site, it is necessary to forecast the cash flows involved. Initially these will be negative whilst the site is being prepared and engineered, and possibly being purchased—although a lease or licence is plainly preferable for the organization acquiring the site, as it enables the payments to be spread over a longer period. Furthermore, there are taxation advantages in leasing (or preferably licensing) a site, since rental payments are allowable against tax, while an outright purchase of the freehold is not allowable. A premium on a lease is only allowable at the rate of 82% of the premium and this must be spread over the period of the lease. The site preparatory works are also not allowable for tax if they can be regarded as 'improvements to land', such as access roads, fencing, drainage, etc. These taxation aspects, although frequently regarded as a boring detail by operators, have a most significant bearing on the economic viability of a given site.

Having established the initial outlays it is then necessary to forecast operating profit, based on a forecast of input and revenue. This is difficult but is quite imperative, and must be done by the local management who know the local markets and the competitive situation. In the writer's experience, managers tend to be reluctant to commit themselves to high volumes or high prices, but then are most vociferous if they are told that the site is not worth acquiring. By demonstrating such a reaction they are implicitly saying that revenues for the site could be higher than those to which they have committed themselves. The art at this stage is eliciting the truth from your own management.

Finally, before performing a return on investment calculation, it is necessary to try and forecast the trend of disposal prices compared with the general rate of inflation. In the last few years disposal prices have risen in real terms, largely due to a rise in the value of licensed landfill airspace. The steepness of the rise has reflected the advent of the controls of the Deposit of Poisonous Wastes Act and the Control of Pollution Act, which have restricted the amount of airspace available. Whether the value will continue to rise in real terms once equilibrium is reached remains to be seen. It seems likely that equilibrium has not yet been reached in the vicinity of the major conurbations—GLC, Manchester, Merseyside, West Midlands—although the County Councils are taking aggressive steps to resist further increases, in some cases by the questionable use of compulsory purchase orders.

This approach by some counties against private contractors seems to be rather short-sighted. Their objective is allegedly to reduce their own costs by resisting the prices being charged for airspace. These prices, nevertheless, reflect a value set by balancing supply and demand, and a more realistic way of minimizing cost would be to deliver their waste to sites operated by the private sector which also take industrial waste. This would spread the operating expense over a higher volume, thereby reducing unit cost since, with the exception of airspace consumption which remains constant, unit costs fall substantially with rising volume.

All the factors mentioned above must be considered before computing whether a site will provide a reasonable return on investment. The most difficult figure to estimate, however, is the price it is worth paying for the airspace. This is because each geographic area is effectively a separate market situation of supply and demand. Consequently, airspace in one part of the country can be worth 10p per cubic yard, where in another it may be worth more than £1. Regrettably, this situation is likely to persist, with disparities becoming greater as transportation costs rise. Maybe this is one of the elements which makes our industry so fascinating.

Practical Waste Management
Edited by J.R. Holmes
©1983 John Wiley & Sons Ltd

6

Techniques of sanitary landfill

KEITH J. BRATLEY, CENG, MIMECHE, MISWM

Deputy Waste Disposal Officer, West Midlands County Council

ABSTRACT

Some of the best work in recent years on sanitary landfill, this chapter describes the performance of modern landfill. Every aspect of the art from selection of plant, the consumption of space in landfill operation, safety, day-to-day management, and site administration are set down. A comprehensive set of appendices gives detailed instructions on responsible and cost-effective operations. Mention is made of health and safety as well as final restoration of sites.

INTRODUCTION

The term 'sanitary landfill' is an Americanism. I and others use this term not because we have followed the techniques and practices in America with respect to landfilling of waste but to move away from the term 'controlled tipping' which in the 1950s and 1960s became an unacceptable method of waste disposal. The scientific process remains the same; the techniques and applications are greatly improved. Experience has shown that successful operations will only be achieved with continuing proper and competent planning and control. Many fine examples of completed landfill schemes can be seen in the UK and around the world. However, many of these schemes were the subject of criticism and complaint during the infill operation. It is this area of my experience I hope to be able to pass on to you in this chapter.

HISTORY

Man has always returned his waste products back to the earth. Some are treated before deposit but all known systems of waste disposal require land for the disposal of the final residues. As civilization is developing we are centring more and more into larger and larger urban communities. The waste produced by each person is increasing, placing greater demands on the diminishing availability of land suitable for disposal of waste close to urban areas. George Bernard Shaw had a running battle for many years with his local authority, Wheathamstead in Hertfordshire, over the rotting, burning stench coming from refuse transported from Islington and disposed of near his home.

Disposal of refuse from a large urban community in a controlled way was developed in Bradford in the 1920s. Since then there have been many further developments in the science and techniques of landfill but like many other community services it has taken legislation to bring out the necessary financial investments to bring about acceptable operational conditions.

SCIENCE

The process for the disposal of household or organic waste by controlled tipping has been accepted since the early 1930s. A scientific investigation carried out by Bertram B. Jones and Frederick Owen (*Some Notes on the Scientific Aspects of Controlled Tipping*) in Manchester from 1931 to 1934 is the major work carried out on the subject. Other more recent research work has shown little to disprove their basic conclusions. It can be shown fairly easily that pathogens die when exposed to certain temperatures for certain times. However, the process of bacteriological decomposition, fermentation, and breakdown of animal and vegetable sources is much more complex. Jones and Owen detailed many of the chemical formulas by which elements of the waste would break down. Since that time the constituents of waste have changed little other than that the amounts of each constituent have varied in relation to the whole. The major addition has been plastics. Plastics undergo little change in this process but remain very much in the same state as when deposited. For further information on the science of the process I would advise reading *Notes on the Science and Practice of the Controlled Tipping of Refuse*, by R. E. Bevan. This can be obtained through the good offices of the Institute of Wastes Management.

LANDFILL SITES

Sites suitable for infill with waste are infinitely variable. It does not necessarily have to be a hole in the ground although it is true to say that most sites are the backfilling operation after mineral extraction. However, natural valleys, foreshores, river mud flats, railway cuttings, and many other areas have been successfully infilled giving rise to community benefit by improving the environment and upgrading land for development or agricultural use.

The final use of the site should be determined before tipping commences. This will enable a satisfactory working plan to be drawn up that will ensure no unnecessary movement of material for final restoration works to be completed.

All landfill sites should satisfy the criterion that there will be a resulting benefit to the community.

TOPOGRAPHY OF THE SITE

The design of the landfill site should include one or more topographic maps at a scale of 1 : 250 with 1 metre contour intervals. The map should show: the proposed fill area; any borrow area; access roads; grades for proper drainage of each lift required and a typical cross-section of a lift; special drainage devices if necessary; fencing; garaging; amenity facilities; and all other information which will clearly show the planned development, operations, and completion of the landfill.

GEOLOGY OF THE SITE

The geological characteristics of the proposed site shall be determined, preferably by on-site testing, or from earlier reliable survey data. Geological investigations are of the utmost importance in ensuring the site will not give rise to future pollution problems, particularly with regard to water pollution. Such an investigation will also show whether a suitable quality of cover material is available in the required quantity or whether material will have to be brought in for this purpose.

WATER POLLUTION

The Water Authority will ensure no site is infilled which is likely to give rise to later water pollution problems. Indeed in almost every case it will be the task of the Waste Disposal Authority or private operator to convince the Water Authority that water pollution will not occur. There are many sites suitable for infill where the natural flows of water would give rise to pollution. However, if sufficient capital investment can be made to prevent this pollution then there would be no constraint on water pollution grounds in most areas. It is then an examination as to the viability of the site as compared with other available options for disposal of the proposed input.

In general, landfills should not be located on sites where the surface or ground water will intercept the deposited waste. Waste often contains infectious material and other harmful substances that can cause health hazards or nuisance if permitted to enter surface, or more important groundwater, supplies. Diversion of clean surface water by drainage provision, diking, or dewatering will often be sufficient to prevent water contamination. In particularly suitable geological areas pollution can be prevented by the slow dispersal and resulting dilution using the natural biological processes.

SITE PREPARATORY WORKS

Having gained planning approval and a site licence for a site which is suitable for infill (topography, geology, water pollution, etc. conditions have all been satisfied) I cannot stress too strongly the importance of carrying out the necessary site preparatory works *before* the landfilling operations commence.

Site entrance

(1) A properly constructed roadway, preferably to normal highway standards, is desirable having adequate visibility splays and extending as far as possible into the site.
(2) An adequate lockable gate together with total perimeter fencing or total fencing of the working area of the site is essential.
(3) An identification board showing name of site; name, address, and telephone number of operator; and Waste Disposal Authority responsible for issuing licence. Also incorporated could be trade refuse charges and attention should be drawn to codes of practice displayed on notices within the site.
(4) A weighbridge facility should be provided or an adequate documentation system to maintain a record of the types and quantities of waste deposited.
(5) Adequate amenity facilities for site operatives, including preparation of documentation, display of working plan and telephone facilities (essential).
(6) Garage facilities, particularly for plant which cannot be moved from site (compactors, track machines) which should also incorporate security and first-line servicing provision.
(7) Storage facilities for site equipment including fuel.
(8) Road lighting aiding security and improving safety.
(9) Landscaping to provide pleasing appearance and screening.

Landfill

Landfill sites are often in remote areas. Security of plant and equipment can often be a difficult and costly operation. It is advisable that facilities are placed as close to the main entrance as possible. Full-time security provision may be necessary, particularly for sites handling hazardous wastes.

Access roads

The successful operation of any landfill site is largely dependent upon the provision of adequate access roads.

Main access road

This road leading from the site entrance should be a properly constructed road permitting two-way vehicular traffic. The site control office must be located on this road. The extent of the roadway into the site will depend on the life of the site and the expected usage. Its route should be planned to ensure it can be used to the end of the landfill project (and possibly beyond). Incorporated along its length should be a wheel-cleaning device.

Secondary access roads

These roads will lead off from the main access road and be constructed in such a way as to ensure the easy passage of vehicles in all weather conditions. It should only be necessary to construct one secondary access road for each lift. There are a number of materials available to facilitate the construction of semi-permanent roads to satisfactory standards.

Drainage

In addition to site drainage ensuring no water pollution risk, drainage of access roads is also necessary. It is not always possible for access roads to be above the level of the surrounding area. Consequently water will run towards them and with vehicular movement will produce unacceptable conditions. This can be avoided by suitable access road drainage provision.

Wheel cleaning

Running vehicles on unmade roads must inevitably produce a problem of depositing mud on subsequent made-up roads. This problem is considerably reduced by the construction and maintenance of good secondary access roads. There are a number of mechanical wheel-cleaning arrangements available. I would favour a more simple arrangement of a 'cattle grid', a long drive through water trough, or a trough of pebbles, all of which are effective.

Signs

Regular visitors to the landfill site will be given a copy of the code of practice for tipping vehicles and any other relevant safety information. This code of practice will be displayed at the site control office for other visitors. Adequate directional signs (preferably standard highway signs) will ensure vehicles using the correct route from control to tipping face. A one-way system is ideal. Dangerous areas must be adequately signed (e.g. lagoons, sumps, etc.).

Fencing

The extent of the fencing and security measures required will depend on the size, nature of operation, type of waste being deposited, and location of the site. It is necessary to fence the total working area of the site as a minimal requirement. Natural barriers that may exist should be taken into account. Dangerous areas must be taped-off or fenced.

Tipping area

The tipping area should consist of a number of constructed bays. These bays should have banks on three sides approximately 2 m high. The open end should be as close to the secondary access road as possible. The shape of the bays will depend on conditions prevailing at any particular site. The width of the bay (i.e. width of the tipping face) will be governed by the number of incoming vehicles. Four vehicle width face, say 16 m, would be suitable for handling an input of 300–400 tonnes/day. The length of the bay will depend on the overall shape and size of the site. A single bay could handle a number of weeks input.

EQUIPMENT

The preparation of a landfill site for the receipt of waste is normally a straightforward civil engineering exercise. The equipment required is that necessary for earth-moving; road-making; drainage provision and building. It is unlikely that a Waste Disposal Authority or private waste disposal contractor could find sufficient use in this context to justify the purchase of machinery to carry out this work. Motorized scrapers, drag lines, road laying machinery, etc., would probably be better hired in.

Assuming that site preparation works are completed in the way I have outlined then the equipment and machinery necessary to operate the landfill site would depend in the main on the following four conditions:

(1) input quantity;
(2) types of waste;
(3) cover material available;
(4) location of site.

My recommendation for any site, in this country, receiving domestic waste and light bulky industrial or commercial waste would be the use of a purpose-built steel-wheel landfill compactor. One machine can compact and place up to 800 tonnes of waste each day to a high final density. To achieve this output to the required standard a 'back-up machine' for the transportation of cover material to the face would be necessary. For inputs of 400 tonnes/day or less, in addition to placing the refuse, the compactor would be able to 'win' cover on-site if material was available.

The type of 'back-up machine' would depend on the site availability of cover material. If cover material could be dug on site then a track-loading shovel would probably do the job best. Depending on length of haul a dump-track or tipper vehicle could be necessary. If easily dug cover material is available then a four-wheel-drive rubber-tyred loading shovel would be the ideal 'stable mate' with the compactor. This arrangement is often more flexible as with Tyrefil tyres the rubber-tyred loading shovel can be used as an effective compactor.

There is a very wide choice of machines available, and matching them together to obtain the best arrangement will always be influenced by the individual site conditions.

The choice of machines for many given conditions is a very involved subject which I am not able to cover in this chapter. (See chapter 11).

Other necessary equipment

(1) Moveable screens for quick assembly.
(2) Insecticide-spraying equipment.
(3) Portable lighting equipment.
(4) Machine firefighting equipment.
(5) Machine cleaning equipment.

MODE OF OPERATION

Incoming vehicles

(1) All vehicles must be received at site entrance control office.
(2) Documentation must be checked (trade refuse, Deposit of Poisonous Wastes Act Notifications etc.) and quantity and type of waste recorded.
(3) Attention should be drawn to relevant site notices (Code of Practice for tipping etc.; see Appendices 1 and 2).
(4) All drivers should be instructed to follow the signed route to the tipping face.
(5) Driver and mate only are allowed with the vehicle in the tipping area.

(6) Proceed to tipping position only when instructed to do so by landfill operative.

(7) Tip in the position indicated (this should be at least a vehicle length from the tipping face).

(8) Leave the site via the signed route (this should pass the site control office).

(9) Special loads to be tipped under supervision at all times.

At the tipping face

(1) All vehicles carrying waste will be tipped in the operational bay at the lower level.

(2) Traffic will be controlled by one site operative in clearly marked protective clothing. No other pedestrians allowed in this area.

(3) Vehicles carrying cover material will be tipped at the higher level of the operational bay, preferably using the secondary access road for the next lift.

(4) The tipped loads of waste will be run over by the steel-wheel compactor immediately after the tipping vehicle has left the area.

(5) The number of passes the compactor makes over the waste at this time depends on the size and nature of the waste.

(6) The compacted waste is then pushed forward in 'onion-skin layers' up a 1 in 7 incline.

(7) Each load of waste is treated in this way with the compactor making many passes up and down the incline compacting, chopping, and breaking-up the waste with each pass (see Appendix 3).

(8) There will be no large voids within the compacted waste and in-place density of over 1 tonne/m^3 can be achieved (see Appendix 4).

(9) All bulky waste must be compacted before spreading up the inclined face.

(10) At the end of each working day the cover material deposited at the higher level is spread over the inclined face. For large-input sites cover material will need to be spread over the finished level of the working layer (i.e. top of incline) during the day as the working face progresses.

(11) During most weather conditions, using this method of operation, there will be no deposited waste leave the working area. However, in conditions of high wind screen fences must be erected. As tipping is taking place 2 m below the surrounding area then a 2 m fence on top gives 4 m screen fence for windblown material catchment. Correct positioning of bays will prevent the open end of the bay being downwind of tipping.

(12) Waste compacted to high densities becomes less attractive to vermin. Seagulls lose interest if an adequate layer of cover material is used.

During hot weather the waste may contain flies, etc., before arriving at the site. In these conditions the working face should be sprayed with an insecticide at intervals throughout the day.

HEALTH AND SAFETY

The Health and Safety at Work Act places onerous responsibilities on line managers at all levels. It also places responsibilities on the operatives. A landfill site can, and often is, a very dangerous place. By its very nature it produces a daily changing workplace. Attention to planning, safe systems of work, and above all trained supervision, is of paramount importance if accidents and health risks are to be avoided. Following the simple *modus operandi* outlined in this chapter will assist in the prevention of incidents, but attention must also be given to adequate training of operatives in all work they are asked to carry out. A trained first-aider is a requisite on every site. Limiting access to any working area will prevent personal accident. A detailed weekly safety report is essential (Appendix 5).

FINAL RESTORATION

Final restoration work of any site is a comparatively easy civil engineering project. The possible public pressures applied when the site is operational quickly recede now the community benefit becomes apparent. This happy situation can be reached much earlier on many sites if the initial planning of the operation is geared to progressive restoration. Large areas of many sites can be completed to final profiles long before the whole site is complete. With suitable screening recreational, agricultural, or industrial use can be made of completed areas whilst the remainder of the site continues to function as a disposal facility. Cosmetic treatment can be achieved by completing boundary areas of a site first and continuing landfilling screened by an attractive area.

COMMENTS

(1) Landfill will remain for some considerable time the major waste disposal system for waste which cannot be further utilized. The future must involve a fewer number of much larger sites up to and over 1,000 tonnes/day. This will occur when disposal sites close to urban areas are filled and disposal relies on the more remote mineral extraction sites fed by bulk transfer stations.
(2) Clearly these large landfill sites will require specialist equipment to handle all types of waste in large daily input quantities. The conditions under which these sites will operate will become more and more stringent as the Site Licence Conditions under the Control of Pollution Act become

effective (see Department of the Environment, Waste Management Paper No. 4).

(3) Landfill space will obviously become a more expensive commodity and methods for the best use of the space will become more important. The method outlined in this chapter will produce high in-place density of waste making the best use of available space.

(4) All known methods of waste disposal are in the final analysis landfill systems. There have been many fine examples of completed projects which have been carried out to high operational standards. The method of operation I have presented would not be suitable in every case. I am a firm believer in 'horses for courses'. However, if some of the techniques I have presented you find worthy of selection for use on your landfill sites then I feel my effort has been rewarding.

CONCLUSIONS

It is essential that the student questions established practices. The system for landfill I outline in this chapter is based on my own experience and the questioning of earlier practices. You must satisfy yourself that they are applicable in your circumstances and that you are in a position to recognize and take advantage of new knowledge at all times.

REFERENCES

Bevan, R.E. *Notes on the Science and Practice of the Controlled Tipping of Refuse.* Institute of Wastes Management, London.
Department of the Environment, Waste Management Paper No. 4, *The Licensing of Waste Disposal Sites.*
Bratley, K.J. *Description of Comparative Performance Tests of Mobile Plant on a Major Landfill Site.* Institute of Wastes Management, London.
Townend, W.K. *Procedures for Vehicles Entering Solid Waste Disposal Premises.* Institute of Wastes Management, London.

DISCLAIMER

The opinions expressed in this chapter are those of the author, and do not represent the policy of any Local Authority.

Appendix 1:   Code of practice for tipping of refuse collection vehicles on SYCC sites

(1) On arrival at site all crew members with the exception of the driver and one mate will disembark at the assembly point and remain at the point only until the vehicle returns.
(2) All loose tools and equipment must be unloaded at the assembly point and collected on return.
(3) Vehicles must travel to and from the tipping face on the prepared site roads. Drivers must obey all directional and instruction signs. At no time must the speed of a vehicle exceed 7 mph.
(4) The driver must report to a SYCC employee as soon as possible on entering the site and must comply with all instructions given by him. The vehicle will be manoeuvred into a tipping position about one vehicle length from the tip face. If the driver has any doubts as to the stability of the site he must satisfy himself that it is safe to proceed.
(5) On completion of the tipping operation the vehicle should be moved into a position away from the tip face where it is safe to clean dust trays, body seals, etc., without obstructing other vehicles using the site.
(6) No tipping may proceed unless there is a SYCC employee on site to direct the operation.

Appendix 2: Procedures for vehicles entering solid waste disposal premises

Procedures for vehicles entering solid waste disposal premises

Collection Authority

Issued to

Date

Collection Authority

I..

being a **Driver** for the above mentioned authority
acknowledge receipt of instruction booklet produced by the
Greater London Council of their procedure for Vehicles
entering solid waste disposal premises.

Signed ....

Date ....

Foreword

The Health and Safety at Work etc Act 1974 places responsibilities on employers not only for the safety of their employees but also for other persons who have access to their operations and who may not be employees.

This document is particularly concerned with my department's responsibility for persons who have access to our premises and who are not our employees. It has been written in the light of the requirements of the Act and as a result of experience gained from incidents in which people have been injured and damage has occurred to plant and associated equipment including collecting authorities' vehicles.

I am most anxious that the basic principles set out in this document should be known as widely as possible in an endeavour to discharge the responsibilities in which we have a mutual interest and I am sure that I can rely on your cooperation to this end.

Director of Public Health Engineering
The Greater London Council
November 1978

Whilst every effort has been made to highlight the major problems associated with the movement of traffic on refuse disposal premises, the GLC reserve the right to amend and update this booklet from time to time as may be necessary.

Produced by the Greater London Council
Department of Public Health Engineering
HE/AE/SA/   1.79

Guide for collection authority employees of safety measures which should be observed when visiting GLC solid waste transfer stations and plants.

Contents

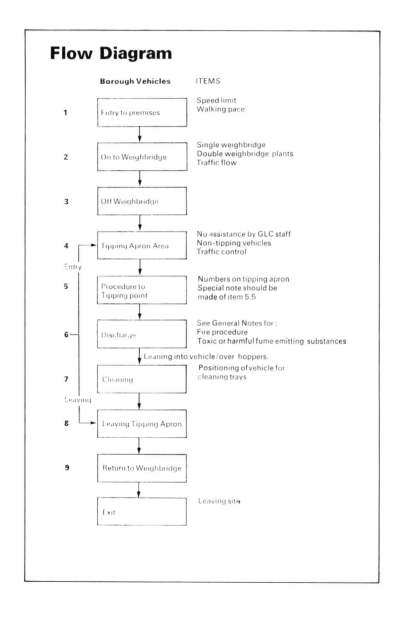

Flow Diagram

Borough Vehicles   ITEMS

1   Entry to premises   Speed limit
Walking pace

2   On to Weighbridge   Single weighbridge
Double weighbridge plants
Traffic flow

3   Off Weighbridge

4   Tipping Apron Area   No assistance by GLC staff
Non-tipping vehicles
Traffic control

Entry

5   Procedure to
Tipping point   Numbers on tipping apron
Special note should be
made of item 5.5

6   Discharge   See General Notes for :
Fire procedure
Toxic or harmful fume emitting substances

Leaning into vehicle/over hoppers.

7   Cleaning   Positioning of vehicle for
cleaning trays

Leaving

8   Leaving Tipping Apron

9   Return to Weighbridge

Exit   Leaving site

1 Entry into Premises

1.1 When approaching main gate take account of vehicles entering or leaving premises.

1.2 Drive at walking pace in restricted depots except where otherwise stated, i.e., where speed limits exist.

1.3 Observe instructions of GLC staff. Beware of pedestrians and other vehicles using the site, i.e., private cars and trade vehicles unfamiliar with depot. Follow the traffic pattern and direction notices at the depot. *Figure 1.*

Figure 1

1.4 When proceeding to weighbridge area in single weighbridge depots, Bulk Carriers have priority when weighing out.

2 On to Weighbridge

2.1 Drive carefully on to weighbridge.

2.2 Declaration of load. Ensure vehicle is safely positioned and all wheels are on bridge.

2.3 Do not allow children or animals in the cab. *Figure 2.*

Figure 2

3 Off Weighbridge

3.1 Do not move off unless advised or signalled to do so, then proceed slowly to tipping area (as 1.2).

3.2 Do not unlatch or unsheet in weighbridge area, this is only permitted on tipping platform.

4 Tipping Apron Area

All Vehicles

4.1 Obey instructions, manual and mechanical given by the traffic control and do not move on to the tipping area until directed.

4.2 GLC staff will determine the number of vehicles allowed on to the tipping apron at any one time.

4.3 GLC staff will not assist in the off loading or discharge of vehicle load.

4.4 Loads discharged manually.

4.4a Ensure that vehicles are positioned as directed, where there is no obstruction or hazard to other vehicles using apron.

4.4b No vehicles must be off-loaded direct into tipping slots. Areas for off-loading will be specially designated.

5 Procedure to Tipping Point

5.1 Driver must wait instructions from traffic control before entering tipping area and position vehicle as instructed.

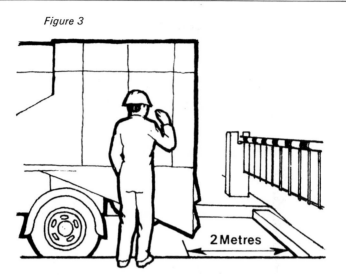

Figure 3

2 Metres

5.2 Position the vehicle clear of any other vehicle and heed all
other instructions while reversing, paying particular
attention to pedestrians in the immediate vicinity. Stop the
vehicle at least 2 metres from the tipping beam. *Figure 3.*

5.3 If the depot has tipping slots or compactor hoppers the
vehicle must not be positioned nearer than 3 metres for
unlatching. In the case of tipping slots under no
circumstances must any person approach within 3
metres of the tipping slot.

5.3a Tipping Slots :—These are large deep holes with a fall of
6 metres approximately.

5.3b Compactor Hoppers :—These are large open topped,
sunken storage bunkers from which waste drops into
compactor.

5.4 Ensure the vehicle is in a safe position with hand brake on and gear in neutral.

5.5 Only the driver may leave the cab of the vehicle to unlatch the back section. Under no circumstances unless specific written instructions are given by the employing authority will any other person or crew member leave the cab of the vehicle whilst it is on the tipping apron. Driver should proceed to unlatch the vehicle heeding the danger from approaching vehicles from either side. In the case of sheeted vehicles the sheets should be removed at this point.

Figure 4

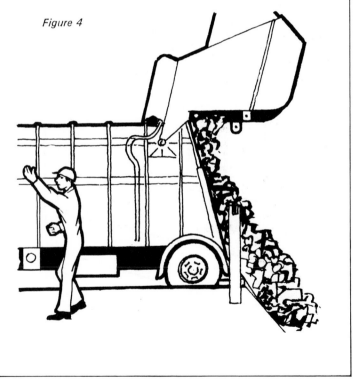

6 Discharge

All mechanical discharge loads.

6.1 In the case of depots with tipping beams the driver should reverse slowly to the tipping beam.

6.2 The driver should proceed to operate controls for discharge of waste. *Figure 4.*

6.3 After discharge, at this stage the driver **will not** lean into or inspect the back of the vehicle or lean out over the hopper or tipping opening.

6.4 Driver after discharging the load will drive forward 3 metres from the tipping beam or to a marked position on the floor. *Figure 5.*

Figure 5

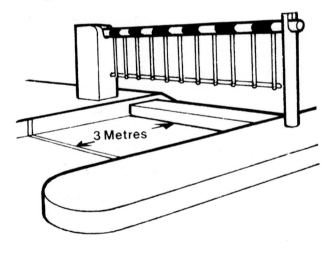

3 Metres

6.5 Skip vehicle drivers must put down their stabiliser jacks before emptying the skip. Before moving forward the driver must ensure that the jacks have been retracted.

6.6 Unstable loads :— If the driver feels or sees the front end of the vehicle starting to lift off the ground or become unstable the driver should immediately reverse the controls and not attempt to continue to discharge.

6.6a In the case of gravity tipping vehicles the body should be lowered. For horizontal discharge vehicles the ram plate should be retracted and the rear end of the body lowered. Skip vehicles should have the skip lowered on the body member. The vehicle must not be moved. The driver should immediately inform the traffic controller of the circumstances, who will then put into operation emergency procedures.

6.7 Sticking loads or Hydraulic failure :— In the case of loads sticking or failing to discharge completely or faulty hydraulics, the tipping kerb must **not** be used to bump out the load. The vehicle should be closed and the traffic controller should be informed immediately.

6.8 Refuse discharged direct on to the ground :—At those depots where bulk vehicles are loaded by power shovels the traffic controller will direct vehicles to a position to tip. Work conditions are extremely hazardous therefore drivers **Must Obey** his instructions at all times.

7 Cleaning

7.1 The driver will pull forward from the discharge position 3 metres (or to another designated area) and park in a safe manner. In this position he may clean the trays and/or clear any other refuse adhering to the vehicle. *Figure 6.*

7.2 Before moving off the driver will ensure that the body is lowered, latches are on and the vehicle is in a roadworthy condition.

Figure 6

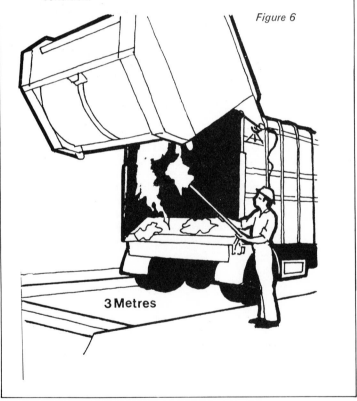

3 Metres

8 Leaving Tipping Apron

If the depot has **No** traffic control, move from the tipping bay if it is safe to proceed. Where there is a traffic controller wait for his signal before proceeding if it is safe to do so.

9 Return to Weighbridge

9.1 The driver will proceed to the weighbridge at a walking pace except where other speed limits are specified.

9.2 If the weighbridge has informed the driver when weighing in that he requires him to weigh out (TARE weight) he should proceed to where there is an out weighbridge.

9.3 If a single weighbridge depot, he should proceed with caution on to the weighbridge taking care to observe any other pedestrian or vehicle in the area.

9.4 Park the vehicle safely with all wheels on the weighbridge.

9.5 On signal from weighbridge drivers will proceed with caution to the exit gate being aware of other vehicles and pedestrians entering the depot.

10 Landfill sites

Where vehicles have access to GLC operated Landfill Sites the following should be observed:

10.1 Drive with extreme caution when entering site. Take account of other vehicles and pedestrians in the entrance and weighbridge area.

10.2 When advised by traffic control, proceed to tip face observing all signs for information and direction.
Figure 7.

Figure 7

10.3 Keep to temporary roads observing climatic conditions and speed limits.

10.4 Manoeuvre vehicle with care on tip face, observing pedestrians and other machines in the working area, particularly when reversing.

10.5 As far as reasonably practicable ensure the vehicle is positioned on stable ground before discharge.

10.6 Unlatch or unsheet vehicle/load. After discharge secure latches.

10.7 Do not drive vehicle on tip face with elevated body.

10.8 During summer and dry weather conditions when leaving tip face take care not to cause any unnecessary dust.

10.9 During wet weather conditions ensure mud is removed from vehicle road wheels before driving on to public highway.

A safe and efficient system should prevail if all reasonable care is taken when observing these instructions.

General notes for further information and guidance

Health and Safety at Work etc Act 1974 Sections 7 and 8

7a   There is now a statutory duty on **everyone** to take reasonable care for the health and safety of himself and of other persons who may be affected by his acts or omissions at work ; and

b   as regards any duty or requirement imposed on his employer or any other person by or under any of the relevant statutory provisions, to co-operate with him so far as is necessary to enable that duty or requirement to be performed or complied with.

8   No person shall intentionally or recklessly interfere with or misuse anything provided in the interests of health, safety or welfare in pursuance of any of the relevant statutory provisions.

Pedestrians

The person on foot is obviously in the most dangerous position and he relies largely on the driver to watch out for him.

Difficult Waste

It is the driver's responsibility to report all suspicious or dangerous wastes in his vehicle, i.e., asbestos, chemicals, drugs, pressure cylinders, paints, oils, radioactive materials, rat infestation, carcasses or burning loads.

Burning Loads

The driver will use headlights and sound horn as a warning and should proceed on instructions from weighbridge clerk to area set aside for dealing with burning loads. Weighbridge will inform other GLC staff and inform the fire brigade if necessary.

Sorting of Refuse (Totting)

It is dangerous to sort over refuse on the tipping apron.
The practice of sorting will not be allowed.

Vehicle Body

Many of the tipping aprons are in enclosed buildings, the
drivers must ensure that the body is lowered before moving
from the tipping area.

Fire Extinguishers

If a fire occurs on a vehicle the traffic control must be
informed immediately. They will know the location of the
nearest fire extinguisher/equipment.

First Aid

If an accident does occur there is on most depots a trained
first aider. The traffic controller must be informed immediately
and he will arrange for a first aider to deal with the accident.

Safety Footwear

Every effort is made to keep loose refuse on the tipping apron
to a minimum. There will still be some loose refuse on the
floor, and therefore to reduce the risk of foot injury, safety
footwear **should** be worn.

Noise

Excessive noise may be generated when tipping. Machines
should only be operated in a manner that keeps any noise
emitted to a minimum level.

Protective Clothing

Every effort should be made to ensure that items of protective
clothing are available and are worn as may be necessary.

Discipline

Where any person who visits a plant or station acts
irresponsibly, his/her behaviour will be noted and reported

Appendix 3:   Landfill operations

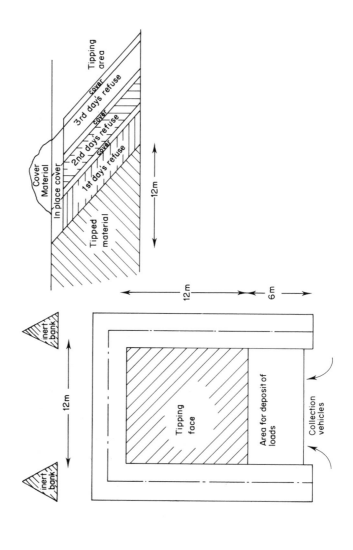

Appendix 4:   Compactor results

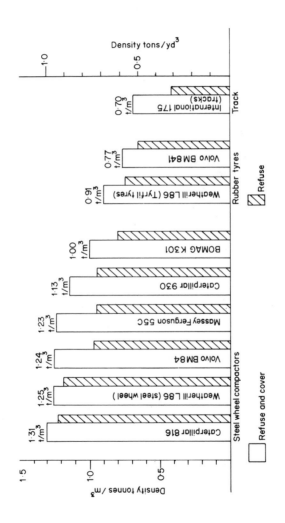

Appendix 5:   Safety report

	Area Superintendent's Weekly Safety Report In respect of sites reference	
	Inspected	Defects Found
Fence and Gates Notice Boards Amenity Unit Garage Site Machines Access Roads Cover Material Refuse Fully Covered Working Faces Excavations Fires Leachates Bonding Culverts/Drains Shafts Stability of Tipped Areas Stability of Original Site Face Proximity of Power Lines Rodents Insects		
Additional Remarks		

Date ....Operations Superintendent ....

Assistant Operations Manager ....

Operations Manager ....

	Inspected	Defects Found
Waste Disposal Treatment Plant .... Weekly Safety Report		
Fence, Gates and crash barriers Notice Boards Access Roads/Parking Areas Road Markings Drain and Inspection Covers Shafts Drains and Culverts Access and Emergency Doors Lighting (Internal, external, emergency) Public Address System/Intercon Fire Alarm Lifts (check alarm) Fire Hoses and Extinguishers Dust Extraction Equipment Machinery and Conveyor Guards Ladders, Stairways and Walkways Castall Key System/Door interlocks Safety Barriers Emergency Generator Heating Amenity Facilities Office Block Rodents Insects Building Fabric (including windows) Hoppers/Refuse Storage Areas		

Additional Remarks

Date ....Operations Superintendent ....

Assistant Operations Manager ....

Operations Manager ....

Practical Waste Management
Edited by J.R. Holmes

7

Behaviour of wastes in landfill—leachate

ALAN PARKER, CCHEM, FRSC, MISWM
Environmental Safety Group, Harwell Laboratory

ABSTRACT

Coming from an executive of Harwell Laboratory's Environmental Safety Group, this chapter describes the behaviour of wastes in landfills with particular emphasis on the generation of leachates and their likely impact on underlying water sources. The methods of treatment, the principal hazardous chemicals commonly found on industrial waste sites, and the design of sites are described. The policies of containment versus dispersal and attenuation of wastes, the use of impervious liners, and assessments of the difficult decisions to be taken complete the contribution.

INTRODUCTION

The Key report[1] which was issued in 1970 gave the types and amounts of industrial toxic wastes generated in the UK. A survey of disposal methods showed that landfill was the principal disposal route. Instances of groundwater pollution were given in the report.

Disclosures about cyanide dumping in 1971 and other happenings led to the Deposit of Poisonous Waste Act 1972. The need for the Department of the Environment to give guidance to Waste Disposal Authorities on site selection and control was instrumental in their setting up a 3-year research programme ending in March 1977. This was carried out by the Harwell Laboratory, the Water Research Centre, and the Institute of Geological Sciences. Twenty landfills were selected for investigation to give information on the behaviour of many industrial wastes when deposited in sites situated in a wide variety of geological strata. This work has subsequently been extended for a further 2 years but the emphasis has changed to study also the behaviour of domestic waste. In particular, investigations are being carried out on the effect on gas and leachate production of various refuse deposition techniques. Data obtained on the density of refuse in these experiments have been summarized by Campbell and Parker.[2] Co-disposal of hazardous and domestic waste is also being studied since reactions within the landfill may minimize the environmental impact of many hazardous wastes by immobilizing such materials or by converting them into less hazardous compounds.

TYPES OF WASTE

Estimates of the quantity of notifiable wastes in the UK vary, but a figure of about 4–5 million tonnes appears reasonable compared to 25–50 million tonnes industrial waste plus 20 million tonnes domestic.

A classification of difficult wastes is given in Appendix 4 of the DoE Waste Management Paper No. 4[3], dividing these into various groups and subgroups. Obviously when faced with disposal of any wastes including hazardous materials, questions such as the following should be asked:

(1) Is landfill the optimum disposal route?
(2) Can a suitable landfill be found within an economic distance of the source of the waste?
(3) If landfill is to be used what should be done to prevent problems due to gas and leachate production?
(4) Can the use of liners make an originally unsuitable site acceptable?
(5) Should a containment site be used or will a 'dilute and attenuate' philosophy be more appropriate?
(6) Is the deposited material likely to react with other wastes on the site or can it affect the health of site workers, e.g. acid wastes, asbestos?
(7) Should the waste be specially treated before landfilling; e.g. chemical fixation or sealing in concrete?

GASES IN LANDFILLS

The main gases to be evolved will be methane, carbon dioxide, and hydrogen resulting from the decomposition of organic material. Traces of many other organic compounds have been identified in landfill gas. Some of these, particularly mercaptans and esters, are responsible for the typical smell associated with landfill sites. Hydrogen sulphide is often present, particularly if significant quantities of sulphate-containing waste, e.g. plasterboard impregnated with gypsum, are present. Smell problems can be minimized by use of adequate cover or by installation of gas interceptor systems followed by burning the methane collected together with combustible impurities.

The relative quantities of the major gases will depend on the age of the landfill. This variation in composition has been investigated by various workers including Farquhar and Rovers,[4] while conditions favouring methane production have been reviewed by Rees.[5] It is found that up to about 20% of hydrogen is present in landfill gas soon after refuse is deposited. This gas, together with any oxygen, quickly disappears and is replaced by nitrogen and carbon dioxide. After some months have elapsed, methanogenic bacteria in the refuse start to multiply so that methane concentrations build up slowly to about 65%. The residual gas is carbon dioxide together with a little nitrogen and traces of impurities. This gas mixture is of interest:

(1) as a hazard, e.g. by lateral diffusion from site with methane possibly causing explosions in nearby buildings;[6]
(2) as an energy source: collection and utilization of methane from large sites, e.g. Palos Verdes, Mountain View and other sites in the United States is being practised.[7]

The rate of gas production will depend on the moisture content of the waste as well as on its age, but the presence of some hazardous materials may inhibit the decomposition processes thus minimizing gas release. Vapours from deposited solvents may cause a potentially hazardous situation due to their possible flammability and toxicity.

LEACHATE PRODUCTION FROM LANDFILL

When rain falls on freshly deposited refuse, the latter will gradually become saturated (at least in the UK) until the field capacity is attained. When this is exceeded then leachate will be formed. Leachate production is shown diagrammatically. Once field capacity is reached then it is estimated, in the southeast of the UK, that 10–20% of the annual rainfall can appear as leachate. However, this can be minimized by good management techniques, e.g. by capping the site after refuse has been quickly built to its final level thus preventing water ingress, and by preventing surface water from entering the

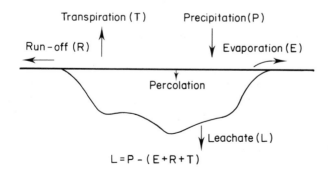

$$L = P - (E + R + T)$$

Figure 1

Table 1   Composition of leachate (mg/ℓ)

	Fresh leachate (0–2 years)	Old leachate (>5 years)
Total organic carbon	13,800	70
Fatty acids		
C_2	1,340	not detected
C_3	660	not detected
C_4	2,000	not detected
C_5	980	not detected
C_6	1,090	not detected
Protein	1,750	16
Carbohydrate	3,245	1.3

landfill. Landfill leachate, especially from industrial sites, is a nuisance since it may pollute surface or underground water sources. The composition of the liquid is often complex and can vary with the age of the landfill with the organic content decreasing with age (see Table 1).

When industrial waste is co-deposited with domestic waste then the complex mixture resulting from domestic refuse will contain an even greater variety of chemicals. For example, from a site in the United States at least 42 organics have been identified including camphor, di-ethyl phthalate, tri-*n*-butyl phosphate from plasticizers, *p*-cresol from creosote, and *o*-xylene used in phthalic anhydride manufacture.[8]

Leachate treatment methods can be divided into several categories (see Table 2).

REACTIONS OF SOME SPECIFIC INDUSTRIAL WASTES IN LANDFILL SITES

Much useful information on behaviour of hazardous materials has been given in the recent DoE report on research in this subject.[9] This work involved a study of 20 landfill sites, and further data were obtained by studies involving lysimeters and monoliths together with experiments in the laboratory on a smaller scale using columns and drums. The main findings may be summarized:

(a) pollution plumes around landfill sites are quite often restricted in extent;
(b) the site geology and hydrogeology, especially the presence of an unsaturated zone, are of great significance in determining the degree of attenuation of leachates;

Table 2   Leachate treatment categories

Method	Notes
1. Spray over nearby land	Spray rate up to about 56000 litres/hectare/ day in the UK. Is suitable land available? Will smell cause problems? Process likely to be more effective during dry weather. Liquid from industrial landfill may contain objectionable chemicals.
2. Spray over landfill surface	Elevated temperature in landfill will help evaporation. Decomposition of waste assisted by increased moisture content. Smaller volume of more polluted leachate obtained.
3. Treat at sewage works	Cost of tankering? Is a foul sewer nearby? Has the works sufficient capacity? Steady hydraulic and BOD loading must be maintained.
4. Aerobic treatment	Store in lagoons and oxidize with air or oxygen. Use of trickling filters—problem with blockages due to precipitation.
5. Anaerobic biological treatment	Conversion to methane and carbon dioxide—only on pilot-plant scale so far.
6. Chemical treatment	Oxidation with hydrogen peroxide or ozone. Use of precipitants and coagulants, e.g. lime, alum. More suitable for removal of colour and suspended solids rather than dissolved organics.
7. Physical treatment	Reverse osmosis. Polishing of effluent with activated carbon.

(c) attenuation mechanisms (defined broadly to include dilution) are available in the landfill and underlying strata, which are extremely beneficial if used with discretion.

The mechanisms of attenuation in the landfill itself are of particular interest since these will tend to reduce the pollution load of any leachate escaping from the site. This will mean that attenuation underneath the site need not be so effective, e.g. the thickness of the unsaturated zone may be reduced. During this experimental programme and associated investigations valuable information was obtained on the behaviour of many potentially hazardous materials within the landfill. For many of these, e.g. mercury, oil and PCBs, co-deposition with domestic waste has significant advantages since this waste has significant retention capacity for these materials. Specific wastes studied included the following:

(a) *Halogenated organics.* A combination of evaporation and adsorption largely prevents their release to groundwater. If oils are present in the refuse then halosolvents tend to remain associated with this phase.

(b) *Cyanide.* Various decomposition and elimination mechanisms have been identified. For instance, conversion to volatile hydrogen cyanide, formation of complex cyanides, hydrolysis to ammonium formate, formation of thiocyanate, and biodegradation may occur. In a pilot scale experiment in which cyanide-containing waste was admixed with domestic waste, after 3 years less than 3% of the added cyanide was found in the leachate. Thus co-disposal enhances decomposition but for cyanide an initial chemical treatment is strongly advocated.

(c) *Heavy metals.* Electroplating sludges have been co-disposed in landfill but have been exhumed virtually unchanged after 2–3 years. Chromium is often found in industrial wastes, e.g. in tannery wastes, and can pose an environmental hazard if present in a soluble form, e.g. as chromate or dichromate. However in landfill such compounds are reduced, in the presence of organic material, to the trivalent state so that at the normal near-neutral pH of the site chromium is precipitated as the insoluble hydroxide.

Mercury can be deposited in batteries, fluorescent tubes, demolition wastes, etc. There is evidence that it is immobilized as the insoluble sulphide under the anaerobic conditions in the fill. In some sites where waste containing clay has also been deposited it may become firmly bound by adsorption or ion exchange. The quantities of mercury-containing wastes which can be landfilled have been suggested by the Department of the Environment.[10] The objective should be not to increase the assumed national average of 2 mg mercury/kg fill by more than a further 2 mg/kg.

(d) *Acids.* Liquid acidic wastes are often delivered to landfill sites for disposal. It is normal practice to neutralize them prior to deposition in trenches or shallow lagoons in the fill. It is essential that the inherent neutralization capacity of the waste is not exceeded, or dissolution of heavy metals may occur and microbial activity be inhibited. It has been found that 1 kg fresh waste can neutralize 22 g sulphuric acid and 1 kg decomposed waste is required for 33 g of this acid.

(e) *Oils.* Adsorption on to fill materials is an important attenuation mechanism. Studies at two sites indicated no free drainage of oil at a concentration of 5% w/w. Simulated landfill experiments also showed that domestic waste had the capacity to absorb and retain cutting oil, amounting to at least 2.5 g/kg dry waste. There is interest also in the disposal of marine oil spillages in landfill and their subsequent behaviour, since this disposal route is probably the one which is most economic and convenient especially in the UK.

(f) *PCBs.* These have been found in certain industrial landfills and are often derived from small capacitors, distillation residues, and filter cakes. Their generally low solubility and poor biodegradability suggests that once deposited in a landfill they are retained within it. Generally there is no evidence that the presence of other organics influences the mobility of such substances although the presence of solvents could have a marked effect. Site investigations show that PCBs tend to be readily adsorbed by suspended matter in aqueous solution. Measurements from two landfill sites in South Wales showed PCB concentrations in leachate ranged from 0.01 to 0.05 $\mu g/\ell$.

(g) *Phenols.* These can be a major problem since the WHO limit for phenols in drinking water is 0.002 mg/ℓ and many industrial wastes contain such compounds. They can also be produced by breakdown of timber and sawdust, and by pyrolysis of rubber and similar materials. Evidence has been found for the escape of phenols from landfill sites although phenols can be biodegraded under both aerobic and anaerobic conditions. Some adsorption occurs on domestic waste but since the process is reversible it is probable that this mechanism would only delay escape from the landfill. Many tarry wastes will also contain phenol as their main potential pollutant.

(h) *Solvents.* These may be lost by evaporation during deposition or may be adsorbed on the refuse where they may undergo biodegradation. Laboratory experiments show that it is difficult to predict the extent to which these processes will occur. The method of deposition appears to be important with volatile, water-miscible solvents evaporating more readily from a pool of liquid than from domestic waste.

In order to obtain the maximum attenuation from co-disposal practice ideally notifiable and domestic waste should be mixed intimately. Solids or sludges should not be deposited so that they form discrete 'pockets' in a domestic waste matrix. In practice, it is normal to spread these over the working faces of the landfill and cover with fresh domestic refuse as the working face advances. Several metres of domestic waste should be left ideally between the notifiable waste and the base of the site. Liquids are pumped into trenches or bunded areas in the landfill but care must be taken not to overload the absorptive capacity of the waste. Co-disposal of drummed waste can pose problems since drums should be opened and their contents removed in order to obtain an intimate mix. However this may present a hazard so that any potentially dangerous drums should be buried unopened in trenches and quickly covered with other waste. Care should be taken not to mix drums whose contents could react together violently. Also 'nests' of drums should not be allowed to accumulate since problems with subsidence may occur. The position of any notifiable wastes should be marked on the site plan.

USE OF LINERS IN LANDFILL SITES

Water Authorities may insist on installation of liners to prevent leachate from reaching groundwaters. Liners form two main groups:

(1) synthetic materials, e.g. polyethylene, PVC, butyl rubber, and ethylene propylene diene monomer (EPDM);
(2) natural materials, e.g. bentonite and other clays, asphalt, and concrete.

Synthetic liners are not cheap and considerable care is required in engineering the site before their installation. They are normally covered with a layer of sand or soil to protect them from sharp objects in the waste and from delivery vehicles. The joining technique for various sections of liners has been much improved during recent years. A choice of synthetic material for site lining is difficult to make, since it is necessary for the liner to remain intact for probably 20–30 years after its installation. Obviously knowledge of the possible interaction between the liner and any chemicals deposited on the site is essential while the work of Haxo[11] should eventually give valuable information on the effect of domestic leachate on synthetic liners.

More information is available on the interaction of leachate with natural materials, e.g. clays. For example, the absorptive capacity for heavy metals of clays, e.g. montmorillonite, is well established. Bentonite clay, which has often been used for lining purposes, has the useful property of swelling in contact with liquids thus tending to seal any imperfections which may appear.

If liners are to be used then operators must realize that these will cause the formation of containment sites, and potential problems with water infiltration must be overcome. Sealing of the top surface with an impermeable material, e.g. clay, followed by grading, covering with soil, and grassing is often used. The use of a synthetic membrane above the waste is another possibility. Research on permeable liners—e.g. limestone chippings, pulverized fuel ash, or builders' rubble—is being carried out in the UK at present. Hopefully such liners will attenuate leachate without the problems associated with a containment site.

'CONTAINMENT' VERSUS 'DISPERSE AND ATTENUATE'

Containment sites will not allow leachate to escape unless the site fills with water, thus causing leachate to flow over the rim of the site. Containment sites initially appear attractive especially for the disposal of hazardous wastes but good management of possible water inflow is essential.

The converse of the 'containment' philosophy is that of 'disperse and attenuate'. In this, leachate is allowed to escape from the landfill with the pollutants contained therein being attenuated by the surrounding strata. Such attenuation mechanisms have been studied both in the laboratory and in lysimeters during the current DoE programme. Of particular relevance is the behaviour of heavy metals such as lead, cadmium, and nickel in strata such as the Lower Greensand which contains a substantial proportion of clays, such as montmorillonite. For instance, lead is strongly adsorbed while nickel is much more mobile. Such research enables predictions of the behaviour of hazardous materials to be made and, together with a knowledge of interactions between wastes within the landfill itself, can prevent overloading of the site. It should be emphasized that, in addition to chemical effects, a knowledge of the hydrogeology of the landfill site is essential. Thus close consultation between interested parties, e.g. Water Authorities and Waste Disposal Authorities, is of paramount importance before any deposition of hazardous materials is permitted. Finally, the problem should be kept in

Figure 2　A well-equipped vehicle used in Harwell Laboratory's Chemical Emergency Service. (Courtesy of Harwell Laboratory)

SITES INVESTIGATED DURING D.O.E. LANDFILL PROGRAMME 1973-7

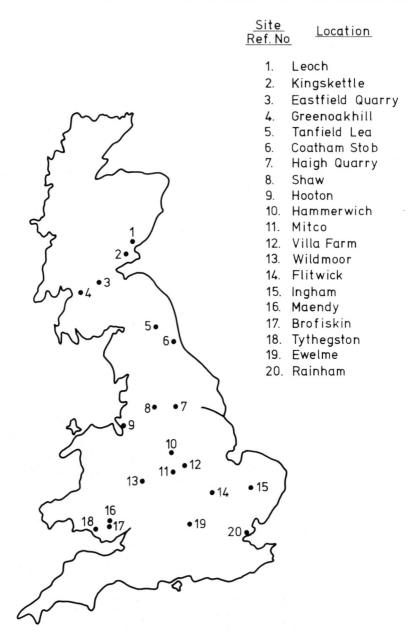

Site Ref. No	Location
1.	Leoch
2.	Kingskettle
3.	Eastfield Quarry
4.	Greenoakhill
5.	Tanfield Lea
6.	Coatham Stob
7.	Haigh Quarry
8.	Shaw
9.	Hooton
10.	Hammerwich
11.	Mitco
12.	Villa Farm
13.	Wildmoor
14.	Flitwick
15.	Ingham
16.	Maendy
17.	Brofiskin
18.	Tythegston
19.	Ewelme
20.	Rainham

Figure 3   Sites investigated during DoE landfill programme, 1973–7

Figure 4 Heavy-duty impermeable linings being placed prior to waste disposal operations. (Courtesy of Harwell Laboratory)

Figure 5 Leachate treatment by spray irrigation over a completed landfill. (Courtesy of Harwell Laboratory)

Figure 6   Landfill Experimental Programme. London Brick Landfill Ltd in conjunction with Harwell Laboratory

perspective since a review carried out in England and Wales in 1972/73 by the
IGS[12] identified about 2,500 landfill sites of which only 51 presented a serious
pollution risk to aquifers. Similarly in the United States a 1977 survey
identified only 190 pollution incidents involving landfills, surface dumps, or
lagoons.

SOME CURRENT RESEARCH IN THE UK

As a continuation of the DoE programme on the behaviour of wastes, two
large-scale experiments have been commissioned. At Edmonton, the Water
Research Centre have constructed a series of large concrete cells lined with
plastic in which toxics may be admixed with domestic waste. Amongst the
materials being studied are solid wastes from gasworks, i.e. spent oxide. The
behaviour of phenolic lime sludges and highly putrescible vegetable waste is
also being investigated. The cells, which contin about 250 tonnes of waste,
have facilities for the collection of gas and leachate samples, and the
monitoring of temperatures.

At Stewartby, Bedfordshire, the DoE have funded an experimental
programme which is being operated by the Harwell Laboratory in conjunction with London Brick Landfill Limited. In a worked-out clay pit eight large

cells have been constructed, each capable of containing about 1,200 tonnes of refuse. Some of the cells have been filled using the same type of waste but with different machinery or deposition techniques. Useful data are being obtained on density, gas, and leachate composition.

Other experiments are being carried out to study the effect of varying the percentage of moisture in refuse in an attempt to optimize methane production. Such information is of value if large landfill sites are to be run in order to obtain methane in an economic fashion. In other cells the co-deposition of acid wastes and oils from different sources is being studied. In addition, work is being continued at various selected landfill sites in order to gain a greater understanding of processes which occur within a landfill site—a subject which, until a few years ago, lacked much basic information.

Finally, information is required at times to predict the behaviour of potentially hazardous materials in landfill sites. For example, Waste Disposal Authorities may require guidance as to whether a certain waste can be landfilled or whether an alternative disposal route, e.g. incineration, should be chosen. Attempts are being made to develop a standard leaching test in which the waste is treated with an appropriate extractant, e.g. dilute buffered acetic acid.[13] Measurement of the pollution load in such an extractant should give an assessment of its potential environmental impact.

REFERENCES

1. *Disposal of Solid Toxic Wastes.* HMSO, London, 1970.
2. Campbell, D. J. V., and Parker, A. Density of refuse after deposition using various landfill techniques. *Solid Wastes* (August 1980), pp. 435–440.
3. Department of the Environment, Waste Management Paper No. 4. *The Licensing of Waste Disposal Sites.* HMSO, London, 1976.
4. Farguhar, G. J., and Rovers, F. A. Gas production during refuse decomposition. *Water, Air and Soil Pollution* 2 (1973) 483–495.
5. Rees, J. F. The fate of carbon compounds in the landfill disposal of organic matter. *Journal of Chemical Technology and Biotechnology* 30 (1980), 161–175.
6. Bromley, J. and Parker, A. Methane from landfill sites. *International Environment and Safety* (August 1979), pp. 9–11.
7. Stearns, R. P. Landfill methane: 23 sites are developing recovery programs. *Solid Waste Management* (June 1980), pp. 56–59.
8. Dunlap, W. K. Organic pollutants contributed to groundwater by a landfill *EPA–600–9–76–004*, pp. 96–110 (1976).
9. Department of the Environment. *Research on the Behaviour of Hazardous Wastes in Landfill Sites.* HMSO, London, 1978.
10. Department of the Environment, Waste Management Paper No. 12. *Mercury-bearing Wastes.* HMSO, London, 1977.
11. Haxo, H. E. Liner materials exposed to municipal solid waste leachate. Third interim report. *Water Pollution* 600–2–79–038 (1979).
12. Gray, D. A., Mather, J. D., and Harrison, I. B. Review of groundwater pollution from waste disposal sites in England and Wales, with provisional guidelines for

future site selection. *Quarterly Journal of Engineering Geology,* **7** (1974), 181–196.

13. Wilson, D. C., and Waring, S. The safe landfilling of hazardous wastes. Paper presented at 3rd International Congress on Industrial Waste Waters and Wastes. Stockholm, 6–8 February 1980 (sponsored by International Union of Pure and Applied Chemistry).

Practical Waste Management
Edited by J.R. Holmes
©1983 John Wiley & Sons Ltd

8

Behaviour of wastes in landfill— methane generation*

ALAN PARKER, CCHEM, FRSC, MISWM
Environmental Safety Group, Harwell Laboratory

ABSTRACT

A review of the theoretical background and practical progress in the generation and extraction of landfill gas from landfill sites. Both a menace and a beneficial source of energy, landfill gas is a feature of the bigger, deeper, landfill sites increasingly a feature of the waste management scene in Europe and the United States.

INTRODUCTION

Waste usually contains organic material of different types, e.g. vegetable material, paper, wood, and plastics. The total quantity of organics will vary with the source of the refuse and can show marked variation from country to country. For example waste which is largely industrial/commercial deposited in a landfill in Hong Kong has a markedly different composition from that of UK domestic waste (see Table 1). The low vegetable/putrescible content of the Hong Kong waste will tend to minimize initial gas production but the overall high organic content should ensure that the total amount of gas produced will be greater than the UK waste although it will be produced over a much longer period of time. This is due to the fact that organic materials vary widely in their decomposition rate. For example, vegetable waste will break down rapidly while, at the other extreme, plastics can be exhumed from landfill sites usually torn but otherwise untouched.

---

* This chapter was first presented as a paper read at the National Association of Waste Disposal Contractors, Practical Waste Management Course, 1981.

Table 1 Comparison of refuse composition, Hong Kong and UK (percentages)

	Hong Kong	UK refuse (average, 1973)
Ash	1.2	19
Vegetable/putrescible	2.6	18
Paper	5.0	33
Textiles	9.0	3.5
Rubber/leather	1.0	—
Plastics	8.7	1.5
Wood	63.2	—
Metals	2.9	10
Glass	0.3	10
Miscellaneous	6.7	5

DECOMPOSITION OF WASTE

When solid waste is deposited in a landfill aerobic decomposition will occur for a short period until trapped air is consumed. When this occurs anaerobic micro-organisms become dominant and these initially break down organic compounds, principally carbohydrates, to form fatty acids—e.g. acetic, propionic, and butyric acids. This production of acids is accompanied by evolution of carbon dioxide, hydrogen, and nitrogen. Typical leachate produced during this phase has been found to contain the following acids:

Acetic	3,800 mg/ℓ	*n*-Valeric	2,100 mg/ℓ
Propionic	1,600 mg/ℓ	iso-Valeric	70 mg/ℓ
n-Butyric	3,500 mg/ℓ	Caproic	3,700 mg/ℓ
iso-Butyric	145 mg/ℓ		

After oxygen has disappeared, then methanegenic bacteria become active; they decompose organic acids into methane, carbon dioxide, and water. These bacteria are strictly anaerobic and even small quantities of oxygen are toxic to them. In order to maximize methane production the system must be in balance, i.e. the bacteria should use the organic acids as rapidly as they are generated. In order to optimize conditions control of alkalinity to about pH 7 (e.g. by the addition of calcium carbonate) may be considered although in a landfill site such sophisticated control is seldom practised. The desirable sequence of events in landfill has been described by John Rees[1] and is shown in Figure 1. As can be seen stage I involves rapid displacement of air from the waste. During stage II a small quantity of hydrogen is evolved accompanied by much carbon dioxide. Methanogenic bacteria become active during stage III so that the methane concentration builds up to about 60–65% with carbon

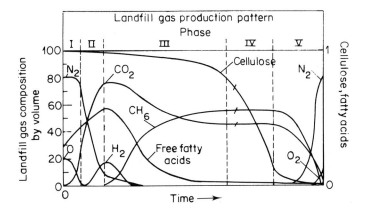

Figure 1 Landfill gas production pattern. I, II, III, IV, and V refer to the landfill gas production pattern phase

dioxide forming the bulk of the residual gas. The time associated with each of these stages will vary from site to site. Factors influencing the rate of gas production include:

(1) size and composition of refuse;
(2) age of the refuse;
(3) moisture content of refuse;
(4) temperature conditions on the landfill;
(5) quantity and quality of nutrients;
(6) pH of liquids in the landfill;
(7) density of refuse.

It is obvious that some of these factors can be controlled by the site operator. For example, input might be adjusted with various operating techniques being used to vary the density. In future it may be desirable to alter various parameters so that reaction rates can be accelerated or retarded, depending on whether the gas is to be released at a steady predetermined rate or released rapidly so that the site may be restored for other uses as quickly as possible. Co-disposal of industrial waste with domestic waste may inhibit methanogenic decomposition. For example carbon tetrachloride or chloroform have been shown to have detrimental effects on this process. Conversely enhancement of the organic content of the system by the addition of sewage sludge or agricultural wastes will increase gas production, as will a high mosture content (60–80%). It has been shown, for example, that gas production rates increase after heavy rainfall. In practice it may be found that while it is desirable to prevent water ingress into landfill to prevent leachate formation this is at variance with a desire to maintain high moisture loadings to enhance gas production. Obviously the optimum operating strategy for each individual site must be clearly thought out.

Once waste has been deposited then there will be an interval, i.e. stages I and II (see Figure 1), until methane starts to be evolved in significant quantities. This usually amounts to about 12–18 months. The time period for stages III and IV will depend on factors given above but it will normally last many years. It has been suggested that 50% of available landfill gas will be evolved within 5–15 years of placement of the waste. Traces of methane may still be detected after many decades, as was shown by measurements at a landfill completed around 1900 where several hundred ppm of gas were still present.

It is possible to calculate the quantity of landfill gas which can be evolved from a knowledge of waste composition. Calculations by Tchobanoglous *et al.*[2] indicate that 6.5 ft^3 of landfill gas (3.3 ft^3 of methane) could theoretically be obtained per pound of solid wastes. This is equivalent to 406 ℓ/kg of landfill gas. Other calculations by Encom Associates suggest gas yields of 3–8 ft^3 of landfill gas per pound of wet refuse, i.e. 1.5–4.3 ft^3 methane. It must be realized that in practice it will be impossible to recover all of this gas since

(1) gas may be lost through the top cover and sides of the landfill;
(2) there may be a delay before gas recovery is started; and
(3) as the waste ages methane production rates will fall until the point is reached when it is no longer economic to recover the gas.

Figures have been given by Encom Associates which show low recovery rates of landfill gas from operating sites in the United States. These range from 0.043 to 0.120 ft^3/lb refuse/year. Such low recovery rates may be due to the generally dry nature of the refuse since the data refer to landfill sites in California. However at a large landfill containing 1 million tons of waste and assuming that 0.1 ft^3 gas/lb of waste can be recovered then an annual landfill gas recovery of 2.24×10^8 ft^3 (5×10^6 m^3) could be possible. Since the energy contained in a cubic foot of landfill gas is approximately 500 BTU—about the same as for coal gas—then the total annual energy output would be about 1.1×10^{11} Btu. A typical figure for the energy requirements for an average house in the UK is estimated to be 5×10^7 Btu, i.e. enough heat for over 2000 homes.

COMPOSITION OF LANDFILL GAS

When anaerobic decomposition is proceeding steadily then the methane concentration varies between about 45% and 65% with the remaining gas being mainly carbon cioxide. Traces of impurities are also present. Ham *et al.*[3] give the composition of a typical landfill gas. In the active sites which we have investigated in the UK we find methane concentrations to be generally higher than indicated above, with 58–65% being common. In at least one site in the UK and one in the US methane concentrations of approximately 80% have been reported. The reason for this is not clear but it is possible that the hydrogeological conditions of the site cause scrubbing and removal of the carbon dioxide from the landfill gas.

Table 2   Typical landfill gas composition and characteristics

Component	Component percentage (dry volume basis)
Methane	47.5
Carbon dioxide	47.0
Nitrogen	3.7
Oxygen	0.8
Paraffin hydrocarbons	0.1
Aromatic and cyclic hydrocarbons	0.2
Hydrogen	0.1
Hydrogen sulphide	0.01
Carbon monoxide	0.1
Trace compounds*	0.5

Characteristic	Value
Temperature (at source)	106 °F
High heating value	476 Btu/scf
Specific gravity	1.04
Moisture content	Saturated (trace compounds in moisture)†

* Trace compounds include sulphur dioxide, benzene, toluene, methylene chloride, perchlorethylene, and carbonyl sulphide in concentrations up to 50 ppm.
† Trace compounds include organic acids 7.06×10^{-3} ppm and ammonia 7.1×10^{-4} ppm.

The smell of the gas is due to the presence of impurities. Work at the Harwell Laboratory has shown the presence of at least 30 such compounds which can be grouped into the following classes:

(1) saturated aliphatic hydrocarbons;
(2) unsaturated aliphatic hydrocarbons;
(3) esters;
(4) alkyl substituted benzenes;
(5) terpenes;

(6) volatile sulphur compounds, e.g. alkyl mercaptans and dialkyl sulphides;
(7) miscellaneous compounds including chlorinated hydrocarbons.

The odour is due to a blend of many of these components with the esters and sulphides being the most pungent. Concentrations of many of the components are about 100 μh/ℓ although concentrations of organo sulphur compounds are probably an order of magnitude lower.

LANDFILL GAS AS A PROBLEM

It is being recognized that under some circumstances landfill gas can pose an environmental hazard due to the following factors:

(1) When the concentration range of methane in air is in the range 5–15% then an explosion can occur if a source of ignition is present.
(2) Landfill gas can contain up to 65% methane and under these circumstances the residual gas is mainly carbon dioxide. Although the methane causes the main hazard, the carbon dioxide occasionally has been found to collect preferentially in cellars of buildings forming a potentially asphyxiating atmosphere.
(3) Gas evolution occurs over many years but even a slow rate of methane production can result in the concentration of the gas gradually building up in confined spaces so that the lower explosive limit is eventually exceeded.
(4) If the strata on which the landfill is situated are permeable, e.g. sands or gravels, or fissured, e.g. chalk or sandstones, then lateral as well as vertical diffusion of landfill gas can occur. Lateral diffusion will be encouraged by covering the landfill surface with impermeable material, e.g. moist clay.

Methane problems were of little consequence when sites were small and, more importantly, situated outside the boundaries of towns or villages. However, in recent years there has been great pressure to acquire land for building so that in many instances filled and disused landfill sites have been incorporated into urban areas. Often such sites are levelled and grassed over to be used for recreational purposes. This is a very satisfactory end-use, especially as any settlement which occurs can be filled in. Very occasionally poor growth of grass is found, due probably to displacement of oxygen from around the root system, but it appears that this can be overcome by compacting well the top cover on the landfill before sowing grass. Problems may arise if sports pavilions are erected on such sites. An example is given in Case 2, below.

Usually more serious trouble arises if there is pressure to erect buildings on disused landfill sites without realizing that landfill gas can be a potential hazard. In order to avoid trouble the safest, cheapest, and easiest option is *not* to erect any buildings on landfill sites at all. However, it must be pointed out that if it is essential to develop such sites then buildings can be designed to overcome gas problems although this will result in increased costs of construction; provision of ventilation systems capable of operating for many years may be necessary. The design of the buildings should be such that settlement problems which are normally associated with landfill sites can be eliminated. For these reasons it is probably more practical and economic to erect large buildings, e.g. factories or blocks of flats, incorporating effective safeguards than to contemplate individual houses.

A few case histories of problems encountered in the UK in buildings erected on or close to a landfill are given below. It should be emphasized that the latter situation involving lateral gas migration is more insidious and its potential for causing an environmental hazard may not be so readily recognized by Planning Authorities and other bodies.

Case 1

A sports complex was erected on a recently completed landfill. Domestic waste had been used to infill marshy land where the water table was close to the surface. An indoor bowling rink had been built with its base slab being placed directly on the waste which was several metres thick. Landfill gas accumulated in a toilet, was ignited, and the resulting explosion caused injury to a woman. Two interesting points were noted. Firstly it was impossible to find the major route of gas ingress although minor leakage paths were found, e.g. around poorly sealed pipes entering the building. Secondly, evidence suggested it was probable that the methane evolution rate had increased with rising temperature at the end of the winter, the so-called 'spring flush'. However the incident resulted in the building being shut for a lengthy period while remedial measures were being carried out.

Case 2

In Essex a sports pavilion was erected on a completed landfill site. Storm water from the roof was drained via vertical pipes into soakaways which were in the landfill itself. Gas therefore was able to travel up the pipes which were partially blocked and collect in the roof void and then gradually diffuse into the pavilion itself. This case shows again that pipes, be they service ducts, drains, etc., can allow ingress of landfill gas into buildings. Care must be taken during the design stage to eliminate such routes.

Case 3

Houses and detached garages had been built on material deposited about 20 years ago in worked-out sand/gravel pits. Here the fill consisted of industrial and domestic waste together with much inert material. Subsidence has been the main problem so that several of the buildings have had to be demolished. However, significant methane concentrations have been detected in manholes and in cracks in pavements; in the houses themselves the methane concentrations found appear to be due mainly to leaking gas appliances rather than to landfill gas.

Case 4

A school was built on the edge of a shallow valley which was filled with domestic waste. Landfill gas diffused laterally into the school playground where it appeared in cracks. In order to overcome the problem, closely spaced boreholes were drilled between the school and the landfill and a PFA/cement grout injected to form an impermeable grout curtain. This remedial work is believed to have cost around £15,000.

Case 5

Lateral landfill gas diffusion is causing a problem in a housing estate in Essex built alongside a partially completed landfill some 20 m deep. The landfill is in a sand/gravel quarry and gas has been found in trenches some 200 m from the edge of the site. The site has been filled in discrete areas with inert and domestic wastes; measurements clearly show that lateral diffusion is confined, as was anticipated, to the ground adjacent to that area which had received domestic waste. Measures to overcome the problem are currently being investigated.

Environmental problems of this type must be kept in perspective. Although incidents have occurred, as far as I am aware there have been no fatalities due to landfill gas explosions in the UK. The message I want to give is that problems can occur and that would-be developers of landfill sites should be aware of these and get expert advice as to the best way of overcoming them. Ignorance may not be bliss!

EXPLOITATION OF LANDFILL GAS AS A FUEL SOURCE

As indicated in the section entitled 'Decomposition of waste', the annual energy output in the gas from a large landfill is considerable. The technology for the extraction of this gas, coupled with its utilization, has been pioneered in the USA although both the UK and Western Germany are starting to exploit it, but both countries are about 4–5 years behind the USA. Interest in the exploitation of landfill gas sprang from the need to install gas interception systems to prevent lateral migration of gas through adjacent porous strata. The most efficient of these systems has perforated pipes inserted into the landfill around the perimeter of the site. These pipes are subjected to reduced pressure, being linked by header pipes, with the landfill gas being flared off in order to prevent odour or fire problems. The next logical step is to use the gas as a source of useful energy as a means of raising steam, heating water, or by the generation of electricity. Since energy costs are rising there is every incentive to use landfill gas providing that the economics are right. Obviously if an anti-gas migration system has to be installed, economic considerations will be different from those in which the gas will be used solely as an energy source. The latter scenario will be considered in more detail. In order to effect an economic recovery several criteria must be met.

(1) The landfill must be of considerable size. Thus in the USA the minimum size is considered to be a landfill having 1 million tonnes of waste in place with further waste being introduced at a rate of at least 500–1,000 tonnes/day.
(2) The landfill should be at least 10–13 m deep. This minimum depth is necessary so that gas may be extracted at a reasonable rate without air being drawn in through the top cover. Ingress of air will result in a decrease in methane concentration accompanied by an increase in carbon dioxide and nitrogen content which will lower the calorific value of the gas.
(3) Careful consideration should be given to the method in which the gas will be used. The distance of potential users from the site will have an important effect on economics.

Several options have been considered in the USA for the most efficient use of the gas. At present these are the following:

(1) Purification of the landfill gas to remove firstly moisture and then impurities, together with the carbon dioxide, so that essentially pure methane remains (high Btu gas). This gas can then be introduced into a natural gas pipeline. At the Palos Verdes and Mountain View landfills in California this purification is carried out by the use of molecular sieve adsorption. At the CID landfill in Illinois and the Operating Industries landfill in California carbon dioxide is absorbed using the dimethyl ether of polyethylene glycol. Recovery of carbon dioxide from the landfill gas may be a worthwhile bonus.
(2) Use the gas with the minimum of purification, i.e. with only the removal of moisture and any particulate material. The dry gas can then be pumped away and used in boilers to generate steam or to heat water which could be used in greenhouses or even in a district heating scheme. In the UK

such gas is being used for the firing of bricks. This type of gas (medium Btu) could be used for the generation of electricity on-site using a steam turbine, gas turbine, or gas engine; this application is being investigated extensively in the USA.

(3) Landfill gas could be used as a feedstock for the production of chemicals such as methanol.

The technology for the extraction of landfill gas has been extensively developed. Typical recovery wells are 12–36 inches (30–90 cm) diameter often drilled to the base of the fill. A well casing (3–8 inches (75–200 cm) diameter) which has perforations in the bottom one to two thirds of its length is inserted with coarse backfill over this distance. The top of the outer well is plugged with bentonite or concrete. The well casing top is fitted with a control valve with individual wells being joined by a manifold to an extraction pump. At present the spacing of wells is decided in a rather haphazard fashion but with the approximate density of 1 well/acre being fairly widely accepted. The American attitude appears to be one of drilling the wells, extracting initially the gas at the maximum rate possible, and then deciding whether additional wells are required. This overcomes the significant variation which can occur between sites but also tends to highlight the lack of detailed knowledge necessary to predict long-term rate of gas generation in landfills.

CURRENT STATUS OF LARGE SCALE LANDFILL GAS SYSTEMS

Table 3 gives some details of the 10 projects in the USA which are on-stream at present. In addition to these, about 30 further sites are either being tested or being considered for gas extraction projects at present. It is also considered that very many more sites in the USA have good potential for economic gas extraction.

The situation in Europe is not nearly so far advanced. In the UK there is only one scheme operating which uses landfill gas as an energy source. This is operated by London Brick Landfill Limited at Stewartby, Bedfordshire and was described by Cheyney and Moss at the recent Harwell landfill gas symposium. At present the gas, which contains about 55% methane, is being extracted at rates up to 500 ℓ/min (17 ft^3/min) and is being pumped to a kiln where it is being used to fire bricks. At the Aveley landfill site in Essex a gas extraction system is being installed by the Greater London Council in conjunction with National Coal Products Ltd. It is planned to pipe the gas to a nearby factory where it will be used as a boiler fuel. Some utilization of landfill gas is practised in Western Germany and Switzerland where it is being used for space heating in greenhouses and offices.

It seems that in the UK the utilization of landfill gas will continue to grow since this is encouraged by the emphasis on fewer but larger landfill sites together with the ever-increasing cost of fuel. The method of most effectively

Table 3 Current US gas abstraction projects (1981)

Landfill	Location	Landfill area (acres)	Project area (acres)	Depth (ft)	Tons in place ($\times 10^6$)	Gas quantity ft^3/day $\times 10^6$	Wells	End-use
Ascon	California	38	38	60	1.8	1.5–1.8	60	Unprocessed as fuel
Azusa	California	325	37	130	6	0.5–4.3	17	Unprocessed as boiler fuel
Bradley East	California	67	67	120	8	2	39	Unprocessed as boiler fuel
CID	Illinois	300	100	130	7	5	14	Upgraded to pipeline
Cinnaminson	New Jersey	64	55	60	2.5	<1	30	Unprocessed; used in steel works
City of Industry	California	120	120	35–40	3.6	0.6	30	Unprocessed, boiler fuel
Mountain View	California	250	25	40	0.55→4	0.7–3	33	Partially purified injected to pipeline
Operating Industries	California	150	150	250+	23	8	56	Purified to pipeline
Palos Verdes	California	176	40	150	20	1.8	12	Purified to pipeline
Sheldon Arleta	California	36	36	120	6	1.4–2.8	14	Unprocessed as boiler fuel

using the gas is still to be decided, and in very many cases local circumstances will be all-important. In my opinion, use of the gas for the on-site generation of electricity appears to be a flexible and attractive option since electricity is cheaper to transport than gas. However this assumes that it is possible to sell the electricity which is generated at an economic price. Legislation controlling the sale of gas and electricity must have an important effect on the way in which landfill gas is used.

Finally, operators of large landfill sites should seriously consider the potential of this new technology since substantial financial savings may be possible by the use of a valuable by-product rather than by allowing it to dissipate into the air where it may cause an environmental hazard.

REFERENCES

1. Rees, J. F. 'Major factors affecting methane production in landfills'. In Proceedings of a landfill gas symposium held at Harwell Laboratory, 6 May 1981.
2. Tchobanoglous, G., Theisen, H., and Eliassen, R. *Solid Wastes: Engineering Principles and Management Issues.* McGraw Hill, 1977.
3. Ham, R. K., Hekimian, K. K. *et al.* 'Recovery, processing and utilization of gas from sanitary landfills'. *EPA–600/2–79–001* (1979).

SUGGESTED FURTHER READING

Encom Associates, *Methane Generation and Recovery from Landfills.* Ann Arbor Science, 1980.
Landfill Methane Utilization Technology Workbook. Prepared by Johns Hopkins University, Applied Physics Laboratory (CPE–8101), 1981.
Proceedings of a landfill gas symposium held at Harwell Laboratory, 6 May 1981.
Stafford, D. A., Hawkes, D. L., and Horton, R. *Methane Production from Waste Organic Matter.* CRC Press, 1980.

Practical Waste Management
Edited by J.R. Holmes
©1983 John Wiley & Sons Ltd

9

Solid waste disposal—problems associated with tipping and the licensing of landfill sites

C. TUNALEY

Waste Disposal Officer, South Yorkshire County Council

ABSTRACT

A Waste Disposal Officer from a Metropolitan County looks at the essence of the function of solid waste disposal, the behaviour of the public, and what is thrown away eventually becoming the responsibility of the waste disposal authorities. The nature of landfill operations, vermin, litter, fire hazards, and an analysis of how the public use waste disposal facilities are descibed. Illustrations are given of site development plans and behavioural analysis in the use of sites.

To an extent the phrase 'solid waste disposal' is a most misleading one to apply to the function, when in fact the material involved need not necessarily be 'solid', may indeed include potentially recoverable 'resources' and in any event cannot be considered as 'disposed of' no matter what we may do to it. The term 'waste management' more accurately describes the practice of handling and treatment of this mixture of household, commercial, and industrial discards.

THE ESSENCE OF THE FUNCTION

Waste management is essentially a materials handling exercise. Admittedly, waste is arguably the most difficult, multi-variable, unpredictable material known and therein lies the source of the many problems associated with waste treatment and tipping alike. Even the most elaborate treatment methods merely seek to alter its form and properties, with the intention of reducing the environmental impact of its ultimate deposit. It must be accepted that all treatment methods will inevitably result in some form of solid residues, which will still require landfilling. It is evident, therefore, that whatever circumstances may exist, or policies prevail, within any authority's area, landfill must always form part of all disposal strategy options, to a greater or lesser extent. It is for these reasons, together with increasing public awareness and demands for improved environmental protection standards, that the Control of Pollution Act first introduced the statutory licensing of the deposit of controlled wastes. This legislation closely followed the establishment of the County Waste Disposal Authorities in 1974 and represented a most significant advance in waste management philosophy and practice. It was unfortunate in the event, that economic pressures prevented its immediate implementation, but this is now being achieved in stages.

In order to discuss the problems associated with the effective licensing and control of landfill operations, it is necessary to identify the properties of waste, the good practice of tipping, and those elements of the environment requiring consideration.

THE MATERIAL

Table 1 shows an example of the analysis of mixed urban domestic waste, but this cannot be called typical, as the characteristics, proportions, and properties of waste from different locations will vary considerably, even from area to area within one town, or from season to season and particularly as a result of socioeconomic differences.

Waste is all things to all men. To the producer it is valueless, uninteresting, and even distasteful and he demands that it be removed for disposal by the authorities, at little or no cost and preferably to someone else's area. To the waste manager it is a heterogeneous mixture of materials, predictable only in its infinite variability. At best it is awkward, and at worst virtually impossible,

Table 1   Analysis of a sample of urban waste (Doncaster, 1980)

Fraction	Percentage of total (by weight)
Paper	25
Plastics	5
Textiles	4
Putrescibles	17
Glass	8
Ferrous metal	7
Non-ferrous metal	1
Miscellaneous (unclassified)	
(a) Combustible	4
(b) Non-combustible	2
Fines (less than 20 mm)	27

to handle. It is the most complex cocktail of opposites imaginable, composed of individual fractions being at one and the same time, large and small, light and heavy, hard and soft, inert and reactive, obnoxious and highly attractive, the conflicting descriptions of its properties are endless. The irony of this situation is that whilst many of these constituent elements may well be of value, the resultant mixture is invariably a considerable liability, requiring disposal at the minimum possible cost to the community. The separation of individual fractions from this mass has not traditionally been possible within the constraints of commercial viability. Indeed, few recovery systems are viable in terms of total energy balance, particularly where they have required commitment to substantial capital investments, high energy consumption, or labour-intensive methods. It is evident, therefore, that the reclamation of land in areas of mineral extraction, industrial activity, and dereliction or unsuitable topography, will undoubtedly continue to represent an attractive end-product of the landfilling of wastes. The creation or return of a valuable land asset for use by industry, or for recreation, agriculture, or re-afforestation, must rank as a worthwhile resource recovery from waste.

Table 2 shows the mix of wastes originating in two typical but very different County authorities and shows the variation which each fraction represents as a proportion of the total. It is appreciated that not all of these wastes are necessarily the statutory disposal responsibility of the Waste Disposal Authority, but their very presence in an area will influence, if not dictate, the disposal strategy and the mix of facilities to be provided. It should be noted

Table 2 Waste generation in two Waste Disposal Authority areas

Type of waste	Type of Authority			
	Shire County (rural with some industry)		Metropolitan County (urban and industrial)	
	Quantity (tonnes)	Percentage of total	Quantity (tonnes)	Percentage of total
Household and commercial	292,000	14	480,000	4
Industrial	452,000	21	404,000	3
Mining and quarrying	1,087,000	52	12,070,000	91
Construction and demolition	280,000	13	245,000	2
Totals	2,111,000		13,199,000	

from these figures just how much of the waste produced in an area is suitable only for landfill.

THE POTENTIAL PROBLEMS

It is the variable properties and unpredictable handling characteristics of waste which results in the majority of problems when the material, particularly domestic and commercial refuse, is deposited in the open air at tipping sites.

Litter

Inevitably, in an uncontrolled situation, this material will give rise to litter with numerous fractions being liable to be liberated by the winds usually common in exposed rural situations. Once airborne, litter will precipitate and collect in surrounding vegetation and is always unsightly, potentially unhealthy, and may be damaging to natural ecosystems in hedgerows and watercourses. In addition the non-degradable elements, such as plastics film, represent a possible danger to wild animals and livestock. The combined answer to this problem is the control of deposit in order to eliminate the initial liberation, together with the effective sealing of the tipped surface throughout the operation and the provision and maintenance of effective litter screens. The attention to visibility screening in the development and preparation of tipping areas is also important in this aspect. The essence of successful litter suppression is the speed with which tipped materials can be deposited in layers; compacted; and blanketed with a covering of soil, inert material, or further layers of other suitable wastes. The old methods of using small tractor-mounted shovels just cannot hope to cope with the volume of the discharge from a modern refuse collection vehicle. It is, therefore, of paramount importance that large-capacity handling machinery be employed. Compaction of tipped materials is an essential prerequisite of a good landfill and the current generation of steel-wheeled landfill compactors have proved their value on many sites throughout the country.

Putrescible materials and vermin

The presence of putrescible materials in waste makes the tipping sites potentially attractive to animal life—particularly to vermin, predatory birds, and insects. In the management of a landfill it must firstly be recognized that these are possible effects, in order to minimize or eliminate them. Here again, the speed of deposit, compaction, and covering are the fundamentals of good successful practice. Densely compacted waste does not provide the habitat desired by vermin such as rats and mice, and well-covered layers prevent

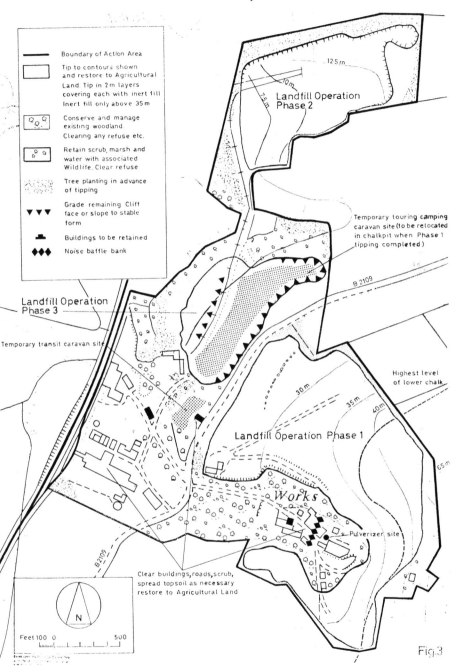

Figure 1   Planning operations on a major landfill site in the South of England. (Courtesy of East Sussex County Council)

disturbance of the surface by birds and animals. The suppression of insects such as flies, wasps, and crickets is simply achieved by regular insecticide spraying and attention to site cleanliness. Deposits to the agreed tipping plan, with the minimization of the working area, is also very effective in the suppression and treatment of vermin. Total eradication of all pests may not be entirely achieved and in any event, re-infection of the site is always a possibility with each new load of refuse, but the essence of the control activity is 'continuity' to prevent infestation. Where a particular problem may exist due to local circumstances, such as proximity to an airfield giving rise to fears of birdstrike possibilities, it is advisable to take special precautionary measures to control the particular pests involved. An authority should be prepared to undertake its own extensive trials to establish the most effective controls applicable to a site, and may find it necessary to employ a number of methods simultaneously to ensure success.

Leachate pollution

The degradation of putrescibles in a landfill can give rise to problems usually collectively referred to as 'pollution', along with the direct leaching of soluble and liquid fractions and the contamination of natural water flows through the tipped materials. The most common pollution in evidence at tipping sites is the discharge of polluted rainwater into surface watercourses surrounding the site. This is almost invariably due to the lack of proper site preparation work to allow for the hydraulic loads imposed upon the sites and in all but a very few cases is certainly avoidable. This particular problem is usually as a result of the overflow of excess water, much of which could have been prevented from entering the site in the first place. It must be emphasized that clean rainwater is infinitely to be preferred and far easier to handle than is polluted water. The important principle is, therefore, prevention rather than cure. The use of site perimeter cut-off drains, to prevent ingress of adjacent run-offs and percolation of surface waters is essential. However, ironically the massive percolation of underground waters through permeable strata beneath a landfill can represent an advantage, in that a high dilution and dispersion of small quantities of percolating leachate is desirable in cases where the pollutant load derives from biodegradation of normal wastes. Conversely, where industrial wastes are to be handled which may contain concentrations of chemical pollutants or toxic elements, the containment of resulting leachate may be necessary and the collected flows subjected to some form of treatment. This may range from simple re-irrigation over the landfill or oxidation in lagoons prior to discharge, right through to specialized treatment, removal, or discharge to a foul sewer. Where such containment is required this may be achieved by the use of naturally occurring formations such as impermeable basins or the forming of non-porous layers beneath the

Figure 2   Public instruction graphics on use of waste containers. (Courtesy of the DoE)

landfill using clay or man-made membranes. Such methods are expensive and should be used only where percolation cannot be allowed for reasons of local water abstraction or major aquifer protection. The final completion of a landfill is also very important to the future flows of rainwater and leachates. It is imperative that a suitably thick layer of natural soils be used to overlay deposited wastes and that the surface be effectively contoured to promote good run-off and vegetation growth. It can be seen, therefore, that pollution problems of landfill are avoidable by the careful selection of sites; the recognition of hydraulic parameters; and the matching of development, operation, and reinstatement works to the wastes proposed.

Scavenging

One of the properties of waste material which has always given rise to problems is that one man's waste can be surprisingly attractive to another. Unauthorized salvage, totting, or scavenging—call it what you will—represents an uncontrolled activity which must inevitably disrupt and jeopardize the good practice of a landfill operation. The material most usually sought after by unauthorized totters is the various metals, particularly the non-ferrous fraction, which surprisingly can often be present in quite large quantities. Where material salvage can be incorporated during the collection, treatment, or transport of waste this should be encouraged, but the sorting of refuse in the open cannot be permitted, on grounds of health and environmental considerations. The partial elimination of this problem comes as a result of the speedy deposit, compaction, and covering of wastes. The sooner it is out of sight the less the temptation, and the continuous control of the site by authorized personnel during working hours is an added safeguard. Security of the site by provision of adequate operational staff and physical installations such as fences, gates, etc., is essential, and good visibility screening at the site perimeter also helps.

Fire hazards

Inevitably a large proportion of waste material is combustible and a landfill can represent a potential danger from accidental or deliberate ignition. However, well-compacted waste is not susceptible to combustion and in the unlikely event of a fire the rate of spread of flame is restrained, enabling the seat of the fire to be isolated and dealt with. The old spectacle of the continuously burning open tip was the inevitable product of the uncontrolled, uncompacted, insecure operation. A modern landfill need represent no more of a fire risk than any other commercial or industrial operation, and the provision of normal precautionary equipment and firefighting facilities will minimize this still further.

Site traffic

The effect of site traffic on the local environment can represent a very real problem, which may be rather more difficult to solve, particularly in the rural areas usually surrounding landfill sites. Even large-capacity facilities cannot normally justify major improvements to community infrastructure, and in practice it is often difficult to restrict traffic use to preferred routes. This factor alone can, in some cases, be the overriding criterion against the development of an otherwise suitable site. There is no simple solution to it, but in considering new major landfills authorities will need increasingly, in the future, to consider making limited improvements to access road networks, well beyond the immediate site entrance.

Operatives and resources

One problem which stems only indirectly from the properties of the material is the generally low level of manning and standard of resources employed in waste disposal, both on landfill sites and in the management structure. In many cases the disposal service has been the 'Cinderella' function, commanding little or no respect and even less resources. This problem can only be solved by the establishment of correct manning levels to achieve the required standards, together with the correct emphasis being given to the duties by management. The rate for the job must be objectively fixed in line with the skills demanded and the value of the service. Whilst it must be accepted that predominantly 'attendance' is the basic requirement of many site operatives, situations can be created where genuine work incentive conditions apply and it is possible to supplement basic pay with a measured bonus scheme. This should be the aim wherever possible, in order to recruit suitable plant operatives and tipping attendants in the competitive labour market.

Visual appearance of sites

In practice the most usual criticism of landfill sites during past years has been their general unsightly appearance, apparent haphazard development, and visual obtrusion upon the environment. This has usually been the sum total effect of the problems previously referred to, with the gross product being totally unacceptable to the community. But it has been amply demonstrated that careful attention to each potential problem area can eliminate these sources of individual complaints and in the development of new sites the Waste Disposal Authorities can incorporate effective visibility screening by the use of existing natural features, the creation of embankments, and liberal planting of trees and vegetation. Phased restoration is also desirable in that it demonstrates awareness of environmental needs.

LICENSING CONDITIONS

The licensing of the deposit of controlled wastes in landfill situations has had the desired effect of centring attention on the problems of the past and the reasons for them. In the evaluation of the physical parameters and problems of existing sites it has been possible readily to identify the individual elements which are desirable, even essential, or otherwise, and to formulate an operational policy for each site. This has resulted in the building up of a clearer picture for the development of new sites, based upon objective assessments of the nature of the wastes and the physical conditions likely to result within the geological situations found. In this respect recent research programmes on the behaviour of wastes in landfill, and the passage and attention of leachates through strata beneath them, have been of considerable value to Waste Disposal Authorities.

The specific conditions applied through the licensing procedures have been aimed at achieving acceptable operational standards, to safeguard the public and the environment, whilst being compatible with disposal needs and

Practical waste management

Figure 3   Site entrance and offices on a modern landfill site, Sheffield

Figure 4   An example of restored landfill operations, Sheffield. (Courtesy of South Yorkshire County Council)

commercial viability. Licences have specified the necessary physical site development requirements such as access needs; void preparation; site security; and general facilities like amenities, buildings, and services. In addition the operational requirements in respect of manning and provision of proper plant have also been covered, along with measures for the identification, litter prevention, visual improvement, and restoration or after-use of sites. But perhaps the most significant conditions imposed by licences are those aimed at securing the management of a planned programme of operations and the following of nationally accepted procedures, together with the monitoring and recording of materials deposited. The real strength of these controls is in the power which the new legislation affords the authorities to ensure compliance. Sensibly and uniformly applied by the County Councils, the licensing of landfill sites has resulted in a considerable improvement in the management and operational standards of waste deposits. Whilst it must be admitted that the enormous physical workload facing the Waste Disposal Authorities, when licensing was first introduced, initially presented problems stemming from the lack of staff resources, and the necessary expertise, the majority have risen to the task and in so doing the benefits to the community have been significant.

Far from being a problem to the Waste Disposal Authorities, the introduction of the licensing of landfill operations will ultimately prove the saviour of this most economic waste disposal method. The proper control of landfills during the next few years should ensure the continued availability of further capacity for future needs and it is to be hoped that the blind opposition to it, shown by the public and authorities alike in the past, will be replaced by a more realistic attitude to objective criteria assessment. The community must ask itself 'what real economic alternative is there?'

Having said that landfill can be acceptable and is undoubtedly economic, we must not be deluded into believing that it is inexpensive. It is patently obvious that the proper provision, development, control, and operations of a landfill site and its ultimate restoration are costly items. The one real problem facing the waste manager at present, and for the future, is the securing of a continuous commitment of labour, staff, plant, materials, and capital expenditure.

THE ULTIMATE IN LANDFILL

The tipping of wastes unfortunately has a poor reputation, which was understandably established with the uncontrolled dumping of both domestic refuse and trade wastes, often practised in the past. This unacceptable face of disposal need not, and indeed must not, be allowed to be perpetuated. It has been demonstrated that properly organized and managed operations can raise standards to the extent that today's controlled landfill sites will satisfy even the most demanding environmental requirements of a modern society. But it must be emphasized that this will demand the allocation of sufficient physical and human resources.

The days of the uncontrolled local dump are gone and no-one mourns their passing. The era of controlled landfill is thankfully now a reality; indeed it is arguably an inevitability, as no other disposal solution offers equivalent physical and financial advantages.

ACKNOWLEDGEMENTS

The author gratefully acknowledges permission to publish this article, the contents of which are a personal opinion and do not necessarily represent the views of the Chief Environment Officer or the policies of the Environment Committee of the South Yorkshire County Council.

The article first appeared in the *Chartered Municipal Engineer* in January 1979 and is reproduced by permission of the Council of the Institution of Municipal Engineers.

BIBLIOGRAPHY

Waste Management Paper No. 1: *An Evaluation of Options.* Department of the Environment. HMSO, London.

Waste Management Paper No. 4: *The Licensing of Waste Disposal Sites.* Department of the Environment. HMSO, London.

Co-operative Programme of Research on the Behaviour of Hazardous Wastes in Landfill Sites. Final Report of the Policy Review Committee, Department of the Environment. HMSO, London.

Incentive Bonus Schemes in Waste Disposal Landfill Operations. Reports of Working Party. No. 1—County Surveyors' Society Committee; No. 4—Leicestershire County Council.

Appendix 1:   An analysis of the behaviour patterns and use of an urban civic amenity site in the City of Sheffield, 1976. (Courtesy South Yorkshire County Council.)

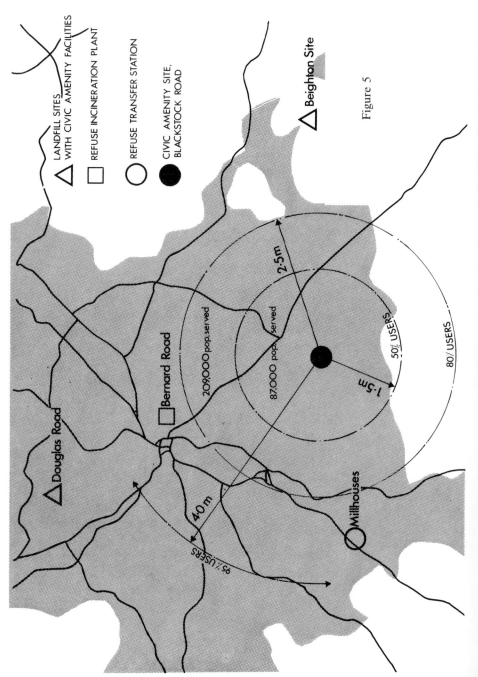

Figure 5

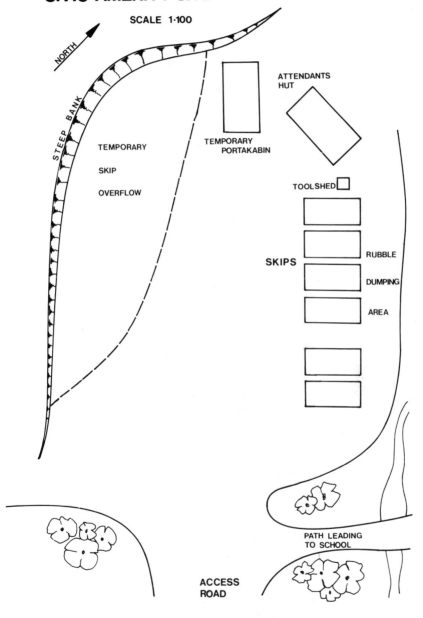

Figure 6

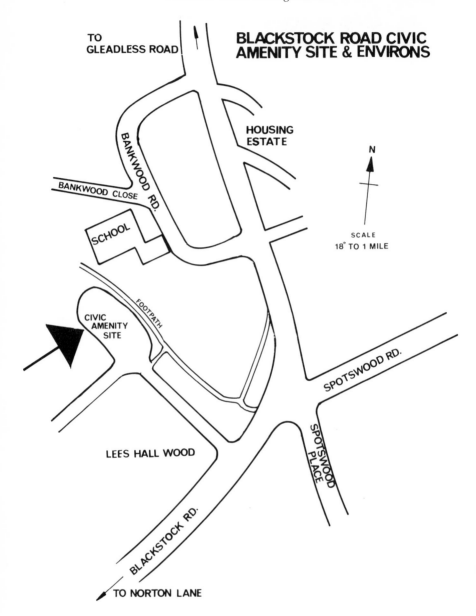

Figure 7

Number of loads of various types of waste

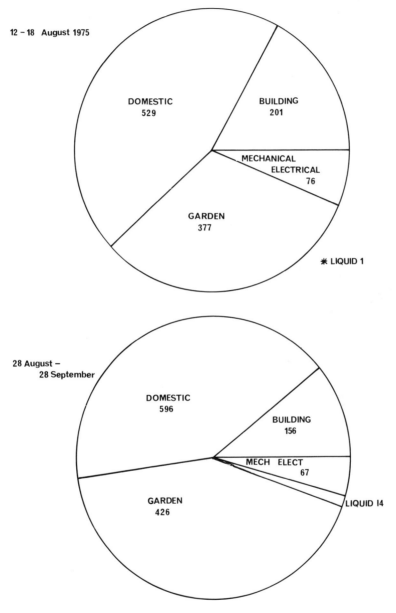

*BASIC INFORMATION FROM STUDENT SURVEY

Figure 8

SIZE OF LOAD

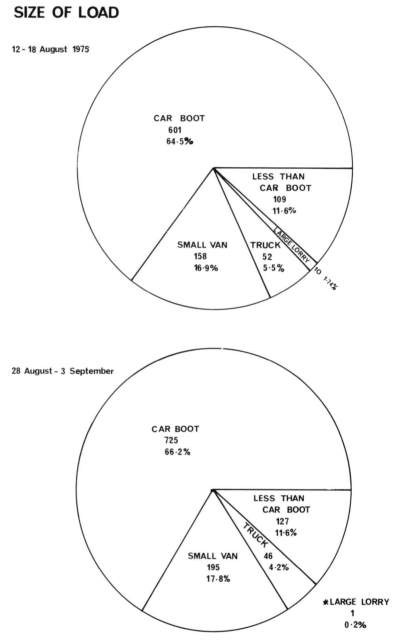

Figure 9

TRANSPORT USED

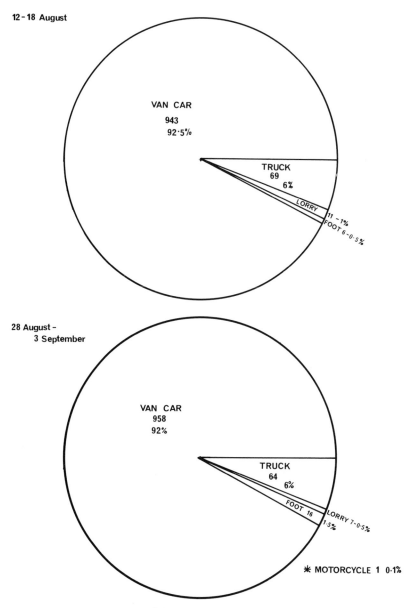

Figure 10

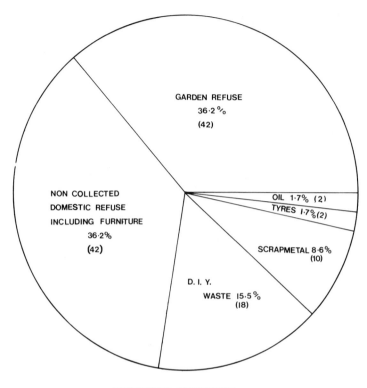

GARDEN REFUSE
36·2 %
(42)

NON COLLECTED
DOMESTIC REFUSE
INCLUDING FURNITURE
36·2%
(42)

OIL 1·7% (2)
TYRES 1·7%(2)

SCRAPMETAL 8·6%
(10)

D. I. Y.
WASTE 15·5%
(18)

CATEGORIES OF RUBBISH

*BASIC INFORMATION FROM STUDENT SURVEY

Figure 11

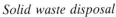

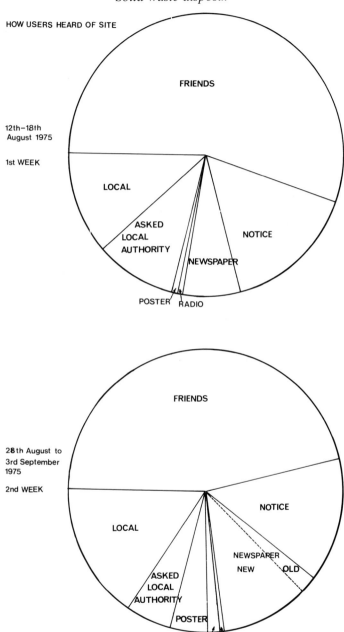

Figure 12

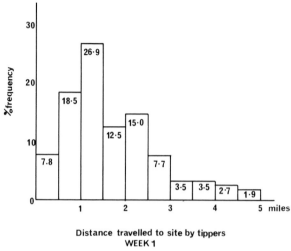

Distance travelled to site by tippers
WEEK 1

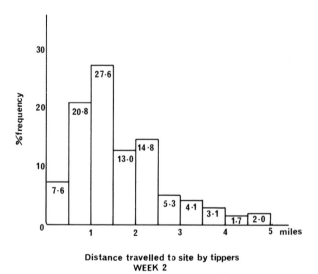

Distance travelled to site by tippers
WEEK 2

Figure 13

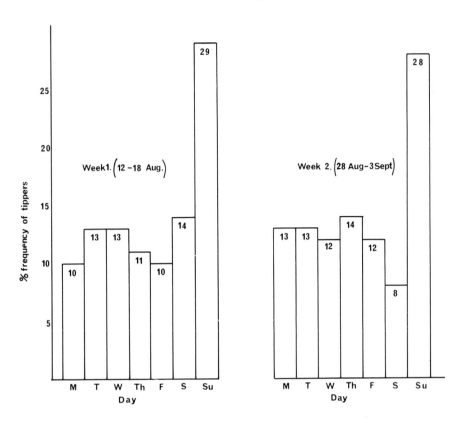

The day to day tip usage during the two weeks of the survey

*BASIC INFORMATION FROM STUDENT SURVEY

Figure 14

Practical Waste Management
Edited by J.R. Holmes
©1983 John Wiley & Sons Ltd

10

Co-disposal of wastes—a positive view of sanitary landfill by a Pollution Control Officer

A. Q. KHAN, MSc, MISWM

ABSTRACT

How to make pollution legislation work in the real world is the theme of this chapter. The Pollution Control Officer of South Yorkshire County Council tries to balance the realities of industrial economies, where industry looks for the most economic disposal solutions, against the zeal and duties of public officials charged with maintaining the law. Co-disposal of domestic and industrial wastes; the granting of licences, the enforcement of conditions, appeals, and planning matters are looked at. Surveys, plans, and guidelines to industrialists and site operators complete the work. Salutory examples of bad operation and hazards to life and the environment are described.

INTRODUCTION

In recent years problems of waste disposal have become critical in all industrialized nations as these nations have been extremely active in the production of more and newer chemicals with newer and difficult waste resulting therefrom. It must, however, be remembered that no nation, no industry, no man will discover or produce these new chemicals and their mixtures unless they are required by the consumers. The consumer requires these chemicals as they perform for him in better, more efficient, and beneficial ways than the existing materials. Several examples of chemicals, particularly those used for agricultural purposes, could be given to prove that without these the production of food, in quality and quantity, could not be achieved. It is not unreasonable to visualize that without these chemicals many people could have died of starvation.

The problem of waste disposal that industrialized nations face is complicated. No nation wants—or could ever be visualized to want—that production of all those chemicals, which are difficult to dispose of, should be completely banned. Such an action would have extremely serious implications.

The choice, therefore, is in ensuring that the disposal of all waste, and particularly hazardous waste, takes place in a well-regulated and controlled method, using not the cheapest tolerable methods but the best environmentally suitable method to do this. The industrialized nations have reached a stage where this is the only choice left. For ages disposal of all waste, particularly industrial waste, has been taking place in an extremely hazardous manner resulting in problem sites such as Love Canal (Niagara, New York) and Valley of the Drums (West Point K.Y.), in the United States of America; Ravenfield Site (South Yorkshire), Malking Banks (Cheshire), Ripley (Derbyshire), in the United Kingdom, and many more in other countries. This has created suspicion in the minds of the public. Unfortunately because of this the present-day problem is not limited to the disposal of hazardous waste. The waste disposal industry has been facing problems of finding acceptable landfill sites for disposal of domestic or even non-hazardous waste. The reason is that in the past not enough resources were devoted to finding hydrogeologically safe sites; in undertaking preliminary preparation; in preparation of proper operational schemes; in carrying out of scientific control and monitoring of these sites, and in quick restoration of the completed areas. Most sites, and the exceptions are very few, were dumps where anything or everything was deposited; no cover, no fence, no security, no equipment or plant was used to operate the sites properly. These sites were free for anyone to pick up things, or for children to play or drive motorbikes on them. Most of the time these sites were on fire. It is also understood that some of the operatives of waste disposal sites were issued matchboxes to fire refuse. (Members of the general public even now remember these old sites and visualize that a waste disposal site in their area today will be no different.) It is unfortunately not realized that during the last 10 years appreciable advance has been made in improving the image of the waste disposal industry. Most regulatory authorities are employing scientific staff to deal with waste disposal in an environmentally safe way. Controls and operational techniques have improved considerably.

HISTORY OF CONTROLS

The first controls on waste disposal in most countries started with the regulation of the discharge of chemical effluents into the drainage system, either because it was causing problems in the sewers and had adverse effects on the biological treatment of sewage or because of pollution of watercourses. This resulted in industries providing treatment of their effluents to make them suitable for acceptance into the drainage system. The resulting sludge from

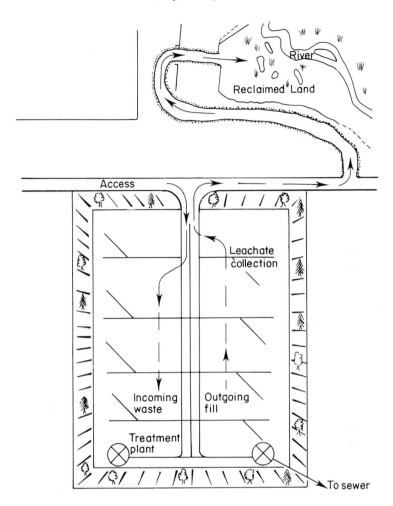

Figure 1   Plan showing a simplified version of a two-stage tipping operation. Top right-hand corner illustrates that land near a river could also be restored by well-degraded waste

such treatment processes had to be removed elsewhere for disposal. At that time most of the industrial effluent treatment plants were designed with absolute minimum capacity for the holding of sludge, as sludge disposal was not expensive. When the sludge had built up to such a level that it overflowed into the sewerage system the firms would employ waste disposal contractors to desludge the settling tanks. This exercise generally involved the emptying

Practical waste management

Cross section

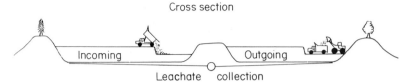

Figure 2   Cross-section of the two-stage tipping scheme. In one area waste is being
deposited while in the other area it is being removed

of the full contents of the tanks. The slurry was then either deposited at a
convenient place within the curtilage of the factory of taken to a local quarry.
There used to be very little, if any, public reaction or opposition to such
disposal.

Similarly when air pollution became a problem demand for the abatement
of air pollution resulted in industries providing filters and scrubbers to limit
emission of pollutants into the atmosphere. The resultant effluents from the
scrubbers were either heated to make them acceptable into the sewerage
system or removed for disposal somewhere else. In both these cases the need
for the disposal of sludge or liquor at a landfill disposal site became inevitable.

The point being made here is that the control and restriction of discharges
into marine environment or emission into atmosphere resulted only in the
pollutants taking a different route. In most cases pollutants cannot be
completely destroyed but can be made to take a different route which is
acceptable at that time. The most convenient route left, in most cases, was the
deposit of these wastes on the land. A stage then arrived when the
contamination of the ground and the resultant environmental and public
health implications became so serious they could no longer be ignored by the
concerned public. They were not satisfied that land disposal of toxic waste
was environmentally acceptable. However at that time planning legislation
was the only means of dealing with landfilling problems. In some countries it
was possible to use some of the provisions of public health or environmental
health regulations to exercise some control. However the complex science of
contamination of lands and the associated implications were too complicated
to be easily understood or dealt with by planning authorities or the majority
of environmental health experts. It required a different approach involving a
multidisciplinary team to understand, control, and deal with the problem.

When all this was happening the members of the public were getting
concerned and pressure was building up against the landfilling of waste. This
was the inevitable result of the past abuse of landfilling as there were no laws
effectively to control the indiscriminate dumping of waste. It soon became
evident that the environment was being threatened, and to protect the
concerned public became the primary motivating force in pressurizing the
Government to provide the necessary means to ensure adequate protection.

Figure 3   Uncontrolled tipping—an example of the hazardous situation caused by an underground fire

The unfortunate fact is that this increased awareness and resultant strong feeling against all waste disposal has tilted the balance too much in the opposite direction. In spite of various legislative controls, regulations, and Codes of Practice the members of the public are not able to notice any significant change in the standard of disposal. Some of the older sites are still in operation and may continue to be operational for some years. These transitional sites are, no doubt, coming under legislative control. However, some of these sites cannot be made to operate at a high enough standard due to inherent problems of site geology and hydrogeology, lack of preparatory engineering work, or lack of scope for the cell system of working. At the same time there is no merit in closing down these sites. This is either due to the fact that legislation does not cater adequately for old but operational sites, or because closing a site at such a stage could result in more problems. Continued use of these old sites has other serious implications. The regulatory authorities are caught in the middle between a concerned member of the

Figure 4   Hollow hot spot where children could walk, causing themselves harm

public, who suddenly becomes 'expert' when a site is planned near him, and public and private concerns seeking to locate a facility. The regulatory authorities are often left with no choice but to allow poorly sited, under-designed disposal areas, which have been upgraded from dumps to landfills to become overstretched to the limit or even overfilled. However this cannot go on for ever. Very few new sites are being allowed to be opened. At the same time there are large areas of land where quarrying or mining has taken place in the past and which are either totally unrestored or partially restored. Newer planning permissions for mineral extraction invariably contain a condition for quick backfilling and restoration of such sites. On the one hand the mineral operator is asked to fill the quarry quickly, but on the other there is opposition to allow him to use waste to fill it.

SUGGESTED APPROACH

To win back the confidence of members of the public it is essential that the problem of public relations involving their education in the problems of disposal of waste and control should be given top priority. Confidence must be created so that members of the public are convinced that both the industry and the regulatory authorities know what is required of them. They should have confidence in the regulations and controls that exist.

Disposal of all waste, and in particular the hazardous waste, should take place under a foolproof system: a foolproof system in the eyes of the public and not only in the eyes of the regulatory authorities is essential.

PUBLIC RELATIONS

It is therefore essential that a balanced approach is taken when dealing with disposal of all waste and in particular hazardous waste. The first and the most important step is to ensure that all concerned are made aware that industry produces wealth, and in doing so produces waste, and that the waste is an inherent by-product which needs to be disposed of as at that point in time there is no further use for it. Once the concerned public is made aware of this fact and the importance of the processes which produce a particular waste, along with the full implications of not producing that product, there is no reason why there should be any opposition to production of such wastes.

When this awareness is achieved then the need for a disposal point becomes obvious. Once this is realized then the opposition to waste facilities reduces.

When opening a disposal site (either landfill, treatment plant, or incineration plant) the best approach is to write an explanation of the options available, their implications on cost of disposal, energy requirements, risk to other road-users from long-distance transport, the damage to the immediate environment, and risk to after-uses of the site in question. There should be a definite recommendation by the regulatory authority's officers of at least one method so that a meaningful decision can be taken by the elected members or the appropriate body. The point being made here is that no application for disposal of waste should be rejected without giving an acceptable alternative

Figure 5   Hot vapour/steam coming out of adjacent land near an operational landfill site

Figure 6   Uncontrolled tipping—advancing underground fire can cause serious
environmental problems to adjoining land

which is acceptable from planning, operational, and environmental aspects. This is essential as by rejecting an application waste cannot be made to disappear; it has to go somewhere. If no economically viable alternative exists then there is every likelihood that the waste will be deposited on a site which may not be suitable. This may also increase the risk of illegal disposal of waste.

ADVISORY SERVICE

Regulating authorities should also take the role of providing advisory services to industry. Authorities could look into the processes which produce waste which is difficult to dispose of. There have been several cases where by change in process or chemicals a different and easy-to-dispose-of waste has been produced without any adverse effect on the product. There are also examples where difficult waste has become a saleable by-product by a change in the pretreatment chemicals.

Most of the difficult wastes are in liquid form. The problems of leachate and water pollution at a landfill site become more serious if liquid wastes are allowed to be deposited. In view of this water authorities, planning authorities, and licensing authorities discourage liquid waste on landfill sites. Conditions are quite common on planning permissions or site licences which prohibit liquid waste.

Regulatory authorities could assist waste producers by suggesting methods of pretreatment. A large number of wastes travel long distances to disposal outlets because they contain over 90% of water. It is comparatively easy to dewater them by on-site filtration. The resultant cakes become suitable for landfill disposal. Similarly it is comparatively easy to neutralize acidic and alkaline waste. Sometimes it is even possible to get two wastes to mutually neutralize each other. Here again neutralized and filtered waste becomes suitable for disposal at local sites. Soluble oils are another example where a preliminary treatment involving cracking and separating oil layers for reclamation and the bottom layer for filtration can made them easy to dispose of.

Waste disposal authorities can also assist by encouraging reclamation, reuse, or recycling of waste or redundant materials. There are several examples where industries have been storing raw chemicals and other usable by-products which were either bought or produced for a purpose but could not be utilized due to changes in circumstances or market forces. Regulatory authorities do have expertise and knowledge to know who in their area are potential users of such materials. They could bring these two industries together for them to negotiate the arrangements for use of such materials. This type of work has been carried out in various places and indicates that substantial quantities of such materials have been saved from destructive disposal.

CO-DISPOSAL OF INDUSTRIAL WASTE

Any void in the ground could be used for disposal of waste. If the void is impervious then it is safe from the groundwater pollution risk; if it is pervious then there is risk to groundwater. Therefore the type of waste that can be deposited on a site depends mainly on geology and the hydrogeology of the site. With the expertise and equipment that is now available any site could be made as impervious as one would like it to be. Once a site has been properly engineered, preliminary work carried out to protect water resources, and operation is on the cell system (the cells are so formed that they do not remain exposed to become saturated with rainwater), the rate of fill is fast enough for quick filling, the waste material is absorbent enough to hold rainwater within

Figure 7   A burning tip advancing and destroying vegetation

itself, then there is very little if any risk from the water pollution point of view irrespective of the type of waste material, whether it be paper waste or a mixture of domestic waste and most industrial waste. It should be noted that a leachate from a landfill site, irrespective of what the waste material is (except when the material in question is so-called inert building rubble), is not acceptable in any water resources. This means that when a landfill site for most types of waste is opened, it becomes essential that no leachate gets out of the site into either ground or surface water. A site chosen on such exacting criteria is suitable to take a number of industrial wastes in conjunction with domestic wastes which are acceptable from an operational and environmental pollution point of view. It is also worth mentioning that recent works that have been carried out by various organizations have indicated that there is very little difference, if any, between the leachate characteristics of purely a domestic waste site and a site operated on a co-disposal basis.

A waste disposal facility operating on a strictly scientific basis and taking domestic waste and a calculated percentage of industrial hazardous and non-hazardous waste, both in liquid and dry solids, can provide a realistic local safe waste disposal outlet for many industries of the region. When industrial wastes do not have to travel long distances one could afford to divert some of the saving from transportation to improvement of standards at landfill sites. Industrial concerns will be attracted to regions where co-disposal is allowed. Co-disposal sites will be able to be operated at higher standards,

Figure 8 Uncontrolled tip continued to burn for a few years and created extremely hazardous conditions

Figure 9 The result of uncontrolled, unauthorized, unscientific, and indiscriminate tipping—acid tars and other wastes were bubbling out of a site

Figure 10   High acidic waste bubbling out of a fenced-off area where children used to play. Many children were reported to have been burnt by the waste

thus improving the image of the waste disposal industry with the resultant reduction in opposition to the landfill sites by the public.

As the number of co-disposal sites will increase, the ratio of hazardous waste in them will decrease. In such an ideal situation one could safely say that leachate that ultimately reaches the substrata will not be different in its properties as far as biodegradation and self-purification is concerned. A survey of refuse tips[1] in the areas of several water undertakings, carried out by the Research Panel of the Society for Water Treatment and Examination and the Institute of Water Engineers (UK) clearly indicated the tremendous bacterial and organic purification capacity of the soil and the dilution capacity of the abstractable groundwater. What is, however, significant in these sites is that they do not provide quick degradation of organic materials due to the operational method which restricts the moisture intake, compaction of waste, thin layers, and intermediate covers resulting in anaerobic conditions. Moreover some of the hazardous industrial wastes slow down the bacterial activities. The advantage of this phenomenon is that the strength of leachate tends to be comparatively weaker (but continues to be produced for longer periods). The disadvantage is that the site remains potentially hazardous for the after-use and development, due to the risk of incomplete settlement and potential gases getting into surrounding properties.

CONTAINMENT SITES

Containment sites apparently appear to be a good system as there are a limited number of such sites needed, thus it is easier to provide effective preliminary work and monitoring during and after the operation closes. The main difficulty is the reluctance of authorities to have a containment site located in their area due to public opposition. Containment sites tend to receive more liquid waste and invariably have a pool of liquid which creates a smell nuisance. They also tend to remain exposed to rainfall, thus creating a potential risk of overflowing and causing water pollution. A containment site will remain a potentially hazardous site for indefinite periods, even when it has been fully filled in and covered up, and there is no means of ensuring such a completed site will not one day overflow and cause a serious problem. Moreover due to its contents being mainly toxic there is very little, if any, biodegradation of materials. This means that the site can be put to very limited after-use for an almost indefinite period.

TWO-STAGE LANDFILL SITE

There always seems to be strong opposition to opening up new landfill facilities even for domestic waste. At the same time voids are being created by mineral extraction that need to be filled in and restored ideally to their

Figure 11   An example of industrial waste damaging the vegetation and the environment

Figure 12   A industrial waste site being decontaminated by dragging out chemicals from a distance and then treating with lime. Treated materials are being spread in layers in the foreground

original level. Both these objectives could be achieved by a novel method. The principle behind this system is similar to building a sewage works which are built on a permanent basis at a convenient place which is acceptable to the community of the area. Once the sewage works is built and operated in such a way that it does not cause any visual or smell problems, no objections are made against it. The treated sewage goes into the water courses and the sludge is taken for disposal (in most cases) somewhere else. Sewage sludge is either taken to a large sludge disposal area away from the residential area, or to smaller sites where disposal takes place for a short time and the site is then restored.

If a suitable central landfill site is chosen to act as a permanent primary landfill which only provides treatment of waste to make it suitable for disposal elsewhere, there is every likelihood that, provided the environmental improvement is carried out and the operational methods are well controlled, the members of the public will not strongly object to such a system. The details of the working of such a system are explained in the following paragraphs.

This will work on the system that the most suitable site, based on its size plus nearness to the area of waste production, be selected. The distance of this site must be such that a collection authority vehicle could economically deliver waste to the site. This could be an existing landfill site or an old domestic waste site awaiting restoration/after-use. If no such site is available then any piece of suitably sized land could be used with a degree of engineering work to make it visually acceptable.

This then acts as the primary tipping area for all domestic waste on a fill-and-draw system, i.e. there must be at least two distinct areas of land at this site. The waste materials would be deposited in one area and conditions would be created for fast and controlled biodegradation of domestic waste by pretreatment to reduce the material size. (There could be provision to reclaim saleable material if required at this stage.) Faster biodegradation would be achieved by creating aerobic conditions and optimum moisture content plus higher temperature. The deposit in this area would be organized in such a way that it would make it convenient for removal of first-come, first-go material for the second stage of tipping elsewhere. The two areas on the primary site would be of such capacity that each could take and store refuse input for a suitable period. The period will depend on the degree of site preparation, type of refuse, and method of operation. After a period the second area of the site would start receiving waste on a similar system as the first one and the removal of stabilized waste from the first area would start at a rate which depends on the demand at the site of secondary deposit. The system of fill-and-draw in areas one and two would constantly be repeated in cycles.

The second stage of tipping for the stabilized material should not create any problem. This material could be made suitable as raw material to restore any appropriate piece of land—large or small—at any reasonable distance.

The sites which are not suitable for depositing crude domestic waste, either because of nearness of houses, being within bird-striking distance of airport, risk to water pollution, small and uneconomical size of site, objection to long-term use of the site, or due to designated after-use, would become available for stabilized material. The rate of restoration of such sites will be purely dependent on the operational need of the disposal authority and not on the constraints imposed by the collection authority. Some of the smaller secondary sites could be restored in weeks rather than in months or years. The product resulting after degradation under optimum conditions should be such that it must not be considered as waste but a raw material for the restoration of derelict lands and unrestored quarries.

Environmental requirement of primary sites

(1) The primary site must have an embankment surrounding it, with a fence at suitable places. As the primary site is a permanent feature it should have suitable trees planted around it to make it environmentally acceptable. Ideally the operation and activities at this site should not be visible from outside the site.

(2) The access roads and the amenity facilities must be of suitable high standards.

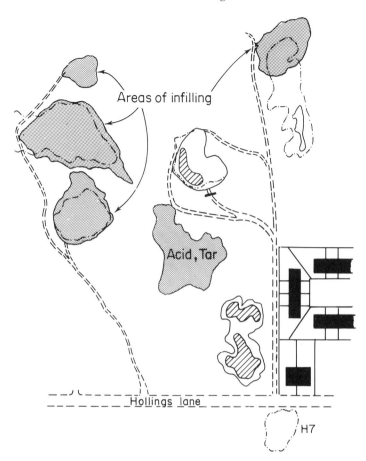

Figure 13   The diagram shows how near to dwellings the highly toxic and hazardous wastes were deposited. The nearest house was only 30 ft (10 m) from the acid tars. Many children were burnt and injured by this

(3) As there should be no need to provide daily cover, except to reduce windblown litter, there must be separate provision for deposit/storage of inert materials brought to the site.
(4) There could be leachate production, particularly if there is a need to seal the bottom of the site for whatever reason plus spraying of water to provide optimum moisture for increased rate of biodegradation. There must be provision of a leachate collection system. Collected leachate could be used to spray back on the site and the surplus must be treated *in-situ* by biological treatment plants to the standard at which it becomes suitable for discharge into the foul sewer or, when acceptable, into

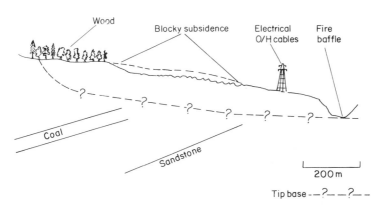

Figure 14   The diagram shows the serious environmental and public health risk due to a site vigorously burning. There is a risk to overhead electricity cables, gas mains, and persons who walk onto the site

watercourses. There should also be ducting to provide aeration of the site to keep it aerobic for quick degradation. Where appropriate to allow anaerobic degradation, provision must be made for collection of methane gas.

In the long term the primary site should be able to provide a more effective reclamation exercise, particularly for ferrous metal and energy either as methane gas (by anaerobic degradation process) or by segregation of combustible material before degradation.

Advantages of the system

(1) The major advantage is that there is no constant necessity to search for a large hole nearer the place of waste production and then go through the process of satisfying members of the concerned public, planning authorities, airport authorities, water authorities, and countless other interested parties. Once all these parties have been satisfied it is essential that resources are available for the cost of preparing the site, requiring a degree of work to obtain connection with various services—electricity, water, drainage, and telephone—and to provide amenity blocks, places for parking, and maintenance of equipment and plant.

(2) Depending on the distance of the site from the collection area there is no need to liaise with the trade union every time a disposal point is changed, as there will be no change of sites. The additional cost of collection vehicles travelling long distances in some cases could, however, be avoided by operation of transfer stations.

(3) No hazardous gases being generated at the second site of deposit. The site could be put to any after-use soon after restoration.
(4) No risk of water pollution at the primary site as economically it should be viable to install a leachate collection and treatment system. Moreover aerobic sites produce weaker leachate. No risk of water pollution problems at the secondary site of deposit as the material must have produced all the polluting leachate at the primary site under aerobic conditions. The risk could be further reduced by quick filling and suitable contouring where appropriate.
(5) No risk of mud at the primary site as it should be economically viable to install a suitable wheel-washing system plus road sweepers. The mud problems are bound to be less severe as the secondary sites could be operated when the weather is suitable.
(6) No risk of fire at the secondary site due to the stabilized nature of the materials. There could be a higher degree of risk at the primary site due to the aerobic degradation of material. Better supervision and control could be exercised to avoid this risk.
(7) Bird population at the primary site could be reduced by spraying of leachate plus higher temperature of the site. (This, however, needs to be studied to see the result.) There are means available to discourage birds at landfill sites. There should be no objection to the secondary sites within the bird-strike zone from the airport authority due to the stabilized nature of secondary material which should not attract birds.
(8) Fewer experienced operatives and less supervision may be required, depending on the size of the county and the number of landfill sites reduced by this system.

Disadvantages

(1) Disposal of asbestos waste will not be acceptable at primary sites and in a large number of shallow secondary sites.
(2) Co-disposal of industrial hazardous waste may not be advisable for some of the difficult wastes.

HAZARDOUS WASTE DISPOSAL

For controlling generation, handling, transport, and disposal of hazardous waste it is essential that the controls are so devised that they do not hinder the industry and at the same time do not cause problems to the regulatory authorities either because of their being too restrictive or too ambiguous.

The most important requirement is to have clear definitions of all relevant and appropriate characters, words, and phrases. To quote an example waste must be defined at the point of the production. It should be the duty of the

Figure 15   An example of progressive restoration of an industrial waste site. The area in the foreground is almost ready for final restoration while the area in the background is undergoing active treatment

person who employs a process to prove to the regulatory authority that the resultant materials are reuseable or reclaimable material and is not a waste, as it is being sent to a named company for a named process and the company producing or receiving it take the responsibility of its safe transport to the point of its further use. The producer of the waste should also be clearly defined, as the company who occupies and/or have control of the building where the process is taking place. This is important in cases where certain maintenance or remedial work is being carried out. Asbestos stripping from buildings or plants is a good example. Some organizations consider that the contractor who actually carried out the stripping is the waste producer and should be responsible for all relevant documentation and safe disposal. At the same time there are organizations who consider that the person who decides and places the order for the stripping work to be carried out is the waste producer. It is difficult, if not impossible, to have a meaningful control if the contractor is termed a waste producer. He, being mobile and subcontracting some of the works, has potential to deal with the waste illegally without much risk of being caught with sufficient and acceptable evidence for legal action. This is particularly true if there is no system under which registration of such contractors is required.

The most effective system in the view of the author is to require by law that any company or person who uses specified processes, or produces specified

Figure 16   An industrial waste site almost ready for top soiling and seeding with grass

types of waste, should register his company with the appropriate regulatory authorities of the area. It should not be difficult to specify the processes or the waste which requires registration. It is advisable to register all firms who produce liquid waste which needs to be transported for disposal.

The potential for illegal disposal of liquid waste by deliberate leakage through valves during transport, discharge into the drainage system, water-courses and spare lands, without the likelihood of being caught, will be lessened. Processes which produce solid waste which contain toxic chemicals should also be candidates for registration.

The registration of firms by regulatory authorities should mean that the regulatory authorities be either made responsible to approve the name of the haulage contractor and also the disposal contractor, or should give the names of contractors who have the expertise, equipment, and facilities to provide transport and disposal. The main advantage of this arrangement to the regulatory authorities will be that only those operators who are bona-fide will provide disposal. The advantage to the waste producers will be that they will know that the contractors have the capability to handle their waste safely and in an environmentally acceptable manner. Regulatory authorities will be under obligation to ensure that facilities exist either in their areas or within reasonable distance for proper disposal of waste. If such facilities do not exist then they will be under public pressure and legal obligation either to encourage others to provide such facilities or make arrangements themselves.

Once a registration has been approved there is no need for waste producers or disposal contractors to issue documents to authorities with each consignment. A copy of the registration document could be held by the producer, transporter, and disposal contractor to ensure that the waste being transported is authorized.

A periodic return drawn up independently by all the parties involved, giving dates and times of removal and time of deposit (if need be, with the name of the driver and vehicle registration number), should provide official check-up to ensure that wastes are being handled and disposed of in an approved manner. If, on top of this, it is possible to require by law that the cost of disposal is paid to the company who provides disposal by the waste producer, this will ensure that waste goes to the correct disposal site.

Disposal in emergencies could also be controlled by a system where approval by telephone could be obtained and then confirmed in writing, giving full details who the contractor is, and who the site operator is, and that they have facility to provide disposal.

One could argue that this sytem of registration should not be limited to the disposal only of certain types of wastes which are toxic or difficult to handle; it should include all waste except domestic waste. The advantage of this system is that there will be a very significant reduction in the cases of fly-tipping, dumping, and littering. The only point against this is that it will require resources to implement it at least initially. In the view of the author this is insignificant when compared to the benefit to the public and the environment.

DISPOSAL OF CONTAINERIZED WASTE

Disposal of waste in drums or containers should be discouraged. This is particularly important as modern methods of disposal are such that the landfill sites attain fairly high temperatures, because of compacting and covering, for long periods. There is also the danger that wastes transported to landfill sites in drums may not be the same as those which are documented. Very few landfill sites have a chemist and fully equipped laboratory to establish the contents of drums; therefore incidents have occurred where the contents have exploded, with very serious consequences. There is also the problem of drums degrading after they have corroded. This could result in voids appearing within sites after a delayed period and this can result in non-uniform settlement, making after-use of the site potentially difficult and expensive. There is also the risk of chemical emissions continuing to exist even after the site is considered stabilized and safe.

In certain cases after-use of the site may require excavation. When such an exercise is undertaken drums will also be excavated. At that time it is very unlikely that the original marking and labelling will exist to warn the excavator of the contents and their properties. This could result in accidents, where precautions are not taken, or expensive analytical cost plus unnecessary precautions where the chemicals were later found to be non-hazardous.

CONCLUSION

A more positive approach is possible if legislation is so drafted that all parties involved in production of waste, its transport, and disposal, have a clear idea of their responsibilities and powers. Regulatory authorities' powers should not be limited to prosecution and remedial work after the incident has taken place. They should be required to provide an advisory service and assistance to industry in encouraging conservation of materials by devising processes which produce the minimum quantity of waste, and once this waste is produced and the only method left for its disposal is destructive disposal then they should ensure that firms know who are bona-fide contractors and where suitable facilities exist.

REFERENCE

The effect of tipped domestic refuse on ground water quality, *Water Treatment and Examination*, **18** (1) (1969), 15–69.

Practical Waste Management
Edited by J.R. Holmes
©1983 John Wiley & Sons Ltd

11

Plant and equipment on modern sanitary landfills

BERNARD McCARTNEY, MInstM, MISWM, AMIQ
Managing Director, Macpactor Limited

ABSTRACT

In an atmosphere of increasing science and technique in managing the erstwhile and humble art of landfill, this chapter reminds us that there is much more to landfill machinery than many at first suppose. Often forgotten, landfill operations require a wide range of mechanical handling duties that include levelling and site preparation, access roads, digging cover, spreading and placing waste, covering, and final levelling. The wide range of machinery available has many equally varied advantages, disadvantages, strengths, and weaknesses. The chapter looks at tracked and wheeled machines and emphasizes the use of modern steel-wheeled compactors. How the industry has responded to the changing nature of landfill operations is set down.

INTRODUCTION

If operational economics is the main criterion it is inescapable that the sanitary landfill of domestic, commercial, and industrial wastes will continue to be the predominant method of waste disposal in the UK.

The landfilling of domestic waste, now dignified by the title 'sanitary landfill', can be a very different operation from the usual paper-littered, rat-infested eyesores that normally come to mind. It is becoming a science of its own; knowledge of the geology and hydrogeology of landfill sites, the civil engineering and drainage work that is necessary, and the vastly improved mechanical and engineering mobile plant are changing the face of this simple system.

This chapter sets out to review the range of plant that is available, mentions something of the historical background of each, and where it may best be used. The chapter gives special emphasis to the newer steel-wheeled compactors and illustrates the development of these machines.

MATERIALS HANDLING TASKS ON A TYPICAL SANITARY LANDFILL SITE

The efficient handling of waste materials and inert cover is at the heart of good landfill operations and it is worth looking at the principal tasks involved. These tasks start at the point at which the site is first prepared to receive refuse, and end with its final restoration to agriculture or some other beneficial use. At every point earth-moving or refuse-handling machinery will be required. No one machine can perform every task although some, as the chapter describes, can perform several roles. These tasks include the following principal duties:

(1) levelling and preparation of the site;
(2) access roads and manoeuvring areas;
(3) digging and moving covering material (if on site);
(4) initial handling and segregation (if required) of incoming waste;
(5) spreading and placing waste in the landfill;
(6) covering the waste;
(7) final levelling and topsoil placement.

At the beginning, if we assume that our landfill is a relatively large void, it will be clear that in present-day conditions its market value will be high. So too will be the attendant development costs. Site preparatin will be a substantial part of these costs and the work must be executed in such a way as to enable the site to operate efficiently and yet conserve as much of the gross void volume for commercial use.

At this point it is essential that the site developer accepts the concept that the void is not just a place to tip waste indiscriminately, as in the past, but that it must be treated as a 'factory' for waste disposal, because a properly constructed landfill site will, in the long term, fulfil almost the same function as a specialized multi-million pound pre-treatment plant. Moreover, it will have a greater capacity and flexibility, and will be far more useful in the future for the purposes of land reclamation in the form of golf courses or recreation land, whereby it may be possible that substantial financial gains can be achieved to set against the cost of the initial landfill operations.

In order to make the best use of this void it is necessary to have the right equipment available. This will be essential to make access roads down to the part of the site where the landfill operations are to commence. The covering material may have to be brought into the site or it may be possible to dig this out on site, so it is advantageous if the road-making machine could be bought, having in mind the possible continual heavy digging, stock-piling, and placing of the covering material on the freshly tipped waste. Once this is established, the next thing is the decision as to which technique is to be used; the two main ones are:

(a) controlled waste disposal, and

(b) sanitary landfill.

With regard to controlled waste disposal, this is generally accepted now as being somewhat uneconomic because of the low rate of compaction achieved, the over-generous use of covering material, and wasteful consumption of space. The generally accepted method is that of sanitary landfill where the refuse is spread out at an angle of approximately 10–14°, i.e. 1 : 4–1 : 6 gradient, in layers of between 12 inches (300 mm) and 18 inches (450 mm). The use of this technique, by the means of greater compaction, saves up to 50% of the space occupied by the waste and, of more significance, up to two-thirds of the cover required.

A machine must be made available with sufficient power, weight, and traction to roll in layers of refuse and compact to an overall density which these days averages out to somewhere in the region of 0.95 tonnes/m^3. (On experimental and test cases figures far in excess of this have been and can be achieved quite readily, but for practical purposes it has been found that 0.95 tonnes/m^3, is a practical figure on which to work.) A steel-wheeled compactor is one of the better solutions here.

The nature of the site may require that excavations have to take place either prior to, or during the course of, the landfill operation, and it may be that some form of impermeable sealing membrane will be required at the base of the site. This all calls for suitable heavy-duty equipment. Therefore, basically, the main tasks to be carried out are of a heavy-duty earth-moving nature where heavy materials may have to be dug out and, most certainly, very regular road maintenance must be catered for which will involve handling variable quantities of stone, brick, rubble, and clay.

Added to this, of course, is the actual covering operation itself and anybody connected closely with waste disposal via sanitary landfill knows that basically if you have good and readily available covering material, you should have a good waste disposal site. Fortunately, the new sanitary landfill techniques, as previously outlined, do save up to two-thirds of the cover previously required, but it is essential that no covering material is wasted during the covering operation. It is therefore advisable to use some form of bucket loader to place the cover in position and back-blade it rather than to doze it from a stockpile at the top or bottom of the tip, whichever method is preferred. Further to this, equipment must be available to landscape the site as sections are finished to enable planning regulations to be met and eventually to leave the site aesthetically acceptable to the statutory authorities and the public. Some landfills, particularly municipal sites, also provide public dump-pit sites under the provisions of the Civic Amenities Act 1967. These facilities are usually located at the entrance to the site and so it may be necessary to move large amounts of 'weekend' waste from this point to the landfill proper. In this case a loader with a grapple scoop is recommended.

"Macpactor" Wheel Fitted to Landfill Compactor

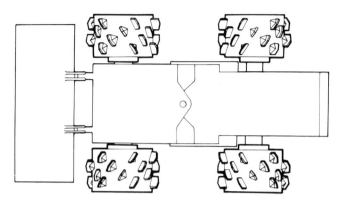

Plan view showing wheels mounted in opposition

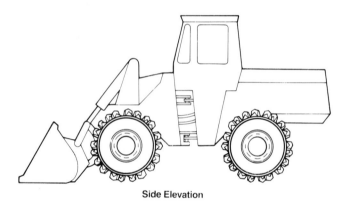

Side Elevation

Figure 1   'Macpactor' wheel fitted to landfill compactor. The McCartney 'Macpactor' steel wheel replacement service provides for the supply of a complete new set of four steel wheels specially designed for landfill waste disposal applications. The steel wheels are of similar or superior construction to the original equipment, matched to the existing machine and are self-cleaning

   To summarize, therefore, envisaging a fairly substantial void to be filled starting from scratch, it is essential to have heavy-duty and if possible multi-functional equipment which will carry out very heavy earth-moving operations from day to day in digging, lifting, dozing, levelling, grading, etc. In fact, the equipment required to handle adequately all the materials involved in a successful landfill operation must virtually be able to cover the whole gamut of earth-moving as related to heavy civil engineering works. Obviously some of the equipment will only be required on a part-time basis, i.e. grading and landscaping, and for this a suitable bulldozer could be hired on a short-term basis. However, the accent should be on multi-purpose machines specially adapted for the varied work on hand.

CHOICE OF PLANT—BROAD PRINCIPLES

In my experience of mobile plant, in choosing one or another type of machine, there are no rigid rules which can be laid down firmly and adhered to. Every waste disposal site has its own individual characteristics and it is, in effect, 'alive' as its needs vary from day to day, dependent on the wind, rain, and general variance of climatic conditions in addition to the nature of refuse being fed into it. In order that all these variances can be taken into acount, what follows is a general set of rules which can be used as a guide towards the selection of the right equipment for the particular job on hand.

Firstly, the machine bought today must not just be capable of coping with existing requirements but must, as near as it is possible to calculate, be well on top of the job at the end of its economic life. The emphasis is on a 5-year replacement period, but I feel that it is more advantageous commercially, and more economical, if mobile waste disposal site machines are replaced after 4 years. During the fourth year equipment begins to require unit replacements and part-overhauls, so that at the end of 5 years the machine has a deal of life left in it, but its commercial value has dropped in relation to its intrinsic value.

Secondly, the management responsible for waste disposal must 'stick by their guns' and purchase the machine they need, not the one which their superiors think they can afford. The size and type of machine purchased must be commensurate to the size and type of work that has to be done and the problems to be overcome. Overworked machines are more costly in the long term—cheapest is seldom best.

Today the purchase of equipment is made far easier than it was in the past by the diverse financial schemes that are available through banks and finance houses. In many cases these schemes have been developed in conjunction with the manufacturers of the plant. These schemes include various forms of hire purchase with graduated payments, simple leasing, and leasing with guaranteed buy-back at the end of 3, 4, and 5 years. It is also possible from some reputable manufacturers to obtain contract hire terms which, although expensive at first glance, do reduce down-time considerably and really make the manufacturer responsible for the efficient operation of the equipment on the site. When one considers the position, equipment is not bought for its intrinsic value but rather for its use. In many cases, ownership of equipment can be a financial hazard and contract hire schemes may be a better proposition. In the private sector particularly, the purchase of equipment enables tax relief and other allowances which, in addition to the guaranteed buy-back schemes, do tend towards the common sense of buying the best for the job, irrespective of price.

THE CHOICE OF PLANT

Bearing in mind the inescapable necessity to compromise in one way or another, these are the principal machines that are available:

(1) *Tracked machines.* These are of two basic types: the bulldozer, which is unable to dig, lift, and carry material but is ideal for site levelling and final grading; and the loader, which can dig, load, and carry but has limitations on levelling, back-blading, and grading. A very common machine in the early days of waste disposal, it is now less popular in dealing with the lighter and less dense nature of modern refuse. Nonetheless it still has numerous applications in materials handling.

(2) *Scrapers.* These are normally used in conjunction with a tracked bulldozer in the preparation of large sites and stripping out and stockpiling material which can be used for cover at a later date. They can also be used to help construct access roads to difficult sites but generally, because of the one-off nature of their work, they are better hired or the work given to a reputable contractor.

(3) *Dragline excavators.* On the larger sites where there is already a usable void, the dragline can be utilized in site preparation and also winning covering material from the site. It comes into its own in places inaccessible to mobile equipment such as tracked machines, i.e. swampland. The dragline, if put on a sound base, can be very useful in excavating 'fingers' of topsoil and subsoil to a suitable depth and stockpiling at one site for use as intermediate and final cover when the landfill operation is completed. Previously this type of equipment had only marginal use, but now certainly it seems its permanent presence is essential to the completion of some jobs. It can also be used satisfactorily for grading the steep sides of an existing hole and can assist in the application of impermeable materials to the sides of the tip prior to tipping.

(4) *Four/wheel drive, rubber-tyred loaders.* A great deal of research has been carried out by manufacturers into the application of this type of machine in many waste disposal duties. This one piece of equipment covers a wide range of work and is extremely versatile in back-up applications, particularly on the larger site, i.e. digging out cover and placing cover into its final position. Its speed and flexibility of operation in this role is a most useful attribute. It is possible on the smaller site up to, say, 200–250 tonnes/day, that this mchine would be capable of handling all aspects of waste disposl on its own, but in the present context of larger, 500–1,000 tonnes/day, sites its role is that of a back-up machine. It is essential that it

be fitted with under-body protection and special puncture-resistant tyres. In the past this machine had its critics because of punctures, but used properly and taking advantage of puncture-resistant tyres, a great deal of advantage can be gained from its use.

(5) *Steel-wheeled mobile compactors.* The steel-wheeled pulverizer/compactor is basically a converted articulated four-wheel drive loader and is an extremely popular machine on many larger sites. The extent of conversion is dependent on the manufacturer. In some cases the manufacturer simply replaces the rubber tyres with steel wheels, fits under-body protection, fire extinguishers, battery isolator switches, redirects air flows, etc. and the machine is basically capable of being quite quickly converted back to a rubber-tyred vehicle at any time during its life or with its secondary owner. Some have gone further than this and replaced the whole front end from the pivot forward. Some have restricted themselves to bulldozer blades only, some to a scoop, but more recently a dual scoop and bulldozer blade has been introduced. Its steel wheels eliminate the risk of punctures and several papers have been published to illustrate its compaction and space-conserving properties on landfills.

AN APPRAISAL OF THE PRINCIPAL FEATURES

Tracked bulldozer with a blade

This was the machine that bore the brunt of waste disposal work in earlier years. It is tending to lose favour as a front-line machine due to its natural limitations. However, very large machines of this type are still in favour on new large sites of up to 1,000 tonnes/day capacity as one-purpose big 'front-runners'. They are extremely useful in the smaller sizes for the landscaping of the finished site, particularly if fitted with grouser blades on the tracks.

Advantages

The ground pressures of 0.4–0.6 kg/cm^2 gives the machine a high flotation factor on soft material, i.e. freshly tipped refuse on soft ground. It is ideal for levelling off and grading the completed site with topsoil, achieving a wonderful finish. It is extremely stable when working across sloping ground and has a high recovery angle either uphill or downhill, particularly with grouser blades. It is not subject to punctures. It is excellent for making access roads and new waste disposal sites. It has the best manoeuvrability in confined spaces. It is not embarrassed by heavy materials, builders' spoil, rubble, etc. on the site. It can be used in conjunction with a scraper.

Disadvantages

Due to low ground pressure, it does not squeeze out sufficient oxygen to make the tip head completely safe from fire risk. It does not give final compaction and tends to bridge soft spots in tipped waste, leading to excessive wear of waste disposal vehicles crossing the site. To achieve level site surfaces, the tracked bulldozer must continually return to fill in the undulations. It will not lift, dig, or carry material. It is very wasteful of covering material which has to be tipped strategically in small heaps and then dozed into position. There is an estimated loss of cover of up to 50%. It is a one-site machine, not flexible in operation and slow in movement. It cannot, under all conditions, effectively 'blind' the top face and, in conjunction with its low ground pressure, it leaves the site face open to infestation by vermin. Tracks running metal to metal have proved on waste disposal sites to have a relatively short life. Normally it has a low clearance and can 'bog' easily. The capital cost is high in relation to its limitations. It has to be garaged on site and all maintenance has to be done on site unless low-loader transport is available, which could result in a lowering of essential maintenance standards. It lacks speed in general operations.

Figure 2   A modern landfill compactor for large sites. (Courtesy of The Caterpillar Company, USA)

Figure 3   A photograph of steel wheels used on mobile landfill compactors. (Courtesy of Macpactor Ltd)

Tracked loaders

The advantages are as for tracked bulldozers, except that this machine can dig, lift, load, and carry materials for short distances. However, levelling and grading of the finished site is not as good as that of the tracked bulldozer. It is useful mainly on smaller applications. It lacks speed in general operations.

Rubber-tyred, four-wheel drive loaders

These machines are capable of doing all the duties on a small controlled waste disposal site, but are not so successful on sanitary landfill as their tractive effort up the ramp essential to high-density landfill is limited and the ramp, in many cases, becomes too long to control and compact with the essentially narrow tyres.

The rubber tyres, if used on normal controlled waste landfill, give a compaction of some 2.1 kg/cm^2 and on the recommended 1.9 to 2.4 m face of tipped refuse, it achieves the correct balance of air, moisture, and heat necessary for the natural decomposition of the material to enable the maximum soil enrichment to take place. The four-wheel drive loader can be used right to the tip edge, and as other vehicles visiting the site will rarely exceed the ground pressure exerted by the rubber-tyred loader, they too can be safely driven quite close to the tip edge.

THE ANATOMY OF A JCB ARTICULATED WHEELED LOADING SHOVEL.

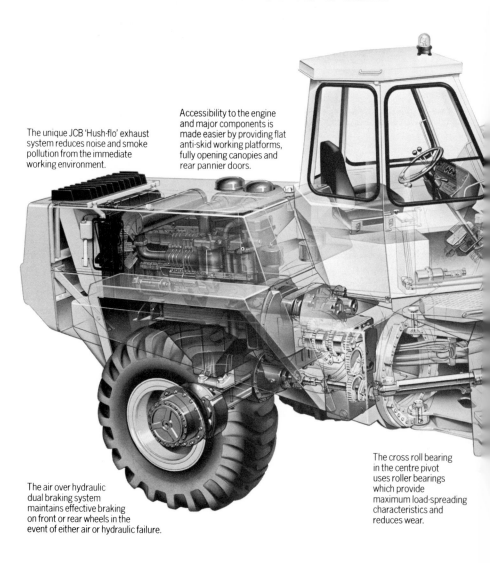

The unique JCB 'Hush-flo' exhaust system reduces noise and smoke pollution from the immediate working environment.

Accessibility to the engine and major components is made easier by providing flat anti-skid working platforms, fully opening canopies and rear pannier doors.

The air over hydraulic dual braking system maintains effective braking on front or rear wheels in the event of either air or hydraulic failure.

The cross roll bearing in the centre pivot uses roller bearings which provide maximum load-spreading characteristics and reduces wear.

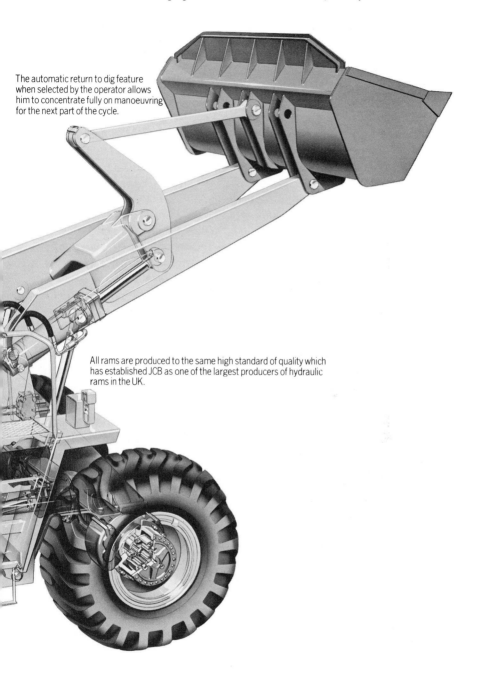

The automatic return to dig feature when selected by the operator allows him to concentrate fully on manoeuvring for the next part of the cycle.

All rams are produced to the same high standard of quality which has established JCB as one of the largest producers of hydraulic rams in the UK.

Figure 4   An illustration of a wheeled mechanical loading shovel as described in this chapter. (Courtesy of JCB Sales Ltd)

The rubber-tyred, four-wheel drive loader is capable of removing covering material from in front of the waste disposal site and bringing it to the top to be used as final cover, thereby carrying out the correct process of land reclamation. It can dig, cover, strip the site, construct access roads, bulldoze, and carry covering materials so that the minimum amount of cover is used for the maximum area. It can be used on high-compaction sanitary landfill sites in conjunction with a compactor unit, mainly as a back-up machine. Models are available in rigid and articulated form, but great care should be taken when considering articulation as this machine, in steering itself, has an infinitely variable centre of gravity and is liable to fall over on soft material where the ground is very uneven, whereas the rigid machine, having a fixed centre of gravity, enables the driver to develop a 'feel' of the stability of the machine and adjust his performance accordingly. The variable stability of the articulated machine does not afford the same advantages to the operator. This relates particularly to the smaller articulated model on rubber tyres. The larger models, mostly fitted with steel wheels, because of the extra width of the steel wheels and generally restricted articulation, do not suffer from this problem.

There are arguments for and against the cab position on the articulated machine as to whether it should be mounted on the front end or the rear end. The general consensus of opinion is that the driver should be situated on the rear end of the machine because then he can see the whole of the machine and it is not possible for him to become disorientated as is sometimes the case with a forward-mounted cab when the driver, on occasions, loses the 'feel' of where the back end of the machine is. In the case of pre-treatment plants, most designers of this type of plant are now incorporating as standard equipment four-wheel drive loaders fitted with a variety of attachments, the most important of which is the large capacity scoop, usually 2 m capacity, with a heavy-duty clamp to contain the refuse. This is mounted by a quick-release mechanism whereby the interchangeability time is between 2 and 3 minutes. Forklifts, extending cranes, bulldozer blades, etc. can also be supplied by the manufacturers interested in waste handling. This is, indeed, a major step forward from the original electro-hydraulic grab crane.

Special puncture-resistant tyres fitted (essential) are, in effect, two tyres in one and virtually eliminate punctures. If required, the tyres can be supplied with Tyrfil or Dunlofil, a light urethane rubber, which makes the tyre completely puncture-proof without affecting road performance. However, there are several schools of thought on this as, although it is theoretically possible to remould the tyres when the tread wears through, due to the action of small particles of glass and other damage caused on waste disposal, it is felt better to stick to the puncture-proof pneumatic tyre rather than bother with the extra expense involved in the special process of remoulding, even if the waste disposal damage does not result in refusal from the tyre remould company to guarantee the process.

Disadvantages

This machine has only limited use across steep slopes. It ruts up the ground in final levelling. It will not grade the finished site as well as a tracked machine. On very soft ground, due to heavy four-point pressure, it can get 'bogged down' more easily than a tracked machine.

The steel-wheeled mobile compactor

In order to appreciate fully the nature of the steel-wheeled machine, one must look back a little way to earlier times when many waste disposal operators stuck to their tracked bulldozers and tracked bucket loaders. This was mainly because they were basically orientated towards civil engineering and, more particularly, just could not equate a pneumatic tyre on their type of work because of punctures and associated down-time. Considering the type of material they had to handle one can readily understand their point of view, but at the same time tremendous advances had been made in the development of the four-wheel drive waste disposal loader and it had, in effect,

Figure 5   Steel-wheeled compactor working on a landfill site. (Courtesy of Bomag Ltd)

reached a position of dominance in local authorities, mainly because of the different type of waste being handled and the generally much smaller landfill sites in operation. The manufacturers of tracked machines were perfectly happy to supply tracked machines to the larger industrial sites, but realized that they were losing a considerable amount of business to the four-wheel drive, rubber-tyred machines. Feeling the need to take some action and try to fill the gap, the four-wheel drive steel-wheeled machine came into being.

The first machine of this type was introduced by Michigan and was called the 'Trashpak' but it was introduced to this country several years before local authorities were ready and, quite frankly, it came on the market at the wrong time because it provided an answer to a question which had not yet been raised. It was designed for large-scale waste disposal operations and the local authorities prior to 1974 were so fragmented that, apart from one or two very large authorities, it was found to be an uneconomic machine and was withdrawn, but the principle was noted and remembered.

Then came the local government reorganization in 1974 and the rationalization previously mentioned with the waste disposal function taking place at county level. Finance was made available to experiment on the larger sites and the stage was indeed set for the introduction of the mobile compactor. However, things were slowed up somewhat because the use of this machine was only seen to benefit the operation which is now called 'high-density landfill' as, if it was used in the normal method of controlled waste disposal, the results were very little different from those using a rubber-tyred loader or tracked machine. Therefore an entirely new technique had to be brought into being, against considerable environmental objections, and I might say that these objections are still taking place. Once these techniques were mastered it became more and more obvious that the mobile compactor was the piece that was missing from the jigsaw. It had its drawbacks certainly, but used correctly on a large landfill site a compaction rate in excess of double the previous compaction rate could be achieved with a reduction of up to 80% in required covering material. So, for the first time, waste disposal personnel had available to them an excellent team of machines from a very small number of proven types, i.e. the tracked machine and four-wheel drive, rubber-tyred loader to the new articulated steel-wheeled pulverizer/compactor. This enabled them, by careful selection, to overcome virtually all the problems they had previously encountered and a spin-off from this was the extended use of the scraper and dragline. Compactors have received a great deal of publicity in recent years and some excellent papers have been published describing their performance on landfills. The decision as to where not to use a compactor depends on several factors, not least the subjective likes and dislikes of individual waste managers. I summarize the position to be:

(1) If the site is big enough and has a long enough life, the specialized bulldozer blade-only compactor will shift tremendous amounts of mate-

rial but must be fed with cover by lorry or ancillary equipment, usually the four-wheel drive, rubber-tyred loader, although tracked machines could do the job but rather more slowly and more expensively. The addition of a dual bucket to this type of machine, if it is not restricted purely to a complete bulldozer action with a very small up-and-down movement, could be an advantage particularly in the final covering of the waste.

(2) If the site is large but of relatively short term, the life of a specialized loader could be limited and it may be that during the course of its usable life it may be transferred to other duties, albeit temporarily, on a different site with a smaller capacity where it would be required to dig, level, doze, and cover. In this case, dependent on conditions, it could be of advantage to change the machine over to rubber tyres to fulfil the necessary flexibility of operation and mobility required, particularly if no back-up machines are available.

A DETAILED LOOK AT STEEL WHEELS

The steel wheel is at the heart of the success of the mobile compactor and it is worth looking, in some detail, at the principal design issues involved.

(1) All compactors, except one, available in this country are of foreign design and manufacture and have been tested in foreign lands whose problems and approach to these problems are somewhat different from our own. With two notable exceptions the design of the steel wheel does not appear to have progressed from the old sheepsfoot roller idea, and here we are going back basically to the civil engineering approach to the compaction of soft earth, etc. No particular effort is made to increase the tractive effort of the machine; in fact no particular regard appears to have been made to the nature of the material on which it will be used. The two exceptions to this rule are:

   (a) one manufacturer who started producing wheels of a semi-chevron type and, indeed, to my knowledge has altered the wheel tooth pattern at least three times in almost as many years;

   (b) another manufacturer who has a somewhat different approach in so far as he has designed a special shape of tooth which is mounted obliquely on the wheel and has incorporated in the tooth pattern a 'screw' type action approach to the material to be handled.

(2) Not much thought appears to have been given to the length of life of the wheel tooth, with two exceptions. Most teeth on steel-wheeled machines are fabricated from varying types of steel and do have a relatively short life of 8–12 months or so before refurbishing is required. The two exceptions to this are two manufacturers who both use solid teeth welded

Figure ·6   Views of wheeled loading shovels fitted with buckets and grapples to enhance refuse-carrying capacity

to the rim of the wheel but who vary drastically as to pattern and tooth configuration.

Certainly, there has been very little development through co-operation between the manufacturer and user. Perhaps this is because the user was not sure of what he wanted and the manufacturer, not having specialized

Figure 7 A tracked bulldozer with specially adapted solid waste handling bucket. (Courtesy of International Harvester Ltd)

in the field to evaluate the situation, worked in the dark and hoped for the best.

(3) From the foreign manufacturers' point of view they seem, for the most part, to go for a frontal assault on the waste material and combine this with speed over the ground. Personal observations, however, have proved (to me at least) that these methods do not necessarily get the best results as I have found that two carefully judged passes over fresh waste at a slow-to-steady speed had a far better result than five or six quick passes over similar material, particularly when the frontal assault sheepsfoot roller type of wheel is used. Possibly these wheels could be better designed to chop, compact, self-clean, and to concentrate on reducing the size of the waste and, thereby, the void, particularly on industrial applications.

(4) The oblique angle of approach to the steel wheel tooth seems to be sensible, as indeed does the cast construction of tooth of special alloy steels, work-hardening, which involves special welding techniques to secure these teeth to the steel wheel rim.

(5) One can also criticize the patterns of steel teeth on the wheel itself. It seems a good idea that these wheels should not work independently but should work together as a four-part unit to produce, wherever possible

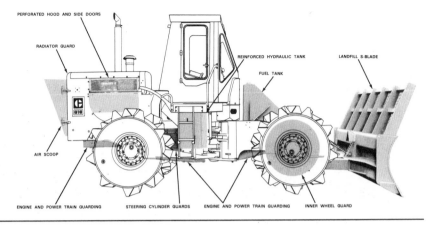

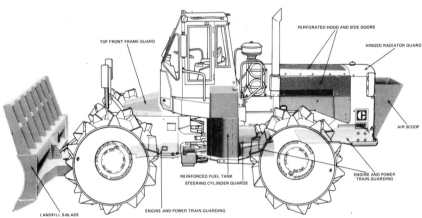

Figure 8 Cross-section illustrations of a modern landfill compactor. (Courtesy of The Caterpillar Company, USA)

(particularly on industrial waste), a tearing and reducing action on the waste at each pass, thereby compacting, crushing, and to some degree pulverizing the material.

Indeed, a great deal more design work must be done in the industry on the steel wheel before an overall successful range of steel wheels will be available for all applications. Even when this is accomplished, with the resultant high standard of operation, these results will only have been reached by extensive practical research.

Figure 9   A traditional bulldozer with blade working on a landfill site

CONCLUSIONS

In this chapter I have tried to show, as impartially as possible, the advantages and disadvantages of the machines that are available and where they may best be used. With the newer machines the chapter has illustrated the link between them, local government reorganization, and all the changes this caused. The newer professionalism in waste management, the development of fewer larger landfill sites, and the changing nature of waste, have all had their effect on these machines. It can never be a precise science; subjective whims and fancies will always influence particular decisions; but at least one can have a better appreciation of the best capabilities of each machine. There is no universal machine 'for all seasons'; in every case the choice must be linked to the job to be tackled.

Practical Waste Management
Edited by J.R. Holmes
©1983 John Wiley & Sons Ltd

12

Transfer and carriage of solid waste

M. C. Harrison, MIWM
Managing Director, Haul-Waste Limited

ABSTRACT

How solid wastes are carried is the theme of this chapter. The principal types of vehicles and containers used, and the operating costs of a number of mobile systems, are described. How a carriage system is chosen, and the capital and revenue implications of that choice, are set down. The use of transfer stations and their costs are mentioned. Illustrations are included of a range of modern equipment for containing and bulk hauling a wide variety of domestic, commercial, and industrial wastes. Examples are given of some rail and river transfer operations in the United Kingdom.

The use of the word transfer in the heading of this chapter does not indicate transfer from the point of origin of the waste to the final disposal; that is denoted by the word carriage. Transfer in this instance is something of a misnomer, as it really implies making the waste ready for carriage, normally by eliminating air space and in so doing improving the efficiency of the carriage element.

The need for movement of waste has built up over the centuries, mainly due to the evolution of the urban civilization which most of us are told we enjoy today. Nomads of old suffered no embarrassment with waste accumulation, as they moved on before such an accumulation reached unmanageable proportions. With mainly agricultural societies the waste was put back to the land or rotted where it lay, but in any case was not of sufficient size to cause nuisance. Town dwelling, in causing the need for a sophisticated network of food supply to the town, caused the parallel need for waste to be disposed of hygienically elsewhere.

Gradually, over the years, a system of rubbish collection and disposal was built up using conveniently handy holes in the ground, old river beds, etc., for the final disposal of the trash. In the UK most Local Authorities were relatively small. Each had its own disposal point and the collection vehicles ran straight to it. With the exception of areas like London there were very few

Figure 1   An industrial waste transfer station in Bristol. (Courtesy of W. Hemmings Waste Disposal Ltd)

Figure 2   A demountable bulk container system for industrial wastes. (Courtesy of Thomas Black Ltd)

Figure 3   A skip lift or 'Load-Lugger' vehicle used for commercial or industrial
wastes. (Courtesy of Biffa Waste Services Ltd)

problems of distance, and in London transfer was effected in a number of
riverside boroughs by water.

Some time after the Second World War the community realized that refuse
disposal troubles were about to descend, if they had not already, on these
small authorities. Whilst previously there had been a private contractor
presence on the disposal scene it was to build up in the 1950s, 1960s and 1970s
in response to a need. The need that was identified, albeit piecemeal, by the
private entrepreneurs, was the need to service the growing industrial and
trade elements of our society. The Local Authorities were well entrenched in
domestic collection by virtue of their statutory duties under Acts of Parliament, but industry and trade had to find their own disposal services. It has
therefore come about in the UK that, for the movement of waste from the
point of origin to the point of disposal we have two systems: the Local
Authority for domestic waste and the private contractor for other waste. This
dichotomy is not nearly so evident in Europe, particularly Northern Europe
where public enterprise is the dominant feature, or in the USA where private

Figure 4   One of Hales Containers' two giant PD Dumpmaster trailers sets off from the refuse transfer station for the disposal point. The trailers—the biggest bulk waste transporters made in Europe—have a capacity of 65 cubic yards. They are 40 ft 2½ in long, 8 ft 2½ in wide and 12 ft 2¾ in high. Discharge is by a hydraulic ejector plate inside the body. Hauling the trailer is a Foden 6AL6/32 six-wheel tractor with a Leyland 680 engine, 12-speed gearbox and double-drive bogie

contractors reign. Both sides of the industry, for that is what it had become, were faced with, and continue to face, the closure of the disposal facilities within easy distance of the point of origin. The story of how the problem is being tackled is varied and interesting.

Before tackling the methods of transfer and onward carriage of waste, let us pause for a moment and review the reasons for transfer. A new awareness of the environment and what we, the human race, were doing to it required us as a nation to pass a set of laws, culminating in 'The Control of Pollution Act 1974' to control our assaults upon that environment and in particular to minimize the effect of man on his habitat. Therefore not only did we find ourselves suffering from a lack of disposal facilities close to the origin of the rubbish; we now found that those disposal facilities had to be monitored and controlled by the operator in a manner far more costly than hitherto. This meant that fewer but larger sites were required, be they landfill or some other form of disposal, to make each site viable. The larger the site the further the geographic distance between each site and the larger the segment of population needed to use the site economically.

The bulk of collection of solid waste is carried out, on the Local Authority side, with refuse collection vehicles built and designed to move slowly through the streets of domestic dwellings, whilst being filled with rubbish by a crew of between three and five operators. These operators' duties may or may not include sorting of saleable items such as iron and rags; it may include distribution of plastic sacks. With crews of this size, with extra duties and with the very construction of the vehicle, running a refuse freighter long-distance to disposal is a very expensive pastime.

Whilst it is true that carriage of waste started with the horse and cart, moving to the tipper lorry via a primitive form of flatbed vehicle, it is also true to say that the private contractor collection of solid waste was started with the four-wheel skip unit; a perfectly adequate machine to move 4–10 m^3 of waste up to about 12 km. At any distance over that the cost starts to escalate because the number of loads one of these vehicles can then clear in a day falls to uneconomic levels. We must always remember, from the customer's or ratepayer's point of view, that the movement and disposal of waste is a money-consuming necessity in which they would not indulge if there was a method of getting the service free of charge—therefore cost is the main force to urge on the use of transfer.

TRANSFER

The transfer of the waste to long-distance carriage can take place:

(a) at the point of origin of the waste,
(b) at a special transfer station.

(a) Transfer at point of origin

If the waste loading and/or the distance from disposal site rise to unacceptably expensive proportions there are three systems which can be used at the point of origin of industrial and trade waste. The first is compaction. In essence nearly all waste loads are made up of discarded material from a process or a trade, and a lot of air. If the air can be eliminated the resultant space can be taken up with more rubbish. One way of achieving this is to press or compact the waste into an enclosed and strengthened container. Two types of compaction at origin are shown:

Figure 5   A 'Big bite' refuse collection vehicle adapted for industrial waste collection.
(Courtesy of Wimpey Waste Management Ltd)

Figure 6   Refuse compaction unit loading unsegregated refuse at the Wembley Conference Centre where full facilities are available for up to 2,700 visiting delegates daily. All waste is handled under a comprehensive scheme drawn up, and serviced on a day-to-day basis, by the contractor. (Courtesy of Biffa Waste Services Ltd)

  (i) the static packer with bin-capacity 2–35 m³;
 (ii) the mobile packer which integrates container and packer unit—these have a capacity range of 10–40 m³.

The amount of compaction available will depend on the type of material to be handled and the efficiency of the compactor. Great care has also to be taken to ensure that in compacting the waste the gross vehicle weight is not exceeded. The normal compaction achieved will be the area 3 : 1–5 : 1, although I have personally witnessed an achievement of 9 : 1 with a specifically very light and flimsy fabric waste. Little compaction is achieved with dense waste such as concrete blocks, brick rubble, etc. Such materials will also tend to ruin the compactor and container. It is therefore not recommended that attempts are made to compact such loads.

   The second popular system of transfer at point of origin is the collection of the waste in a skip capable of being lifted and transported by a skip unit, but transferring the waste at source into a vehicle which itself originated as a

Local Authority refuse freighter, and which has been modified to lift those skips and to compact their contents in its load-carrying compartment. These vehicles are known as rear-end loaders.

Given careful consideration of the loads to be picked up, the cost of disposing of the individual customer's loads can be kept in reasonable line with other costs with this system when the disposal point is moved well away from the point of waste origin. The reason is that depending on the density of the waste, anything from 9 to 14 bins can be emptied on one round. The vehicle and its containers is, for a small company, a large investment of around £100,000, but due to its productivity it is a possible answer to the problem of increasing distance.

The third system is a more well-tried one than the rear loader—that is the front loader, epitomized in this country for many years by the Dempster Dumpmaster. In this sytem the bins are loaded over the top of the cab into the body of the vehicle in which there is a ram plate. This ram compresses the load against the back door. When full the vehicle runs to tip and the ram is used to eject the load. The storage bins for the waste at the customer's premises are very specialized and are not capable of being picked up by any other machine. This is a disadvantage where only one vehicle is available in case of breakdown. On the other hand it has been well proven over many years and most of the 'bugs' have been eliminated.

The above methods of transferring the waste at the point of origin prior to carriage to disposal have the advantage of not requiring a static site run by the operator with all its attendant costs. The disadvantage is that extensive re-equipping is required. Using a transfer station the original collection equipment can be used and the station itself can attract further custom. Opportunities occur at a transfer station for starting a waste reclamation scheme, a subject which is bound to become more important as time goes by.

(b) Transfer stations

Transfer station (compaction)

The words of Mr F. W. Stokes, lately Managing Director of Powell Duffryn, in his paper 'The Transfer Station' places in a nutshell the argument for transfer stations.

> A transfer station is so designed and equipped that it can accept loads from whole fleets of collection vehicles, transfer these loads by means of large and powerful hydraulic packers into high volume steel containers or trailers, which then may be hauled to the disposal site.

The first transfer station to be built in this country was at Fishers Green, North London, for Hales Containers Ltd, and the equipment used was made

Figure 7   Industrial waste bin handling system and hydraulic compactors. (Courtesy of Easidispose Ltd)

Figure 8   A close view of 1.6 m^3 bulk waste container-emptying mechanism into a hydraulic compactor—note safety guards. (Courtesy of Easidispose Ltd)

Figure 9   An industrial waste transfer station in the London area. (Courtesy of S. Grundons Waste Ltd)

Figure 10   A Telehoist CH501 Load Lugger on a Volvo F.86.34 with a Crane Fruehauf 13-ton draw-bar trailer. (Courtesy of Cleanaway Ltd)

by Powell Duffryn. It is of interest in so much as the carriage of the waste to final disposal is undertaken, not in containers, but in special 'push-out' trailers, articulated, or 50 m^2 capacity each.

The Greater London Council operate, at Brentford, a rail transfer station, once again using Powell Duffryn compactors. The waste is carried to Oxfordshire and disposal is undertaken by Associated Roadstone Corporation at Sutton Courtenay. Both the above stations use the Transpack II which has a nominal capacity of 511 m^3 per hour. The working of a compactor is well known. Suffice to say that I believe at the present, and in the near future at least, that compaction is the best method of preparing refuse for most medium- to long-distance journeys.

London Brick Landfill Ltd are now operating a transfer station at Hendon which has a capacity of 800 tonnes/day. The refuse is transported by rail to their landfill site in Bedfordshire. This transfer station uses a Vapsco compactor. The Stewartby landfill also receives, by rail, waste from the Greater London transfer station at Hillingdon, and by road further loads from Northampton.

Smaller stations than the large ones above can be tailor-made for different circumstances. My own company indeed run a station in Plymouth with compactors of a nominal 200 m^3/hour capacity. The bulk haulage is effected by a Rolonoff, single-arm lever vehicle using 27 m^3 containers. In this case the distance to the disposal point is not great, (20 km) each way, but here environmental considerations come into the calculations as well as cost. The whole of the journey is through a city, from one side to the other. It is obvious sense for one vehicle to journey five to six times a day than for smaller vehicles to do fifty journeys plus—this is the type of information the anti-big-lorry lobby does not like to know.

Transfer station (baling)

High- and medium-pressure baling of waste is now being tried in the UK and the USA. In essence the baler squashes the waste into an easily haulable cube which at medium pressure is then bound with wire and at high pressure needs no further binding. Easy transfer by fork-lift trucks from baler to transport, and ease of placement in the disposal point, are claimed for this method. It is thought that this method of transfer in the future will be the most economical to use when the method of carriage is rail, due to the ease with which loading and unloading of the rail trucks can be effected. Baling hs been more successful in the USA, to date, than elsewhere. In the UK maybe it is lack of experience; maybe it is that our waste constituents are different or in different proportions, but success has not yet really been achieved. A major problem is the expansion of the bales—a problem which it is confidently expected will be overcome in the near future.

Transfer station (pulverization)

Pulverization prior to carriage to landfill provides a method of transfer by virtue of eliminating voids in the waste. It enjoys little favour due to the expense of the operation.

Siting of transfer stations

This is problematical, if only because in the UK nobody wants a site processing rubbish at the end of their particular road. In the main this problem is overcome by siting the station in an industrial estate. The Greater London Council Brentford scheme, and the private Grundons scheme, are situated on railway land specifically to take advantage of that method of carriage. The Hemmings station at Bristol, the largest privately operated transfer station in Europe, is built as near to the city centre as possible to shorten the collection vehicle rounds. One of the most spectacular sitings is that in Stockholm, which is in a rock cavern (13 m) below the city centre.

CARRIAGE

Here we have a number of alternatives, not every alternative being viable for every situation, of course: road, rail, or water.

Road

The road method of carriage is an everyday occurrence to us and is normally effected in the UK by a single-arm lever vehicle such as the Rolonoff. These vehicles will normally be six- or eight-wheel rigid-chassis machines. The rigid chassis being preferable for these applications as the non-articulated system is more rugged than the articulated for driving over landfill sites and when discharging the loads by tipping. There is a weight disadvantage against the articulated machine but this is not considered to outweigh the advantage of stability.

Dorset County Council at Poole, and Cornwall County Council at Hayle, amongst others, have purchased the Multilift equipment. The Multilift has a weight advantage of about a tonne a load over the other vehicles and, whilst it is not a favoured vehicle for waste disposal operation in the UK, it enjoys great popularity in Europe and Scandinavia.

The Dumpmaster trailer of the 'push-out' system, as used at Fishers Green, has the advantage of carrying capacity over the single-arm level unit; however it still has to make its mark in a big way here. From the operation point of view, a further advantage is that if the prime mover fails, another can easily be brought in to take over and keep the 'push-out' equipment moving. The

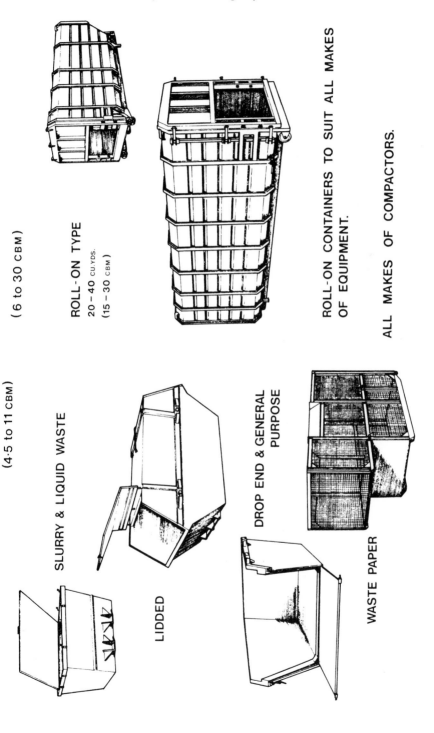

Figure 11   A range of containers used on industrial and commercial wastes. (Courtesy of The Boughton Group)

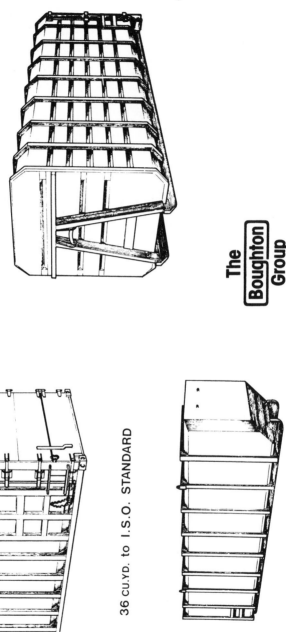

36 CU.YD. to I.S.O. STANDARD

41 CU.YD.
DINOSAUR
18 Ton.

Figure 12   A range of larger waste disposal and reclamation containers. (Courtesy of The Boughton Group)

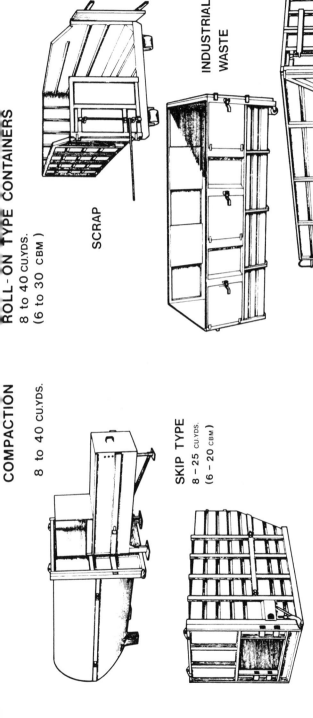

ROLL-ON TYPE CONTAINERS
8 to 40 CU.YDS.
(6 to 30 CBM)

SCRAP

INDUSTRIAL WASTE

MUNICIPAL WASTE

COMPACTION
8 to 40 CU.YDS.

SKIP TYPE
8 – 25 CU.YDS.
(6 – 20 CBM)

DOOR ARRANGEMENTS TO SUIT

Figure 13   Range of lift-on containers for various types of solid waste. (Courtesy of The Boughton Group)

Handling system	Model	capacity
SKIP LOADER 	**GG 1** **GG 2** **GG 3** Single or three phase supply	**7.5^{3}yd** **10^{3}yd** **13^{3}yd**
MULTILIFT 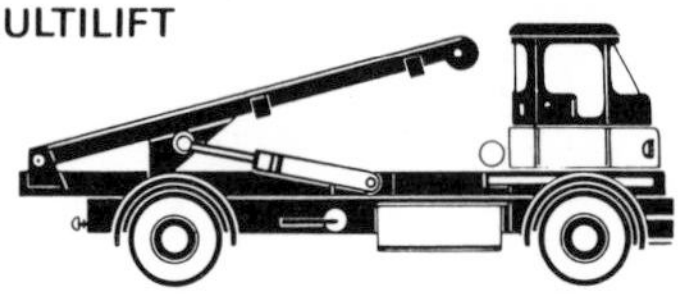	**GG20ML** **GG25ML**	**20^{3}yd** **25^{3}yd**
DINOSAUR 	**GG25DS** **GG30DS**	**25^{3}yd** **30^{3}yd**
ROLONOF 	**GG20RO** **GG25RO**	**20^{3}yd** **25^{3}yd**

Figure 14   Technical descriptions of a range of waste container handling vehicles. (Courtesy of Portable Compactors Ltd)

'push-out' road trailer is favoured on the continent as vehicles may run at 38–40 tonnes capacity.

Rail

We are now, in 1980, looking at movement of bulk rubbish over distances up to and beyond 100 miles (160 km). It is under these circumstances that the use of rail haulage starts to show dividends. Providing a large enough volume of waste is generated in an area that itself is small enough to allow collection vehicles to run to one transfer station, and providing that volume will fill a train of 500 tonnes plus per day, you can start your calculation. A railhead is required at the transfer station and one at the disposal site. The disposal site needs to be able to absorb the above volumes of waste for at least 10, and in

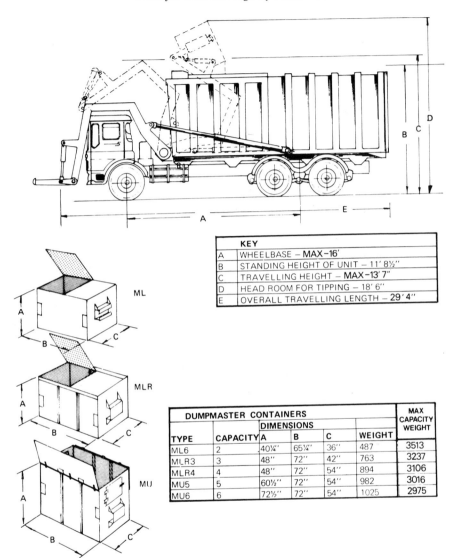

KEY	
A	WHEELBASE – **MAX–16'**
B	STANDING HEIGHT OF UNIT – 11' 8½"
C	TRAVELLING HEIGHT – **MAX–13' 7"**
D	HEAD ROOM FOR TIPPING – 18' 6"
E	OVERALL TRAVELLING LENGTH – **29' 4"**

DUMPMASTER CONTAINERS		DIMENSIONS				MAX CAPACITY WEIGHT
TYPE	CAPACITY	A	B	C	WEIGHT	
ML6	2	40¼"	65¼"	36"	487	3513
MLR3	3	48"	72"	42"	763	3237
MLR4	4	48"	72"	54"	894	3106
MU5	5	60½"	72"	54"	982	3016
MU6	6	72½"	72"	54"	1025	2975

Figure 15   Technical drawings for 'Dumpmaster' front-end loading vehicles and container systems. (Courtesy of Powell Duffryn Engineering Ltd)

reality nearer to 20 years to ensure full economic use of the plant. A reasonably simple rail journey is also an advantage—too many switches across the general thrust of the main rail routes will attract serious operational difficulties. As mentioned before the compacted rubbish in the ISO container is favoured at the moment; but baled waste could well come to the fore in rail transport.

Water

The rubbish moved down-river by the London Boroughs is merely shot into barges and tamped down. It is grabbed out of the barges by grab crane at the disposal site. Nothing very spectacular in this, but it works. Mr. Jaakkoo Paatero, DSc, in his paper to the ISWM on 16 January 1980, in London, reported the use of small boats to collect rubbish from island homes in Sweden and Finland prior to depositing the rubbish in a compactor for onward transmission to disposal.

CHOOSING A SYSTEM

The economic argument is perhaps the basic argument in the choice of any system of transfer and carriage. To illustrate the method of making a choice I must necessarily use a hypothetical station with hypothetical figures. I appreciate that some may feel the cost per mile, for example, of the skip unit is too high or too low, that the tonnage per skip is not correct. If there is one thing that I have learnt in the National Association of Waste Disposal Contractors, it is that methods of costing are all different; that average load tonnages vary; that, in short, no-one works from the same base.

Assumption

Tonnage per skip unit, 10 yd^3 (light rubbish)	1 tonne
Rental per week	
skip unit skip	£1.50
rear-loader skip	£3.00
static packer (4 : 1)	£28.00
static packer bin (14)	£5.00
mobile packer (4 : 1)	£35.00
transfer station	£385.00
transfer station bins	£12.00
Rate per mile	
skip unit	95p
single-arm lever unit	£1.01
rear loader	£189
Old disposal point	5 miles
New disposal point	20 miles

We will assume each plain skip moved three times a week and that the rate assumptions given above provide the necessary profit for all operations shown (see Table 1.)

Table 1   Cost comparisons of hypothetical station

Type of service	Costs					
	A	B	C	D	E	F
Skip unit	£1.50	£9.50	£38.00	—	50p	£38.50
Skip unit + static packer	£33.00	£9.50	£38.00	4 : 1	£61.60	£17.70
Skip unit + mobile packer	£35.00	£9.50	£38.00	4 : 1	£46.60	£21.15
Rear loader	£3.00	£18.90	£75.60	—	£4.00	£11.56

A = rental
B = movement cost to old disposal point
C = movement cost to new disposal point
D = packing ratio
E = rental per movement
F = cost per tonne

Table 2

Cost per week	£385.00
Throughput per week—6,000 m^3 600 tonnes	
Operating cost of station	£0.64/tonne
Skip unit costs, assuming 2-mile (3.2 km) trip complete with bin rental	£2.40/tonne
Single-arm lever unit on long haul together with the bin rental	£2.97/tonne
Load per 35 yd^3 bin 14 tonne	£6.01/tonne

Transfer station

The running and standing costs for this station are assumed as shown in Table 2. The use of even a small transfer station as shown can give very valuable savings in costs per tonne in a long-haul situation, but it needs a large volume of compactable waste to warrant the capital involvement. The rear-loader and front-loader figures, being of the same order, show an advantage over the small packers but it is not always able to get into the customer's premises. The cost of these vehicles and their bins is about half that of the transfer station.

The small packers are invaluable where the approach is too difficult for the biggest machines, they do not require such a big capital outlay, and obviously

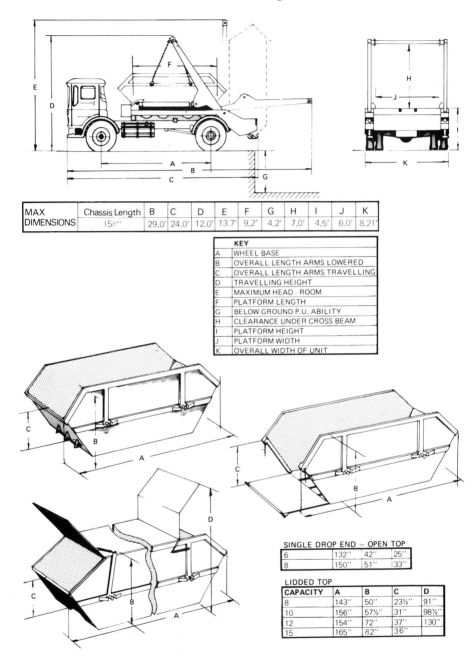

MAX DIMENSIONS	Chassis Length	B	C	D	E	F	G	H	I	J	K
	158"	29.0'	24.0'	12.0'	13.7'	9.2'	4.2'	7.0'	4.5'	6.0'	8.21'

KEY	
A	WHEEL BASE
B	OVERALL LENGTH ARMS LOWERED
C	OVERALL LENGTH ARMS TRAVELLING
D	TRAVELLING HEIGHT
E	MAXIMUM HEAD - ROOM
F	PLATFORM LENGTH
G	BELOW GROUND P.U. ABILITY
H	CLEARANCE UNDER CROSS BEAM
I	PLATFORM HEIGHT
J	PLATFORM WIDTH
K	OVERALL WIDTH OF UNIT

SINGLE DROP END — OPEN TOP

6	132"	42"	25"
8	150"	51"	33"

LIDDED TOP

CAPACITY	A	B	C	D
8	143"	50"	23½"	91"
10	156"	57½"	31"	98½"
12	154"	72"	37"	130"
15	165"	82"	36"	

Figure 16   Technical drawings for skip lift 'Load Lugger' vehicles and container systems. (Courtesy of Powell Duffryn Engineering Ltd)

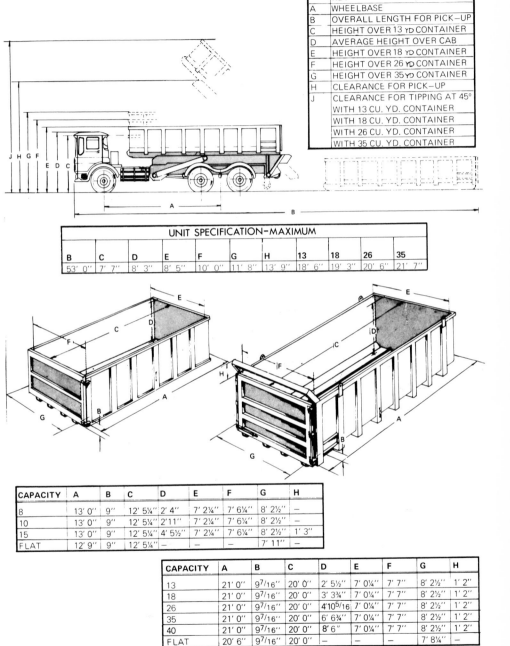

KEY

A	WHEELBASE
B	OVERALL LENGTH FOR PICK-UP
C	HEIGHT OVER 13 YD CONTAINER
D	AVERAGE HEIGHT OVER CAB
E	HEIGHT OVER 18 YD CONTAINER
F	HEIGHT OVER 26 YD CONTAINER
G	HEIGHT OVER 35 YD CONTAINER
H	CLEARANCE FOR PICK-UP
J	CLEARANCE FOR TIPPING AT 45°
	WITH 13 CU. YD. CONTAINER
	WITH 18 CU. YD. CONTAINER
	WITH 26 CU. YD. CONTAINER
	WITH 35 CU. YD. CONTAINER

UNIT SPECIFICATION–MAXIMUM

B	C	D	E	F	G	H	13	18	26	35
53' 0"	7' 7"	8' 3"	8' 5"	10' 0"	11' 8"	13' 9"	18' 6"	19' 3"	20' 6"	21' 7"

CAPACITY	A	B	C	D	E	F	G	H
8	13' 0"	9"	12' 5¼"	2' 4"	7' 2¼"	7' 6¼"	8' 2½"	–
10	13' 0"	9"	12' 5¼"	2'11"	7' 2¼"	7' 6¼"	8' 2½"	–
15	13' 0"	9"	12' 5¼"	4' 5½"	7' 2¼"	7' 6¼"	8' 2½"	1' 3"
FLAT	12' 9"	9"	12' 5¼"	–	–	–	7' 11"	–

CAPACITY	A	B	C	D	E	F	G	H
13	21' 0"	$9^{7}/_{16}$"	20' 0"	2' 5½"	7' 0¼"	7' 7"	8' 2½"	1' 2"
18	21' 0"	$9^{7}/_{16}$"	20' 0"	3' 3¾"	7' 0¼"	7' 7"	8' 2½"	1' 2"
26	21' 0"	$9^{7}/_{16}$"	20' 0"	$4'10^{5}/_{16}$	7' 0¼"	7' 7"	8' 2½"	1' 2"
35	21' 0"	$9^{7}/_{16}$"	20' 0"	6' 6¾"	7' 0¼"	7' 7"	8' 2½"	1' 2"
40	21' 0"	$9^{7}/_{16}$"	20' 0"	8' 6"	7' 0¼"	7' 7"	8' 2½"	1' 2"
FLAT	20' 6"	$9^{7}/_{16}$"	20' 0"	–	–	–	7' 8¼"	–

Figure 17   Technical drawings for 'Dinosaur' demountable container vehicle systems.
(Courtesy of Powell Duffryn Engineering Ltd)

Practical waste management

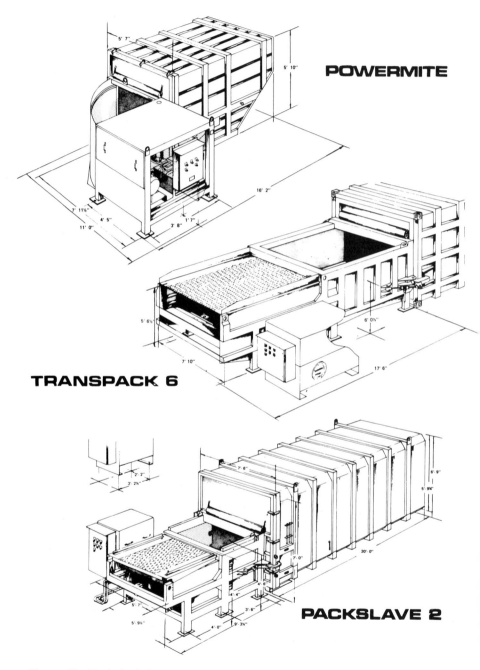

Figure 18 Technical illustrations for industrial and commercial waste hydraulic compactors. (Courtesy of Powell Duffryn Engineering Ltd)

Table 3   Suggested operating costs, September 1980 (£)

	Skip unit	2,000 gallon tanker	Rear-end loader	Front-end loader
Fixed Costs				
Depreciation C/Cab + unit	4,380	4,439	8,600	6,284
Depreciation containers	750	—	—	—
Licence and insurance	1,035	929	2,300	2,625
Overheads—expenses	4,575	4,927	7,000	6,511
Driver's wages	5,637	8,477	6,700	7,352
Total fixed costs	16,377	18,772	24,600	22,772
Variable Costs				
Repair labour	1,260	1,575	2,250	2,234
Spares	690	720	2,400	1,440
Tyres and tubes	759	720	1,500	1,612
Fuel and oil	3,040	2,846	4,458	4,499
Total variable costs	5,749	5,861	10,608	9,785
Total operating costs	22,126	24,633	35,208	32,557
Interest on capital Employed at 25%	7,126	5,428	12,000	11,875
Total costs	29,252	30,061	47,208	44,432

Figure 19   Roll-on–off container vehicle and container at industrial waste transfer station, Bristol. (Courtesy of W. Hemmings Waste Disposal Ltd)

Table 4   Suggested costs per working day, hour, and mile (£)

	Skip unit	2,000 gallon tanker	Rear- end loader	Front- end loader
Per working day				
Fixed costs	71.20	81.62	107.00	99.00
Variable costs	25.00	25.48	46.12	42.54
Total operating costs	96.20	107.10	153.07	141.55
Interest costs	30.98	23.60	52.17	51.63
Total costs	127.18	130.70	205.25	193.18
Per Working Hour				
Fixed costs	7.12	8.16	10.69	9.90
Variable costs	2.50	2.55	4.61	4.25
Total operating costs	9.62	10.71	15.30	14.15
Interest costs	3.10	2.36	5.22	5.16
Total costs	12.72	13.07	20.52	19.31
Per Mile				
Fixed costs	0.65	0.75	0.98	0.91
Variable costs	0.23	0.23	0.42	0.39
Total operating costs	0.88	0.98	1.60	1.30
Interest costs	0.28	0.22	0.48	0.47
Total costs	1.17	1.20	1.89	1.77
Working days	230	230	230	230
Working hours	2,300	2,300	2,300	2,300
Miles per year	25,000	25,000	25,000	25,000

show savings over the plain skip. A similar build-up of figures can be made for large transfer stations: for rail transport, for water transport, for high-pressure or low-pressure baling.

SAFETY

This is an important constituent of the transfer and carriage of waste: at the transfer station, on route, and at discharge. On all these occasions there are potential dangers and the possibility of accidents. The possibility of an accident happening in my opinion is not greater with transfer and carriage than with carriage alone, but it is there and should be recognized.

Apart from the normal training a driver experiences when training for his HGV licence the drivers and operators of the equipment used in this work should be trained to use the equipment—an obvious comment, but one I

Figure 20   Factory-installed hydraulic compactor and bulk waste container. (Courtesy of Anchorpac Ltd)

think with equally obvious merit. Accidents in and around the equipment should be discussed with the manufacturers to attempt to eliminate the occurrence in the future and if necessary to modify the machinery, training, or operating practices. The manufacturers of compaction equipment have banded together in an association, named Container Handling Equipment Manufacturers (CHEM), to ensure high standards of safety and to improve them where possible.

CONCLUSION

The methods of transfer and carriage are many and varied. Whether you use one or more, or indeed any of them, will be determined mainly in the end by economics, but also the equipment you presently operate and the environmental impact of your schemes. Transfer is bound to increase in the UK. We are not going to be able to obtain the full value of our investments in vehicles if too many are running empty long-distance. Good driver–operators are

Figure 21　A large hydraulic 'push-out' refuse container discharging at a landfill site

wasted on long empty return journeys—they are better employed lifting out bins at the customers' premises. Lastly, all fuel will have to be conserved and transfer will do just that. Probably the biggest achievement that will be brought about by the transfer and carriage will be the realization of the Local Authority and private contractor sides of the industry that co-operation at this point will bring great financial benefit to both parties.

ACKNOWLEDGEMENTS

I wish to thank the following for the help they have given in the preparation of this paper: Tom Boughton and Bill Macbeth, Boughton Group Ltd; Anthony Armitage and Brian Rollings, Powell Duffryn Ltd; Vic Beckwith, Shelvoke & Drury Ltd; J. D. Clowes, Jack Allen Ltd; R. C. H. Mead, David MacKrill Eng. Ltd. The opinions and errors are mine, and do not reflect on the policies of the Association, my company, or those who have been of assistance.

BIBLIOGRAPHY

Stokes, F. W. *The Transfer Station.* (1980). Institute of Highways Technicians, London.
Paatero, J. *Refuse Collection in Scandinavia.* (1980). Institute of Wastes Management, London.
Holmes, J. R. *Municipal Waste Processing.* (1979). Developments in Environmental Control and Public Health, Applied Science Publishers, Barking.

Practical Waste Management
Edited by J.R. Holmes
©1983 John Wiley & Sons Ltd

13

The carriage and transfer of liquid waste

H. W. LUTHER
Waste Management Limited

ABSTRACT

A comprehensive review of the safe carriage of hazardous wastes, liquids, and sludges is the theme of this chapter. Mention is made of current legislation, the key aspects of vacuum tanker design, training, and the philosophy of the system of hazard warning panels on tankers. The requirements of the Health and Safety at Work etc. Act 1974 are considered.

INTRODUCTION

In the past waste disposal has been one of those industries which has largely been ignored by the legislators. However, with the advent of the Deposit of Poisonous Waste Act 1972 and subsequent legislation waste disposal tended to become 'a recognized' industry and the various respective governments have begun to recognize that the industry needs, and indeed demands, greater consideration—particularly where transport legislation is concerned. It must be said that NAWDC has played a major part in obtaining this recognition and as an association is now represented on all the legislative committees which affect the industry. This chapter outlines the current legislation regarding the carriage and transfer of liquid waste and also tries to anticipate changes in that legislaton.

CURRENT LEGISLATION

Apart from the general requirements under the Health and Safety at Work etc. Act 1974 the only two pieces of legislation mainly concerned with the transportation of liquid waste are the Petroleum (Consolidation) Act 1928 and the Hazardous Substances (Labelling of Road Tankers) Regulations 1978.

The former dealt basically with petroleum spirit and provided powers by which regulations were made for conveyance of that substance by road. Other substances were brought under control by making orders in council naming these as ones to which the Petroleum Act applied and up to present date, a total of 428 substances have now been named.

The Hazardous Substances (Labelling of Road Tankers) Regulations 1978 dealt mainly with the display of hazard warning panels and for the first time, apart from the Deposit of Poisonous Waste Act 1972, a piece of transport legislation actually dealt specifically with the waste disposal industry. It is now worthwhile considering in detail the requirements of this piece of legislation.

HAZARD WARNING PANELS

The Hazardous Substances (Labelling of Road Tankers) Regulations 1978 came into effect on the 28th March 1979, and from that date all road tankers carrying hazardous substances or, in our case, hazardous waste, were required to display a hazard warning panel on both sides and the rear of the vehicle.

Figure 1   Jubilee special limited edition. Finished in metallic silver, with Union Jack motifs, the attractive Jubilee year sludge tanker from Whale was limited to a total production of only 100 units, offered to collectors at a slightly increased price to cover the extra finishing costs. Completed just before 7 June, the first Jubilee Whale was bought by municipal hire specialists Dassett Hire Ltd, and worked most of the summer for a Hertfordshire District Council. Chassis is a Ford D1614, the 2,000 gal tank (9,000 litre) is evacuated by a Drum Engineering Ltd Gemini 130 exhauster with a swept volume of 120 cfm (204 m^3/h)

One of the first problems encountered was in the definition of a hazardous substance. Unlike previous legislation however, i.e. The Deposit of Poisonous Waste Act 1972, these Regulations actually tried to define on an inclusive basis substances which the Regulations covered. Unfortunately, whilst these were relatively straightforward for hazardous products, the categories for hazardous waste still cause some confusion.

Paragraph 4 of the above Regulation states that:

> Where a road tanker is being used for the conveyance by road of a single load...the *operator* of that road tanker shall ensure so far as is reasonably practicable that it is provided with and displays hazard warning panels—and those panels shall show the following particulars:
> (a) the emergency action code for the substance which constitutes the load;
> (b) the substance identification number;
> (c) the appropriate hazard warning sign;
> (d) the telephone number indicating where specialist advice can be obtained at all times.

Appendix 1 shows the layout of the hazard warning panel; Appendix 2 gives details of the substance identification number, emergency action code, and hazard warning sign.

As mentioned previously, the categories of substances have given rise to some confusion. For instance, substance identification number 7006 is for a liquid containing acid, whereas identification number 7007 is a sludge containing acid. The difference and balance between a liquid and a sludge is very much open to interpretation. Identification number 7015 covers a hazardous waste liquid not otherwise specified. This category has caused a great deal of concern in that unfortunately a number of contractors have seen fit to use it for all their waste removals. The substance identification numbers enable the emergency services to identify the category of waste in the event of an accident, and the action code gives them advice on the action to be taken. The explanation of these action codes is given in Appendix 3.

Although these Regulations seem relatively straightforward, a closer look at the small print reveals a number of problems for waste disposal operators. Section 4 of the Regulations states that the *operator* of a road tanker must ensure that the appropriate warning panel is displayed. Therefore, if the vehicle you are operating is programmed to undertake more than one load in any one day and the wastes which are removed are of different composition, then, before loading, the hazard warning panels must be changed and if the next load is still different, they must be changed again. The size of the plates makes this a very cumbersome operation. Therefore a number of operators have devised systems whereby the substance identification number and the emergency action code can be changed without changing the whole panel.

Figure 2   A modern industrial waste tanker—notice the hazardous-waste identification board. (Courtesy of Thomas Black Ltd)

However, even this requires the vehicle to carry a number of plates and more importantly requires the driver of the vehicle to change the panels after each load. This can be a problem because the Regulations state quite clearly that it is the operator of the road tanker who shall ensure that the vehicle is provided with and displays hazard warning panels, and although there are certain defences in the Regulations, it is necessary for, and encumbrant upon, all operators to ensure that their drivers are fully aware and educated in the use of the hazard warning panels. It cannot be stressed too strongly that clear, precise instructions on which panels to be displayed must be given to the drivers. In my own company this is done by giving details to drivers in writing on a daily basis with their daily worksheets.

Other provisions of the Regulations state that:

(1) the warning panels shall be weather-resistant and indelibly marked on only one side;
(2) the panels shall be either rigid or fixed to the vehicle;

(3) side panels shall be securely attached in a vertical plane with:
   (a) forward edge as close as is practicable to the front of the tank;
   (b) lower edge at least 1 metre from the ground;
(4) rear panel shall be securely attached:
   (a) in vertical plane;
   (b) with lower edge at least 1 metre from the ground;
(5) hazard warning panels shall be kept clean and free from obstruction except that a rear panel may be mounted behind a ladder which does not prevent the information from being easily read;
(6) when a road tanker has had its tank emptied and cleaned or purged, the operator shall ensure that the hazard warning panels are either:
   (a) completely covered or completely removed;
   (b) partly covered or partly removed so as to leave visible only the telephone number.

The Regulations also provide that the consignor of any prescribed hazardous substance shall ensure that the operator has such information as is necessary to enable him to prepare the hazard warning panel. Therefore, again a defence exists if wrong information is supplied by our customers.

Provision is made in S.12 of the Regulations for certain materials to be exempt from the labelling requirements. NAWDC, through its Technical Committee, has taken up this clause with the Health and Safety Executive and generally speaking it appears that for waste disposal contractors the exemption clauses cover most materials which are currently exempt from notification under the Deposit of Poisonous Waste Act 1972. However, it should be noted that the above is only a guideline and an advisory statement, and that the actual interpretation of the Regulations can only be judged by the Courts.

PROPOSALS FOR DANGEROUS SUBSTANCES (CONVEYANCE BY ROAD) REGULATIONS

The above consultative document was published in June 1979 and hopefully it is the forerunner of legislation which will deal with all aspects of the transportation of dangerous substances. The Regulations attempt to cover such things as:

(a) the packaging of dangerous substances;
(b) the construction, the examination, and testing of vehicles and tanks;
(c) the form, specification, and colour for packaged labels;
(d) loading and unloading;
(e) driver training;
(f) the classification of dangerous substances.

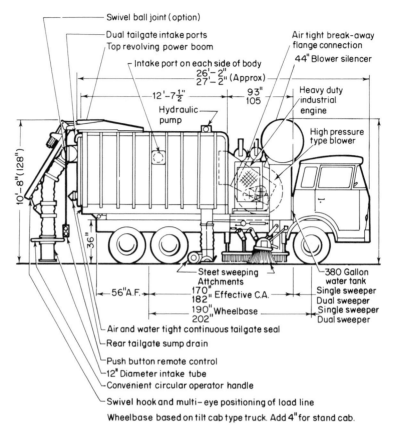

Figure 3   Drawing illustrating multipurpose vacuum suction tanker with street
            sweeping attachment. (Courtesy of Vacall Ltd)

Two of the above categories are of particular interest to waste disposal
operators. The first concerns the provision of instructions in writing to be
available during the conveyance of hazardous substances. This clause states
that an operator of a vehicle being used for the conveyance by road of
dangerous substances shall ensure that information in writing shall include the
following:

(a) the name of the substance or category of substance;
(b) the substance identification number;
(c) the description of the hazard associated with the substance;

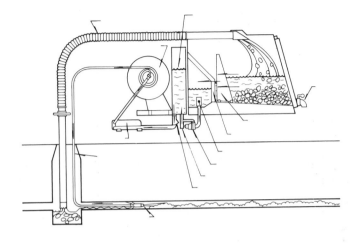

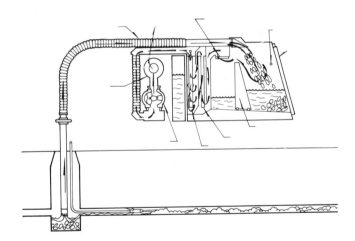

Figure 4   Drawing illustrating sewer cleaning and filtration techniques. (Courtesy of Vacall Ltd)

(d) the nature of the dangers to which the substance to be conveyed may give rise, and the measures necessary to avoid or reduce those dangers during conveyance;

(e) the action to be taken, and the treatment to be given, in the event of any person coming into contact with the substance, including the first-aid measures to be taken;

Figure 5   A view of a 4,500 litre capacity vacuum sludge tanker used by a Regional Water Authority. A Whale 4 × 4 Bedford vehicle with vacuum filling and low-pressure spray discharge capability for processed sludge disposal to land. (Courtesy of Whale Tankers Ltd)

(f) the action to be taken in the case of fire and, in particular, the type of firefighting equipment to be used;

(g) the action to be taken in case of breakage or deterioration of the tank where the substance spills or leaks onto the road.

All the above information will be required to be kept in the cab of the vehicle whilst the substance is being conveyed. To simplify the provision of this information the Chemical Industry Association devised a series of transport emergency cards or Tremcards as they are now commonly termed. NAWDC has therefore produced a series of these cards to cover the categories of waste moved by the waste disposal industry and an example is given in Appendix 4. It is important to note that it is not necessary to provide instructions in writing in the form of Tremcards and indeed various larger independent companies have produced their own set of cards for this purpose. However, it is strongly urged that all members of NAWDC use the cards available from the Association so as to achieve some degree of uniformity throughout the industry. Another important point to note is that although there is no current legislation which enforces the use of the Tremcards, it is strongly recommended that operators start to introduce them into their fleet so that their drivers will become fully conversant with them when the Regulations come into force.

CONSTRUCTION OF VEHICLES

Another important aspect of the new proposals is that construction of road tankers will in fact be governed by Regulations. Realizing the tremendous task it would be to legislate on all aspects of tanker construction, the Health and Safety Commission have stated that codes of practice will be issued on the construction of road tankers. NAWDC, for its part, intends to submit to the Health and Safety Executive detailed proposals on how it feels the tanker vehicles for the industry should be constructed. It is obvious that there are a number of problems to be overcome on this issue and it must inevitably be accepted that there will be a number of recommended designs for the carriage of the varying materials.

TRAINING

No-one, I would hope, would deny that there is a need for drivers of road tanker vehicles which are carrying hazardous waste to be adequately trained to do the task that falls to them, and indeed, bearing in mind the potential hazards that could occur in the event of an accident. It is somewhat surprising that, at this moment in time—apart from general Regulations under the Health and Safety at Work Act—no specific legislation exists relating to the training of tanker drivers. Therefore the new proposed Regulations seem to rectify this matter by making it encumbent upon the operator of a vehicle that he shall ensure that the driver of the vehicle has received adequate training to enable him to do the following:

(a) understand the significance of any hazard warning signs or panels;
(b) understand the requirements of the Regulations for displaying such panels;
(c) understand the requirements of the Regulations as to the loading and unloading of vehicles;
(d) understand the risks associated with the substance he is required to carry, the precautions to be observed, and the action to be taken in the event of an emergency;
(e) understand the operation of the vehicle and where appropriate the proper use and operation of any valves, pumps, hoses, etc.

(f) use any protective clothing and where appropriate respiratory protective equipment;
(g) use the firefighting and first-aid equipment carried on the vehicle.

Figure 6 Remote control 1,200 gal (5,500 litre) Whale gully emptier, with a new design of hose box, as shown by Vauxhall Motors Ltd at the Public Works Exhibition. A fully equipped representative of the Whale remote control series, this unit is one of a batch of four built for the London Borough of Bromley. Based on a Bedford 138 in (350 cm) KGM chassis, it has both channel flushing and sewer flushing equipment, hand hose and hand-washing basin

It will of course be necessary for the training of drivers to be undertaken very much from a practical standpoint, and no doubt many operators are horrified at the prospect of trying to train all their personnel. However, it cannot be doubted that it is reasonable to expect drivers to be able to take all the necessary action in the event of an incident. For some time now NAWDC has been organizing training courses for tanker drivers. Support for these courses has at times been somewhat disappointing, but again I would urge all operators to reconsider their position in this matter in the knowledge that eventually legislation will enforce them to carry out such training programmes.

It is perhaps appropriate that this chapter should end on the training of drivers, because in the final analysis it is these people around whom the carriage and transfer of liquid waste revolves. Automation may come to many manufacturing industries but I have yet to hear the suggestion that robots will eventually drive road tankers. There is no doubt that the industry is going

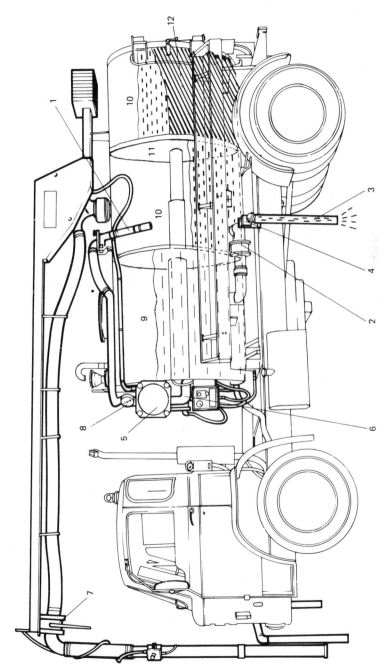

1. Rear level gauge
2. Interconnecting valve
3. Sludge water discharge line
4. Sludge water discharge line valve
5. Sludge trap
6. Sludge trap drain valve
7. Gully arm valve
8. Relief valve
9. Clean Water Compartment
10. Sludge Compartment
11. Compactor Plate
12. Compacted Sludge

Figure 7 The vehicle operates on a vacuum principle and uses an air-cooled vane-type exhauster pump delivering approximately 120 ft³/min and driven by a power take-off on the chassis gearbox. 'Blowback' may be achieved by selecting the pressure setting on the pump thus forcing water down the gully arm to agitate the gully contents. Servo changeover from vacuum to pressure is fitted as standard and is operated from a control lever on the gulley downpipe

Figure 8   Waste tanker discharging at sewage treatment works

through a tremendous technological change both in its disposal methods and its transportation methods. It will be necessary for management and drivers alike to change their ideas to adapt to this new legislation. It cannot be stressed too strongly that it will need a good deal of patience, flexibility, and effort on both sides.

Appendix 1: Specification for hazard warning panels and compartments labels

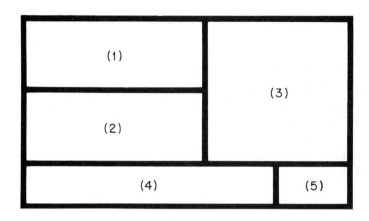

1   Form of hazard warning panel

(1) Space for emergency action code.
(2) Space for substance identification number and, if included, the substance name.
(3) Space for diamond-shaped hazard warning sign.
(4) Space for telephone number for specialist advice.
(5) Space for optional manufacturer's or company house name or symbol.

The background colour for the panel will be orange except for space (3) which shall be white. The hazard diamond warning signs are as shown in Part 3.

A completed hazard warning panel shall be of the style and dimensions shown in paragraph 2.

2   Diagram of hazard warning panel (all dimensions in centimetres)

3   Chromaticity co-ordinates of colour for the hazard warning panels

The colours of the various parts of the panel in conditions of normal use should have chromaticity co-ordinates lying within the area formed by joining the following co-ordinates on the Commission Internationale de l'Eclairage (CIE 45.15.200) chromaticity diagram:

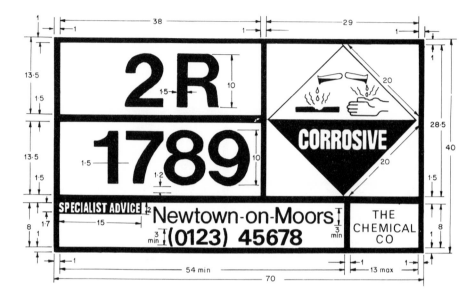

Colour		Chromaticity co-ordinates of colour points determining the permitted colour area illuminant: Standard illuminant D65*				Luminance factor B for non-retro reflective materials†	Typical permissible colours (Gloss)	
		1	2	3	4		Munsell Reference	British standard 381C Colours for specific purposes
Red	x	0.658	0.576	0.605	0.690	⩾0.07	7.5 R max	No. 537 Signal Red
	y	0.342	0.339	0.310	0.310			
Orange	x	0.520	0.520	0.578	0.618	⩾0.22	2.5 YR 6/16	No. 557 Light Orange
	y	0.380	0.400	0.422	0.380			
Yellow	x	0.481	0.439	0.477	0.531	⩾0.50	2.5 Y 8/14	No. 355 Lemon
	y	0.518	0.471	0.433	0.468			
Green	x	0.007	0.248	0.286	0.201	⩾0.15	2.5 G 5/12	No. 228 Emerald Green
	y	0.703	0.409	0.435	0.776			
Blue	x	0.078	0.198	0.240	0.137	⩾0.05	10 B 4/10	No. 166 French Blue
	y	0.171	0.252	0.210	0.038			

*CIE 45.15.145 †CIE 45.60.425
Angle of illumination: 45° with the normal to the surface and viewed in the direction of the normal.

Appendix 2

Substance identification number	Name of substance	Emergency action code	Hazard warning sign
7006	Hazardous waste, liquid containing acid	2WE	Other hazardous substances
7007	Hazardous waste, solid or sludge containing acid	2WE	Other hazardous substances
7008	Hazardous waste, liquid containing alkali	2WE	Other hazardous substances
7009	Hazardous waste, solid or sludge containing alkali	2WE	Other hazardous substances
7010	Hazardous waste, inflammable liquid flash point <23 °C	3WE	Other hazardous substances
7011	Hazardous waste, inflammable liquid flash point 23 °C to 61 °C	3W	Other hazardous substances
7012	Hazardous waste, inflammable solid or sludge, n.o.s.	3WE	Other hazardous substances
7014	Hazardous waste, solid or sludge n.o.s.	2X	Other hazardous substances
7015	Hazardous waste, liquid n.o.s.—	2X	Other hazardous substances
7016	Hazardous waste, solid or sludge, toxic n.o.s.	2X	Other hazardous substances
7017	Hazardous waste, liquid, toxic, n.o.s.	2X	Other hazardous substances
7019	Hazardous waste, liquid, containing inorganic cyanides	4X	Other hazardous substances
7020	Hazardous waste, solid or sludge, agrochemicals, toxic n.o.s.	4WE	Other hazardous substances
7021	Hazardous waste, liquid, agrochemicals toxic n.o.s.	4WE	Other hazardous substances
7022	Hazardous waste containing isocyanates, n.o.s.	4WE	Other hazardous substances
7023	Hazardous waste, containing organo-lead compounds n.o.s.	4WE	Other hazardous substances

Note: n.o.s. = not otherwise specified

Appendix 3

Emergency Action
Code Scale
FOR FIRE OR SPILLAGE

1   JETS
2   FOG
3   FOAM
4   DRY AGENT

P	∨	FULL	
R		BA	DILUTE
S	∨	BA for FIRE only	
S		BA	
T		BA for FIRE only	
T	∨	FULL	
W		BA	CONTAIN
X	∨	BA for FIRE only	
Y		BA	
Y		BA for FIRE only	
Z			
Z			

E	CONSIDER EVACUATION

Notes for Guidance

FOG

In the absence of fog equipment a fine spray must be used.

DRY AGENT

Water **must not** be allowed to come into contact with the substance at risk.

V

Can be violently or even explosively reactive.

FULL

Full body protective clothing with BA.

BA

Breathing apparatus plus protective gloves.

DILUTE

May be washed to drain with large quantities of water.

CONTAIN

Prevent, by any means available, spillage from entering drains or water course.

Appendix 4

Cargo

HAZARDOUS WASTE
Solid or Sludge containing Alkali

Nature of Hazard
Corrosive
Contact with solid, liquid or vapour may cause skin burns, severe damage to eyes and air passages
May contain dissolved or suspended toxic constituents.

Protective Devices
Suitable respiratory protective device
Goggles giving complete protection to eyes
Plastic or synthetic rubber gloves, boots and suit with hood giving complete protection to head, face and neck
Eyewash bottle with clean water

EMERGENCY ACTION

- Stop the engine
- No naked lights. No smoking
- Mark roads and warn other road users
- Keep public away from danger area
- Keep upwind
- Put on protective clothing

Spillage

- Contain leaking liquid with sand or earth and consult an expert
- Shut off leaks if without risk
- Prevent liquid entering sewers, basements or workpits
- If substance has entered a water course of sewer or contaminated soil or vegetation, advise police
- Warn everybody—evacuate if necessary

Fire

- Keep containers cool by spraying with water if exposed to fire
- Extinguish with water spray, dry chemical or foam

First aid

- If substance has got into the eyes, immediately wash out for several minutes
- Remove contaminated clothing immediately and wash affected skin with plenty of water
- Seek medical treatment when anyone has symptoms due to inhalation, swallowing, contact with skin or eyes or fumes produced in a fire
- Even if there are no symptoms sent to a doctor and show this card

Additional information provided by manufacturer or sender.

Practical Waste Management
Edited by J.R. Holmes
©1983 John Wiley & Sons Ltd

14

Commercial landfill management

PETER R. SPENCER, MISWM
London Brick Landfill Limited

ABSTRACT

This chapter looks at commercial landfill management from the point of view of a major mineral extraction and waste disposal company in the United Kingdom. Operational plans, client relationships, management, and leachate control are considered. Specific precautions to control insects, odour, dust, and litter are looked at together with services, fencing, landscaping, and area maintenance. Types of machinery, the handling of waste, and illustration of key aspects of operations complete the text.

INTRODUCTION

We have available today the knowledge and experience needed for the successful practice of managed landfill operations. There is probably no field where the gap between the knowledge and performance of the leaders and the knowledge and performance of the accepted average has been wider or more intractable. That gap still exists, although not so wide as in the past, but our aim as managers must be to narrow the gap between what can be done and what is being done between the leaders and the average.

Management, its competence, its integrity, and its performance is decisive in the achievement of the proper standards in an effective and economic way. I place the emphasis on 'economic' because management can only justify its existence and its authority in the commercial world by the economic results it produces. It has failed if it does not supply waste disposal services desired by the consumer at a price the consumer is willing to pay. There may be great non-economic results in terms of contribution to the comunity as a whole, but the first dimension of management is a specifically economic one.

Landfill sites are major cost centres. Failure to calculate costs makes it impossible to calculate prices or profit. Every landfill site attracts costs in various forms and systems should be set up to identify and allocate them.

Figure 1   Crop growing on completed and restored landfill sites. (Courtesy of London Brick Landfill Ltd)

Many situations arise where knowledge of the costs involved is vital if a realistic decision is to be made. Whilst you may not be directly concerned in operating the costing system you must be aware of the costs involved in operating a site and use the information to make decisions.

Having established the manager's role and the necessity for information relating to the costs of running a landfill site I will turn to the specific operational problems which will concern you as individuals responsible for the day-to-day management of landfill sites.

One fundamental principle of management is to plan so consequently the phrase 'day-to-day' is ill advised if applied at its face value to site management. If an operation plan is devised, however, then day-to-day management within the overall operational plan is a possibility.

In any well-run disposal site the operation plan is essential and serves many purposes. Firstly an efficient site design is very closely related to the method of operation. The plan will spell out the details of operation, ensuring that the site design is consistent with the proposed method of operation. Secondly the operation plan serves as the site operator's guide to construct and operate the site in accordance with the given planning permission and site licence, it translates the technical jargon into instructions and guidance for the manager, supervisor, plant operator, and other employees. Finally the operation plan is a requirement of the Waste Disposal Authority, without which a site licence is unlikely to be granted.

OUTLINE OPERATION PLAN

Entrance area: preparation and maintenance.
Internal road system: construction and maintenance.
Cover material: usage, stockpile, excavation, external supplies.
Surface water control and drainage.
Leachate control and treatment.
Monitoring system, water sampling, boreholes, etc.
Filling process: method, direction, depth etc.

Office-weighbridge, maintenance and employee facilities.
Refuse weighing/classification, charging system.
Discharge method: wet weather, wind.

Accident prevention and safety procedures.
Fire prevention and action procedure.
Vermin control.

Insect control.
Odour control.
Dust and litter control.
Salvage and reclamation procedures.

Services: electricity, heat, water, sewage, telephone.
Fencing: perimeter, entrance.
Gates and signs.
Landscaping and area maintenance.

Plant and equipment: servicing and repair.

This is not an exhaustive list, but one that indicates the necessity for defining the operational requirements before endeavouring to start operations.

Landfill can be defined as an engineering method. This implies a well-thought-out procedure based on technical principles. It means designing and constructing facilities with attention to natural conditions, and using methods that overcome difficulties in a way that maintains the protection of the environment.

Three components to satisfy the requirements of a landfill site are: spreading waste in thin layers, compaction to the smallest practical volume, and daily cover of the deposited refuse. These three steps are a necessary part of the operation to reduce settlement problems, fires, insect, and vermin

Figure 2   Final levelling and topsoil covering on a completed landfill site. (Courtesy
of London Brick Company Ltd)

breeding and to conserve space by compacting to the smallest possible volume. The equipment used to achieve these three components falls into three categories:

(1) refuse movement and compaction;
(2) cover excavation and transportation;
(3) auxiliary equipment.

The choice of equipment to satisfy category 1 is limited effectively to:

(1) tracked loader or dozer;
(2) wheeled loader or dozer;
(3) steel-wheeled compactor;

although a number of manufacturers produce such equipment. The number and type of equipment needed depends primarily on the amount of refuse handled and the degree of compaction required, although on the smaller sites the versatility of certain equipment, able to perform a number of functions, affords the optimum choice.

    To give some guidance, my own experience has shown that for sites dealing with up to 250 tonnes per day of mixed refuse and incoming cover materials, a

Figure 3   Landfill in operation with a landfill compactor in use. Note inert cover on in-place refuse. (Courtesy of London Brick Company Ltd)

track loading shovel of about 60 kW with a 1.0 m³ bucket is adequate; for sites accepting up to 500 tonnes per day on the same basis, a track loading shovel of about 100 kW with a 1.5 m³ bucket is adequate. When dealing with larger quantities the more specialized steel-wheeled compactor machines come into their own; a 125 kW machine with a 3.6 m dozer blade being capable of dealing with up to 1000 tonnes per day.

However, many other factors besides the incoming quantities of refuse must be taken account of before making the final selection.

What degree of compaction is required?

Tests have been carried out which indicate that steel-wheeled compactors compact better than rubber-tyred machines, which in turn compact better than tracked machines. My own experience confirms the advantage of the steel-wheeled compactor in this respect but with the qualification that to utilize this compaction capability to the full is very time-consuming. When the full capacity of the machine is utilized and the performance compared with a tracked machine also working at full capacity my own experience shows that the advantage is small, both types achieving a density of about 0.7 tonne/m³.

Figure 4   Pest and rodent precautions on a landfill site. (Courtesy of London Brick Company Ltd)

The productive capacity of the steel-wheeled compactor can be very high, at the expense of high compaction ratios, or it can produce high refuse density figures of up 1.1 tonne/m^3 but productive capacity is reduced.

If cover material has to be excavated and transported on site, a tracked or wheeled loader may be able to deal with both cover and refuse but if the cover is quite distant, say more than 60 m away, or the volume of refuse is high, then additional equipment will be necessary.

The accepted cover requirements specified in the majority of site licences are that by the end of the working day all exposed surfaces, including the flanks and working face, shall be covered to a depth of not less than 150 mm. Having established your daily working area it is then a straightforward matter to work out the theoretical volume of covering material required daily. The theoretical volume required is in practice a minimum figure, as it can be very difficult to spread the covering material to an even 150 mm layer and fine-grain covering material will penetrate down into the refuse rather than remain entirely on the surface. As a result actual volumes of cover needed to achieve the required standard can be nearly double the theoretical volume. A useful guide is that cover requirements are at least 18% by volume up to 35% by volume of the incoming refuse, that is to say 100 m^3 of refuse will require 18/35 m^3 of cover material. Incoming cover can fluctuate to a large degree, so it is therefore a sensible practice to stockpile material which is in excess of the daily requirements.

LEACHATE

The earlier definition of landfill describing the three components as spreading waste in thin layers, compaction to the smallest possible volume, and daily cover avoids a significant potential problem—leachate generation and subsequent pollution. Rain or surface water passing through deposited refuse emerges as leachate and adequate provision must be made to deal with it. Depending on the formation of the site, leachate can migrate both laterally and downward. It can pollute groundwater, streams, or lakes or seep out of the ground on to the surface. The volume of leachate produced varies according to the amount of water passing through refuse, and as refuse has a considerable capacity for absorbing and retaining moisture, not all the water will pass through and emerge as leachate. Leachate quantities are highly variable, being dependent on the variable factors of rainfall, water ingress, volume and nature of the refuse, and transpiration.

The critical first step in controlling leachate is to reduce the volume of water passing through the refuse. This may be achieved in a number of ways. Surface water entering the site from surrounding land can be reduced by digging ditches on the perimeter of the site and diverting the water away. Deposited refuse should, if possible, be covered with material with a high clay content rather than sandy or gravelly material as less water will penetrate. Refuse should be deposited to a pre-determined gradient encouraging water to run off rather than penetrate. A minimum gradient of 1 in 50 is necessary to achieve satisfactory run-off. Vegetation planted on completed parts of the site can draw water out of the ground, reducing the amount available for leachate production. Grass, wheat, and oats consume more water than trees. Leachate control is dependent on local circumstances but the first objective must be to direct and accumulate leachate in one place separate from any fresh water in the site. Having achieved this, pumping out is feasible, preferably direct to sewer if this is possible. If discharge to sewer is not available, alternatives are to spray irrigate on to surrounding land; this will absorb a considerable volume per acre without damage if carefully controlled. Pump back into the refuse utilizing its absorptive capacity, or treat the leachate so that it becomes acceptable for discharge to stream. The analysis of treated leachate required before a discharge consent is given by the Water Authority varies according to local circumstances and it is essential to consult them before any plan to discharge treated leachate to a watercourse can be finalized.

RAINFALL

Having pointed out the importance of eliminating water ingress as far as possible, we still have to contend with the normal rainfall. When one considers that 1 inch of rain equals 250 m^3 per hectare and that an average

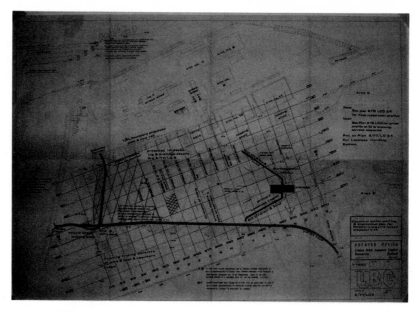

Figure 5 Operating site plans on a major landfill. (Courtesy of London Brick Company Ltd)

rainfall of 55 cm per annum on a 4 hectare site will deposit 22,700 m^3 of water within the site each year, then the importance of the problem can be realized. As previously mentioned, the refuse should be deposited with a designed fall or gradient to encourage rainfall run-off to be directed towards a drainage system. Areas of the site which are not yet in use can be separated from the working area by small bunds, to prevent the migration of leachate into clean water, and ditches dug within the unworked site area, to direct rainwater to clean-water lagoons from which it may be pumped to stream. The important factors are to direct rainwater to a place where it can be handled, using the natural contours and fall of the pit bottom, combined with ditching and piping to achieve this and to keep it separated from any leachate that may be present on the site.

Despite all efforts to drain a landfill site and keep it dry, at particular times of the year the site is going to be wet to some degree and good stable access roads are essential. Site roads can be classified as permanent and temporary.

ROADS

Permanent roads giving access to the site and its office and other permanent buildings should be constructed to as high a standrd as can be afforded. Costs are high but maintenance is straightforward and the value of a good site entrance road from a public relations point of view is worth emphasizing.

Less important permanent roads can be constructed to a lesser standard but nevertheless will have to stand up to everyday wear and tear. The use of a man-made fibre stabilizing mat placed on the graded surface prior to topping with hardcore/brickbats, is well worth the extra initial cost. Such mats, readily available in convenient-sized rolls, prevent the hardcore being compressed into the ground. Roads remain at the graded level and do not need repair so frequently. The costs saved on maintenance will normally outweigh the initial cost of the stabilizing mat within a year. Temporary roads can be constructed in the same way dependent on their expected life. For short-term roads, refuse itself is quite stable and can form a good basis for hardcore topping.

DUST, FIRE, AND LITTER

Having considered a number of problems associated with water and rainfall, we must not forget that dry weather also presents certain problems. The majority of site licences require that, in dry weather, site roads be sprayed with water to suppress dust. Dust is obviously a nuisance both for the people working at the landfill site and for the occupiers of adjacent land and the general public.

Water wagons can be used to control dust, and can range from custom-made vehicles to salvaged tanks towed behind a tractor. A minimum tank size of 1,500 gallons is recommended. Such equipment can be used for fire control, as fire is another hazard of dry weather. Provision should be made for dealing with the initial stages of a fire before it gets out of control. The local Fire Brigade should be contacted, the problems associated with dealing with a fire on a particular site discussed, and a procedure formulated. Fire drills should be conducted periodically.

In the event of a fire, action must be taken after it has been extinguished. Refuse fires are difficult to fully extinguish and can continue smouldering underneath, finally breaking out again weeks, months, or years later. It is essential to dig out a channel around affected refuse and then spread, level, and leave exposed, maintaining a watch on it until you are certain that no resurgence of the fire is possible.

Litter can be the most persistent problem on a landfill site. Windblown refuse festooning the perimeter fence or blowing into a nearby housing estate can consume many man-hours in attempting to rectify the bad public image it creates. The point of discharge, orientation of the working face in relation to the prevailing wind, the existence or absence of nearby wind-shielding features, and the type of waste, all play a role in the litter control problem. Tipping at the toe of the working face is a major help, as the wind cannot pick up the refuse as easily as when refuse is discharged at the top of the face.

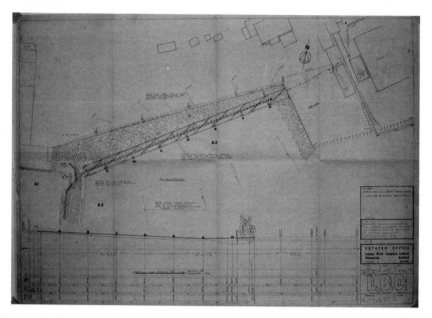

Figure 6　Operating site plans on a major landfill. (Courtesy of London Brick Company Ltd)

Bunds placed at strategic points can provide windbreaks but the most effective method is to provide portable nets or fences to catch wind-blown litter before it can travel any distance.

PESTS

Pests of various kinds can, unless controlled, also become a persistent problem of landfill sites. Insects associated with landfill sites are the housefly, the bluebottle or greenbottle, and the cricket. Crickets do not constitute a health danger but are not desirable company, particularly in the vicinity of housing, because of the incessant noise they make. Food or warmth available near the tip surface, particularly in tins or small crevasses, provide conditions very attractive to crickets. Flies present a potential major health hazard because of the apparent association as carriers of disease affecting man. They may be brought to the site in one of their lifecycle forms, domestic refuse being a major source in this respect. Adult flies may also be attracted from surrounding areas and deposit their eggs on the refuse. Many of the eggs, larvae, and pupae will be buried too deeply to survive, but larvae will tend to emerge as the heat becomes unsuitable for them and pass successfully into the next stages of development. This, as you can see, is one of the reasons for placing so much emphasis on the adequate covering of refuse. Despite good covering it is fairly certain that some adult flies will develop and others will be attracted to the site. The most effective counteraction is by spraying, using an insecticide with a residual toxicity which will knock down adult flies and also destroy them in the development stage. Occasional spraying, particularly in the warmer weather from the beginning of spring to late summer, should be carried out as a routine, apart from any daily attention which is given to any freshly deposited material or where you are forced to carry out emergency treatments. Pressure spraying or fogging machines are readily available from a number of manufacturers.

Rats can also present a problem and are looked upon as a public health hazard. The onus of responsibility for their destruction is upon the owner or occupier of the premises and land infected. It is therefore important that those responsible for site management should be able to recognize the signs of rat infestation or visitation.

The easiest and most satisfactory way to counter the potential danger from rats is either to arrange a system of routine inspections by the Local Authority rodent control officer or a pest control contractor, or maintain a frequent inspection at least once per month by site management calling in specialist advice at the first signs of infestation. The habit of the common rat on landfill sites is to burrow into suitable material on the site or in adjoining banks or hedgerows or to nest in any hollow container offering warmth and shelter. Where the tip surface and faces are well covered, rat-holes and runs will be seen easily. Where holes and burrows are used frequently a path or run is worn smooth by the rats in their nocturnal search for food and water. The holes may be as small as 1 inch in diameter but with a large rat population will be bigger. Holes and runs in use are indicated by the presence of rat droppings which are about 10 mm long and spindle-shaped.

SAFETY

One cannot discuss site management without placing emphasis on safety. Apart from our responsibilities under current legislation, accidents are expensive with the hidden costs often up to five times the costs that are readily apparent. Site personnel work in all weather conditions with many different types of heavy equipment, and with a variety of materials with diverse hazards. The basic types of accident to be guarded against are those resulting from:

(1) falls on rough suraces;
(2) falls from vehicles;
(3) traffic accidents;

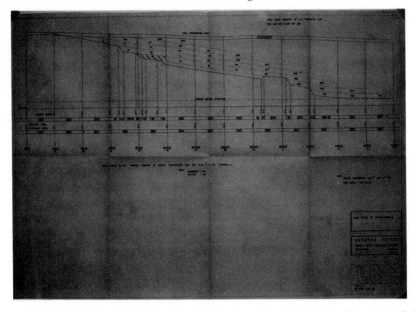

Figure 7   Operating site plans on a major landfill. (Courtesy of London Brick Company Ltd)

(4) attempts to repair equipment while still running or connected to a power supply;
(5) fire;
(6) inhalation of contaminants and dust;
(7) incorrect operation of plant and equipment;
(8) operation of heavy plant;
(9) contact with the waste handled.

Besides the concern for human safety and the economic cost of accidents, Managers must be concerned about safety because the law mandates that safety is part of carrying out work. Carry out regular safety inspections, provide safe equipment and safety equipment, devise safe operational procedures, train your employees and yourselves.

SITE RESTORATION

Finally there is the question of the restoration of the site. This should form part of the initial site design as in practice many planning permissions and site licences specify the final restoration levels and the form the reclaimed site will take. The restoration form actually used is a matter for agreement with the County Planning Authorities and may be subject to a County policy or structure plan. The cost of restoration must be considered in the initial plan as restoring land to a standard suitable for agricultural use, for instance, can well exceed the land price value. However, good planning can minimize these costs and maximize the usefulness of the site in its final form. A basic factor of the landfill site reclamation exercise is that it should be carried out progressively and not left until the site is closed. In this way maximum use can be made of suitable incoming covering materials such as topsoil and subsoils.

Most site licences specify a final capping of at least 1 metre free of materials likely to interfere with final restoration or subsequent cultivation. 5,100 m^3 of material per acre will be required to establish this final cap, add another 2,500 m^3 of topsoil per acre and the reasons for progressive reclamation need no further explanation.

One metre capping is considered unnecessarily thick by some, as is any attempt at landscaping, particularly when the costs involved are considered. There are, however, a number of good reasons which are incidentally long-term sound economic ones for doing it.

Many sites are planted with trees and other vegetation as part of the final restoration process. Frequently, after a period of between 3 and 6 years, these start to die off—usually because the roots are being swamped by anaerobic water. The reasons for this are numerous but the result, whatever the reason, is expensive to cure.

Insufficient or very porous cover material may dry out sufficiently to permit cracks to develop through the cover. Once cracks develop completely through the cover in this way, it can take some 6–8 weeks of wet weather before the cracks seal again; in this time the site is absorbing large quantities of water in the form of rainfall. One metre thickness of capping is considered reasonable for the prevention of this problem, provided that the cap is planted with grass as soon as possible. The grass should be kept cut to promote plenty of root growth to help keep the surface bound together. This in turn helps both to shed rainwater and to consume that which penetrates the ground more quickly.

If it can be avoided do not finish the site in a level plane; do not forget to allow for the effects of subsidence, and create a domed or sloped finish which will throw off unwanted surface water. Too steep a slope should be avoided otherwise rainwater may scour the surface. Such scouring would lead to a thinning of the cap and help to promote crack growth. My own observations are that landfill sites with smell problems are those with inadequate cover. The presence of a 1 metre of cover being sufficient to remove the active smell components from any gas which passes through it to atmosphere.

RESOURCE RECOVERY

Resource recovery must be considered seriously by all who operate landfill sites. Land in itself is a resource and the positive action of reclaiming and restoring unutilized or previously derelict land is resource recovery in the full sense. Beyond that reclamation of materials coming to the site should be examined periodically to establish the viability of such operations.

Raw materials become more scarce and expensive, and in certain cases justification for recovery on site can be established. In certain large sites the production of methane gas, previously regarded as a nuisance—even in some cases a dangerous nuisance—should be further examined. The potential production of methane can be estimated to establish whether commercial production is possible. This valuable fuel derived from landfill sites is already being supplied to industry in the USA.

Positive achievement in the area of resource recovery will do much to improve the public image of our industry.

Practical Waste Management
Edited by J.R. Holmes
©1983 John Wiley & Sons Ltd

15

Liquid waste disposal techniques

STEPHEN L. WILLETTS, BSc, PhD, MISWM
Steetley Construction Materials Limited

ABSTRACT

A review of liquid waste disposal techniques, the types of waste legislative and economic constraints are looked at. Biological and physical treatments, filtration, ventralization encapsulation techniques, incineration, and recovery are considered. Landfill, disposal at sea, mineshaft disposal, and an assessment of these options complete the work. Examples are given of plants and facilities in Western Europe.

INTRODUCTION

This chapter confines itself to industrially generated wastes. Modern industrial activity basically involves the separation and isolation of various substances and their subsequent manipulation. All natural materials exist as an amalgam with other substances and the isolation and purification of any one component necessarily generates waste. Also, the manipulation of materials and their use in various manufacturing processes generates waste.

Such wastes exist in all three of the physical states and it is probably generally accepted that the liquid wastes pose the greatest problems as far as disposal is concerned. Atmospheric emissions and gaseous wastes in general are discharged with the benefit of extremely large dilution to air: solid wastes are easy to contain, store, and handle and relatively easy to dispose of. Liquid wastes, however, are far from easy to contain and safely dispose of.

There are various methods now available for liquid waste treatment and disposal depending on the type of waste in question. These are briefly reviewed.

LEGISLATIVE/ECONOMIC CONSTRAINTS

It is important to recognize that two factors govern the choice of disposal method.

Firstly, there are legislative constraints administered by various statutory authorities. For liquid wastes various Acts relating to public health and water pollution exist on the Statute Book and there will be implemented the relevant and remaining sections of the Control of Pollution Act, 1974, in due course. This legislation is covered in a previous chapter.

Economic constraints are equally important to the industrialist. Direct costs for waste disposal are easy to obtain and interpret as it merely involves the obtaining of a quotation from a contractor. The indirect costs associated with the chosen method are, however, not easy to compute but these are beginning to play an increasing role in how the industrialist assesses his waste disposal route. For example, many waste producers take time to make themselves aware of the environmental consequences of a particular disposal option and are becoming anxious to reduce or obviate any risk of liability rebounding on them after the disposal event. Current reporting of public enquiries into planning and site licensing appeals and incidents such as 'Love Canal' reach the popular press, and industrialists are conscious of the changing situation. As such, waste producers do not now always choose the lowest quotation.

TYPES OF WASTE

The categorization of wastes is complex. For treatment and/or disposal, the broad properties and classifications of importance are:

(1) aqueous or non-aqueous;
(2) acid, neutral, or alkaline;
(3) solution, emulsion, colloidal, or suspension;
(4) inorganic or organic;
(5) inflammability;
(6) viscosity;
(7) interactions with natural environment;
(8) special storage, handling, and transport requirements;
(9) special treatment and disposal requirements;
(10) general health and safety requirements, etc.

An important consideration for liquid wastes is the fact that they normally exist as mixtures of substances. Moreover, the nature of any particular waste may change from hour to hour or week to week, depending upon the process from which it is produced. Quality control of product manufacture can also affect the nature of a waste, and systems malfunctions can totally change its character. Variations in nature can be large and should be taken into account when considering disposal.

Another factor of importance is the possibility of stratification on storage either at a producer's premises or at the contractor's disposal site. Sampling of a waste prior to deciding upon disposal method is a science and an art, and the goal of 'representativeness' is difficult to achieve. Nevertheless, realistic samples must be obtained to allow the contractor properly to assess the waste. It is often imperative to take top, middle, and bottom samples to enable reasonable screening of a waste.

WASTE REDUCTION/RECOVERY/IN-HOUSE TREATMENT

Before considering off-site disposal a waste producer must always look to ways for recycling, recovery, or reduction of the waste in question. The costs of abstraction and use of virgin resources are increasing and possibilities for waste re-use or reduction are becoming more favourable and economical.

Likewise, installation of in-house treatment plant is becoming more competitive with off-site disposal and serious consideration must be given to these possibilities. Such treatment of liquid wastes usually results in a clarified effluent for sewer disposal and concentrated residues for re-use or, if too badly contaminated, off-site disposal.

Bearing these factors in mind, the disposal of liquid wastes through the factory gate is still a popular method. The handling and transportation of liquid wastes has been covered in another chapter. What the disposal contractor does with the waste is discussed here.

BIOLOGICAL TREATMENT/DISPOSAL

Biological methods of treatment rely upon bacteria and other micro-organisms to abstract various organic pollutants from an aqueous carrier medium and metabolize these to inorganic by-products. The aqueous effluent from a biological plant is thereby purified to an extent depending upon the efficiency of the biological system. There is always produced a sludge of debris comprising dead and decaying biological matter in need of separate disposal. This can be used agriculturally as a fertilizer, digested to produce methane and then land-spread, incinerated, tipped on land, or dumped at sea.

Being biological, this system of treatment is prone to upsets from toxic constituents and such things as heavy metals and exotic organic molecules must be restricted.

Sewer discharge

Any waste producer may apply to a Regional Water Authority (RWA) for consent to discharge an aqueous effluent to sewer. If the RWA has hydraulic capacity and ability for treatment at the sewage plant, then consent may be granted: consents are always subject to restrictions. Such restrictions are becoming more severe in the face of the legislative and moral desires to abate pollution of our watercourses, the final recipients of effluent from sewage plant. Payments to an RWA for reception and treatment of an effluent are, of course, rising. Payments and penalties for non-compliance with consent conditions are also increasing.

Sewage plant

Some waste disposal contractors, some private organizations, and some RWAs will accept liquid wastes for treatment other than via the sewerage system: for example, by road tanker. Many wastes are welcomed for disposal using this method; a caustic waste, for instance, can be used to help neutralize an acidic influent to a sewage plant that is serving an engineering orientated area producing steel pickling swills, etc. for sewer disposal. Gas-tar liquors and coke-oven liquors are sometimes accepted at British Steel Corporation plants that have their own sewage plant specifically designed for this type of effluent that they themselves generate.

PHYSICAL TREATMENT/DISPOSAL

The use of physical forces for wastes treatment principally involves the separation of phases as a purification step; this is basically filtration.

Coarse filtration

Bar-screens, rakes, coarse meshes, and booms can all be inserted in an effluent stream to effect removal of detritus. Pyramidal settling chambers and baffled weirs serve the same purpose for heavy debris.

Medium filtration

Separation of suspended solids can be achieved using cloth filter presses, centrifuges, hydrocyclones, rotary vacuum filters, disc filters, etc. Pretreatment using inorganic coagulants such as various calcium, iron, or aluminium salts or polyelectrolyte flocculants is often required.

Fine filtration

Colloidal suspensions and emulsions can be separated by plate separators, reverse osmosis, electrodialysis, air flotation using either compressed air or electrolytically generated gas bubbles, etc.

Aqueous oil mixtures and emulsions are separated at the following facilities:

Croda, Wolverhampton: gravity settlement
Hales, Birmingham: ultrafiltration.
Lancashire Tar Distillers, Manchester: gravity settlement.
Neston Tank Cleaning, Queensferry: gravity settlement plus heat.
Polymeric Treatments, Killamarsh: gravity settlement plus heat plus acid
    cracking.
Polymeric Treatments, Empire: centrifugation plus heat.
Surface Control, Glasgow: ultrafiltration plus tilted plate.
West Wales Tank Cleaning, Swansea: gravity settlement.

The aqueous phase is led away to sewer or estuary and the separated oil is
usually used as a fuel or fuel supplement.

CHEMICAL TREATMENT/DISPOSAL

Chemical methods for waste treatment and disposal are many and varied,
using several different principals.

Neutralization

Aqueous acidic effluents can be neutralized to precipitate various metal
hydroxides from solution. These can then be coagulated and separated from
the effluent. Purified liquors are then suitable for sewer disposal and the filter
cake may be acceptable on landfill.

This form of chemical treatment is operated by many contractors and can
be considered as the provision of a central treatment plant in lieu of several
in-house plants. The principle is the same: aqueous effluents are purified to
render them suitable for biological treatment and the separated solids are
disposed of elsewhere.
Neutralization plants are operated in the UK by:

Contract Gully Cleansing, St Albans.
D.&P. Effluents, Runcorn.
Hales, Birmingham.
Hargreaves, Wakefield.
Lancashire Tar Distillers, Cadishead, Manchester.
ReChem International Ltd., Pontypool and Southampton, and Bonnyb-
    ridge.
Rigby's, St Helens.
Safeway, Birmingham.
Surface Control, Glasgow.
Whelan, Birmingham.
Wimpey, Bladon, Newcastle-upon-Tyne.

Incineration

Many organic effluents can be incinerated as they are self-supporting in combustion or even exothermic, and the heat generated can be used to evaporate water from aqueous wastes and incinerate the residues. The use of supplemental fuel is often required during start-up but its continued use for assisting waste incineration is obviously costly.

Wastes for incineration are broadly classified as chlorinated and non-chlorinated. The significance is that chlorinated wastes liberate hydrogen chloride gas on combustion which attacks the refractory linings and which forms hydrochloric acid in the atmosphere. The general public do not like acid rainfall; thus, incinerators capable of burning chlorinated compounds have to be equipped with water scrubbers to remove the acid gas. The scrubbers also remove sulphur oxides and phosphorus oxides with the same result. This system, of course, produces an acidic scrubber liquor in need of disposal which also contains suspended particulates and other materials in solution. It is not suitable for direct sewer discharge and has to be considered as a waste product itself.

Handling considerations also effect the type of waste that an incinerator can accommodate. Most incinerators are suitable only for bulk liquid feed by direct injection. Drummed wastes can be handled at these facilities by decantation of contents into the bulk tanks. A few incinerators can be fed with whole drums which is the only way to burn out viscous resins, varnishes, and the like, and solid wastes such as capacitors.

The inland incinerators operated in the UK and the types of waste that they can receive are:

Berridge, Nottingham: bulk liquids; chlorinated
Croda, Wolverhampton: oil/water; non-chlorinated
Hargreaves, Wakefield: bulk liquids; non-chlorinated
Lancashire Tar Distillers, Manchester: bulk liquids; non-chlorinated
Polymeric Treatments Ltd., Killamarsh: oil/water; non-chlorinated
ReChem, Pontypool: bulk and drums; chlorinated
ReChem, Southampton: bulk and drums; chlorinated
ReChem, Bonnybridge: bulk and drums; chlorinated
Redland, Ellesmere Port: bulk liquids; chlorinated
Redland, Rainham: bulk and drums; chlorinated
Waste Incineration, Chorley: bulk liquids; non-chlorinated

Marine-based incineration overcomes objections to acid fall-out and two operations at least offer this service.

Ocean Combustion Service B.V. of Rotterdam operate the *Vulcanus* ship equipped with two stern-mounted open-bowl Saacke incinerators. Incineration efficiency is claimed to be 100% and temperature is claimed to be 1,500

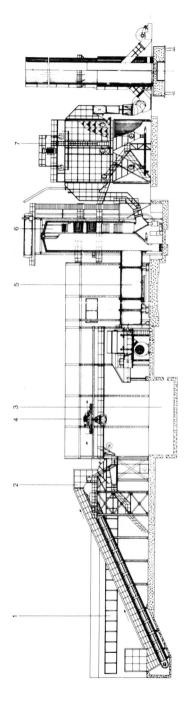

Key 1 – Dry waste storage hall.   5 – Rotary kiln (900–1000 °C)

    2 – Waste shredder   6 – Steam raising boiler

    3 – Wet and dry mixing bunker   7 – Gas cleaning precipitator

    4 – Overhead travelling crane

Figure 1   An industrial incinerator plant in Hamburg. (Courtesy of AVG, Hamburg)

°C. Each of the two incinerators operates at 20–30 tonnes/hour, being fed from one of 13 separate cargo tanks. The incinerator bowls are 4.5 m diameter and 8 m high. The principal wastes handled by *Vulcanus* are the chlorinated hydrocarbon by-product from PVC manufacture in the Netherlands.

T.R. International Ltd of Manchester act as agents for the *Matthias II* incinerator vessel which is equipped with one Stahl-und-Blech-Bau burner. The vessel entered service in 1972 and was a larger version of the original *Matthias I* ship of 1969. *Matthias III* has now been built, but is not yet in service, and is designed to accommodate solid wastes in addition to liquids. This latest vesel is capable of incinerating 15,000 tonnes of liquid and 1,500 tonnes of solid waste during each voyage of 21 days duration.

All marine incinerator ships have to comply with IMCO regulations which specify double-hulled vessel construction, isolation of engine room, and other safety requirements. Additionally, all control equipment has to be remotely photographed by sealed apparatus a 30-minute intervals with the film being removed on return by the authorities. The incineration location is strictly defined.

Solidification

Two, possibly three, systems operate in the UK under the names of Sealosafe, Chemfix and Soliroc (Petrifix). Sealosafe is now firmly established and has three plants operating in the UK and more overseas. Two are located at Brownhills and operated by Polymeric Treatments Ltd, and one is situated at West Thurrock, Essex, operated by Stablex Ltd. A plant scheduled for Yorkshire is at an advanced stage. The Stablex group of companies is responsible for the extension of the Sealosafe system to countries outside the UK.

Liquid wastes can be accommodated in bulk, in drums, or in retail containers at a Sealosafe plant. The types of wastes suitable are those that are principally of aqueous nature but organic contamination of several per cent can be safely handled. The patented process adds various reagents to the wastes and the product, termed Stablex, sets in about 3 days to a hard, impermeable solid. Installed capacity is around 500,000 tonnes per annum.

Much published work is available on Stablex, and several independent assessments have been conducted, both in the UK and abroad. The results thus far confirm the claims for Stablex, viz:

(1) permeability of around 10^{-7} cm/s—that is, less than concrete;
(2) leaching of toxic metals by standard tests less than 1 mg/ℓ;
(3) non-flammable;
(4) compressive strength of 1.4–5.5 MN/m^2—that is, comparable with compressed clays and grouts;

(5) non-biodegradable;
(6) unattractive to vectors of disease (flies, gulls, rats, etc.);
(7) self-compacting, self-seeking, self-grouting;
(8) environmentally inert.

At present, Stablex produced at Brownhills is being used to reclaim a 30 m deep spent-clay quarry for immediate release after filling for light industrial usage. Normal infilling with domestic refuse or general factory wastes would sterilize the land area for 20 or 30 years while biodegradation and settlement occurs. At West Thurrock, the Stablex will be used for reclamation of old chalk quarries previously worked for cement manufacture and considered prohibited for normal waste disposal due to the underlying potable supply aquifer of major importance. Also, there is much work being undertaken on

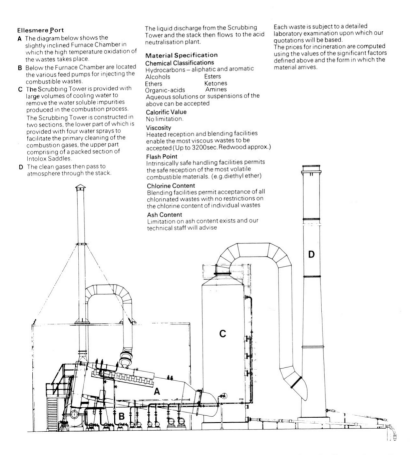

Figure 2   Another example of an industrial liquid and sludge incineration plant in England. (Courtesy of Cleanaway Ltd)

the re-use capabilities of Stablex for void and cavity filling in limestone areas, foundation engineering, block manufacture, etc.

Chemfix was introduced from the USA by P.D. originally but has now been taken into the Wimpey organization following their acquisition of P.D. Beatwaste. The single Chemfix plant is at Bladon, Newcastle-upon-Tyne. The plant is a mobile, trailer-mounted system and is designed to solidify wastes at various customer premises. This is contrary to the Sealosafe philosophy and operation of regional treatment centres. A Sealosafe plant takes in mixed wastes from various customers and is able to use beneficial interactions as part of the treatment process: this is, of course, reflected in the lower price to the customer.

Soliroc and Petrifix are the same system operated by CdF Chimie, Paris, in France and in Belgium. There is rumoured to be a small plant operating on refinery sludge in South Wales.

Encapsulation

The solidification processes described previously are based upon silicate chemistry and the formation of inorganic polymers. Organic wastes are not usually compatible with this system and various organic polymer-based solidification and encapsulation techniques do exist. Most are based on epoxy resins and are expensive. None operates in the UK.

Glassification, and silicate fusion in general, is a form of encapsulation that is usually reserved for radioactive wastes. There has been extensive research into the suitability of Sealosafe for radioactive wastes solidification and results to date have been promising.

COMBINED TECHNIQUES

Various disposal outlets are available that make use of biological/chemical/ physical techniques in combination.

Landfill

Landfill is, of course, the traditional method for waste disposal, and liquid wastes are still disposed of by this system. There are two basic operations for liquid landfill: firstly, there is the system of absorbing the liquid onto domestic refuse and dry industrial wastes; and secondly, there is the system of the 'balanced lagoon' in which liquid seeps away.

Research conducted by Harwell and the Water Research Centre implies that 'dilute and disperse' is an acceptable method for waste disposal. This is the philosophy adopted for liquid landfill. The principle is that liquids are absorbed onto other waste materials and underlying minerals and undergo

Figure 3 Ship used for the controlled sea disposal of industrial wastes

biological, chemical, and physical filtration during their inevitable downward gravity-caused movement. Upon entering aquifers or watercourses, there is hopefully sufficient filtration of pollutant species and sufficient diluting water to make the resultant pollution environmentally insignificant.

Liquid disposal onto refuse is operated by Cheshire County Council at Witton Lime Beds; Derby Waste at Derby; Grundons at Aylesbury; Manchester Tankers at St Helens; Pattersons at Glasgow; Redland-Purle at Pitsea; and Wimpey at Denton, Manchester, and Risley, Warrington. Liquid disposal into leaking lagoons is operated by Haul-Waste at Brampton, Devon; Hutchinson Dock and Estate at Widnes; and White at Villa Farm, Coventry.

In spite of Harwell and Water Research Centre assertions that liquid landfill can be a safe disposal method, there is much controversy over the research work and interpretation of results. Many people, this author included, consider the concept to be suspect.

Figure 4   Sea incineration of industrial wastes

Sea disposal

Sea disposal of wastes in bulk form is reserved for large-volume, aqueous industrial wastes that are readily biodegradable. For example, many wastes disposed of at sea are suitable for normal sewer discharge but may be produced in an area where sewage purification works are already overloaded or where none exist. Sea dumping is one solution for the disposal of these wastes. For such wastes the disposal method relies upon physical dilution into the marine environment, chemical degradation, photochemical degradation, and biochemical oxidation to effect dispersion and destruction of the waste. Vessels can release their loads over the side, into the wake, or by bottom discharge.

In 1979 some 1,145,000 tonnes of industrial wastes were licensed for sea disposal. The main contractor is Effluents Services Ltd operating from the ports of Birkenhead, Barrow-in-Furness, Tyneside, Teesside, Goole, Heysham, and Colchester. Other contractors operate from Sharpness (Bristol) and Southampton, and some contractors undertake the servicing of a single customer as their sole business (e.g. Nypro, CEGB).

Effluents Services have storage tankage at their Tyneside, Barrow, and Birkenhead operations and use multiple boats with direct road tanker loading from their other ports. The shore-based installations are licensed under the Control of Pollution Act, 1974. Their fleet of nine vessels ranges from 300 to 1,000 tonnes payload and includes two rubber-lined vessels for acid work.

The Regional Water Authorities are the largest sea dumpers and in 1979 some 8,800,000 tonnes of sewage sludge were licensed for disposal at sea. These sludges are normally so contaminated by industrial pollutants, such as heavy metals, that they are precluded from agricultural use. The RWAs and their ports of departure are:

Thames: London (5 million tonnes)
Southern: Southampton
South West: Exeter
Wessex: Bristol (also load industrial wastes)
Welsh: Newport
North-West: Manchester and Salford
Northumbrian: Tyneside
Anglian: Ipswich, Colchester
Severn-Trent: Bristol
Yorkshire: Goole (serviced by Effluents Services Ltd. who operate eight
    barges of 250–300 tonnes on inland canals to Goole)

The Dumping at Sea Act 1974 is the vehicle of control and the Ministry of Agriculture, Fisheries, and Food administer the Act. Applications to dump waste are submitted to MAFF with a sample of the waste, and MAFF undertake biotoxicity testing on the sample at their Burnham-on-Crouch laboratory. The usual standard is that a waste must have 96-hour LC50 of more than 3,000 ppm unless very readily degradable when the limit may be relaxed to 1,000 ppm. Some wastes with a 96-hour LC50 of less than 1,000 ppm may be permitted for sea disposal at very slow rates of release but MAFF are obliged to take into account any land-based alternatives for disposal that may exist.

Disposal into the wake of a vessel is the normal method for off-loading industrial wastes, and the rate of discharge is usually specified by MAFF so as to ensure at least a one-thousandfold dilution into the sea. MAFF also specify the dumping ground having regard to navigation, fishing, and amenity interests and the tides. The licence registration fee is currently £45 and the biotoxicity test £450. The dumping of drummed wastes is also permitted. This relies upon initial containment and slow release at seabed level. Cyanide, arsenic, and antimony wastes have all been disposed of to sea.

MAFF, of course, are also responsible for the enforcement of the regulations of the Act and arrive unannounced at ports of departure to take samples, are empowered to travel with a vessel to observe the dumping, and

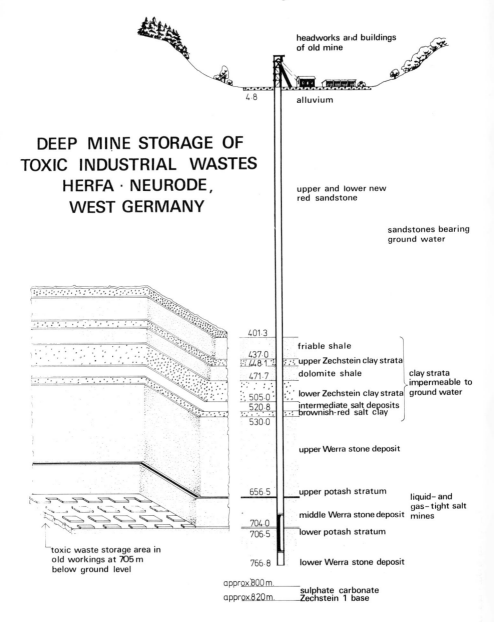

headworks and buildings
of old mine

4·8   alluvium

DEEP MINE STORAGE OF
TOXIC INDUSTRIAL WASTES
HERFA · NEURODE,
WEST GERMANY

upper and lower new
red sandstone

sandstones bearing
ground water

401·3

437·0   friable shale
448·1   upper Zechstein clay strata
471·7   dolomite shale   clay strata
                                                  impermeable to
505·0   lower Zechstein clay strata ground water
520·8   intermediate salt deposits
530·0   brownish-red salt clay

upper Werra stone deposit

656·5   upper potash stratum   liquid- and
                                              gas-tight salt
          middle Werra stone deposit mines
704·0
706·5   lower potash stratum

toxic waste storage area in
old workings at 705 m   766·8   lower Werra stone deposit
below ground level

approx 800 m.
approx 820 m.   sulphate carbonate
                  Zechstein 1 base

Figure 5   Deep mine storage of toxic industrial wastes, Herfa Neurode, West
Germany

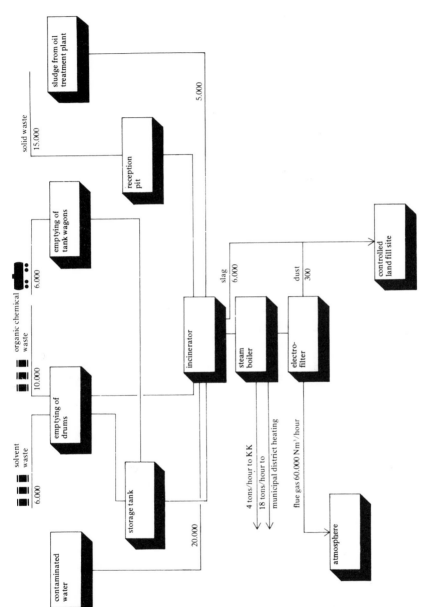

Figure 6 Flowchart of the incineration stream, Kommunekemi waste treatment complex, Nyborg, Denmark. (Courtesy of Kommunekemi A.S)

OIL TREATMENT PLANT (tons/year)

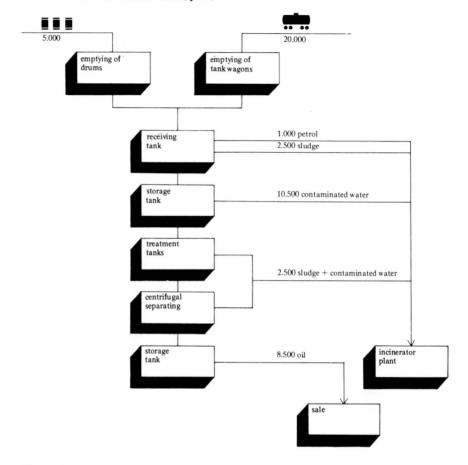

Figure 7   Oil treatment stream at the Kommunekemi waste treatment complex,
Nyborg, Denmark. (Courtesy of Kommunekemi A.S)

have their own patrol vessels. Various International Conventions have been
signed by most industrial nations and these seek to prohibit the dumping of
certain wastes at sea and to restrict others.

Mine disposal

There exist several premises that have private access to underground workings for disposal of their own wastes via shafts, boreholes, wells or pipes.

INORGANIC PLANT (tons/year)

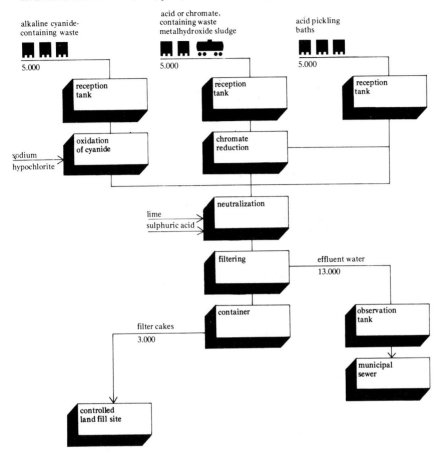

Figure 8   Inorganic wastes treatment stream at the Kommunekemi treatment complex, Nyborg, Denmark. (Courtesy of Kommunekemi A.S)

AgLand, a subsidiary of Leigh Interests, operates a disused coal mine at Chatterley Whitfield, Stoke-on-Trent, for disposal of pottery slip and other clay-based aqueous slurries. The mineworkings are flooded and equipped with a single, valved outlet for liquid discharge. The facility is used as a gigantic settlement and decantation facility in which the clay fines settle out by reason of gravity and fill the voids from the base upwards. Clarified water exits through the valve and is pumped to the surface for monitored and consented discharge to watercourse.

① Storage bin
② Crane
③ Feeder
④ Tanks for used oil and residue oil
⑤ Combustion chamber
⑥ Rotary kiln
⑦ Secondary combustion chamber with flue gas cooler
⑧ Electric precipitator
⑨ Stack
⑩ Ash handling devices

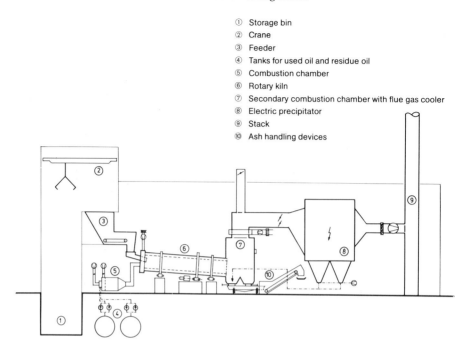

Figure 9   Incinerator plant for mixtures of waste oils and solid industrial wastes. (Courtesy of VKW, West Germany)

AgLand also have planning consent, and intend to re-open, the underground workings at Walsall Wood formerly used for liquid waste disposal by Effluent Disposal Ltd, another Leigh Interests subsidiary. This facility was successfully used for many years gaining access to the workings via a 2 m diameter shaft. The shaft was of brick construction and several decades old, and eventually collapsed during a grouting operaton. Use of the facility was suspended and it is intended to drill a borehole into the same workings from the Polymeric Treatments Ltd Empire complex.

The Walsall Wood mine is unique in that it does not make water, being isolated by massive geological faults that have sealed the worked-out seams by low permeability clays. It is, in effect, a bottle from which liquids cannot escape. Indeed, water had to be pumped into the workings when active for firefighting purposes. The workings are not extant as voidspace but as 'goafs' which is a collapsed state: the land level at Walsall Wood has sunk by over 7 m since coal abstraction. Research work has demonstrated that the montmorillonite clay sequences of the workings effectively absorb ionic pollutant species and macro-molecules are physically filtered by the collapsed ground.

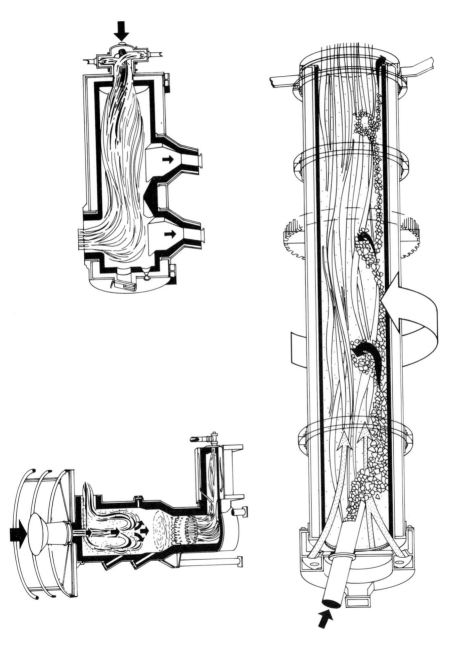

Figure 10 Sketches illustrating various methods of incineration used for liquids and sludges

It is envisaged that as the workings become filled it will be possible to abstract purified liquid from down-dip locations and that this will be suitable for simple biological treatment and discharge into the aquatic environment.

CONVERSION TECHNIQUES

Novel developments in conversion of liquid industrial wastes to a usable fuel offer a disposal outlet for certain wastes.

WDF

Waste-derived fuels are being manufactured at pilot plant incorporating oily wastes. Redcliffe Engineering are using oil-soaked pulverized domestic refuse, for instance, in trials in the north-east.

Leigh fuel

Leigh Analytical Services, a Leigh Interests subsidiary, has developed a process which converts organic wastes such as tar liquors, paints, latex residues, resins, and heavy oils into a solid fuel with a calorific value similar to that of 'brown' coal. The fuel can be pulverized, briquetted, or granulated and full-scale burning trials at CEGB's Walsall power station are most promising.

A plant operated by Polymeric Treatments Ltd is now producing fuel at the rate of 50,000 tonnes per year at Killamarsh on a commercial basis and more plants are planned.

One facet of this system is that it actually requires water for the conversion process and so it is very suitable for aqueous oils and even oil spill clean-ups from fresh and sea water.

SUMMARY

As mentioned at the outset, liquid wastes are by far the most difficult to dispose of. Hence the variety of techniques and methods that exist.

Practical Waste Management
Edited by J.R. Holmes
©1983 John Wiley & Sons Ltd

16

A review of four refuse treatment and reclamation projects in the United Kingdom*

G. L. VIZARD
County Surveyor, Dorset County Council

INTRODUCTION

Solid waste disposal in the United Kingdom is under the legislative control of public sector Waste Disposal Authorities. In England these authorities are the County Councils and they have the duty to dispose of the waste collected by the public sector Collection Authorities (District Councils) which includes all household and some commercial and industrial wastes. The balance of commercial waste and the majority of industrial wastes are collected and disposed of by private sector operators whose disposal activities are subject to statutory control by the Waste Disposal Authorities. The mineral extraction industry is becoming more involved as agents to the County Councils in the utilization of worked-out quarries for landfilling. The estimated arisings of various classes of waste are shown in Table 1.

The majority of waste arising is inert or relatively innocuous and in view of the enormous quantities involved there is no realistic alternative to disposal by landfilling. Those wastes which are not innocuous include the household and commercial waste and a relatively small proportion of the industrial waste which will be toxic. These toxic industrial wastes will need specialist treatment in the form of incineration, chemical neutralization, or specialist landfill. Thus the household and commercial waste which is largely the responsibility of the public sector is the waste that is the most difficult to satisfactorily dispose of, and is under the control of the Municipal Engineer.

Within the United Kingdom the great majority of household and commercial waste has traditionally been disposed of to controlled (sanitary) landfill.

---

* A version of this chapter was first published at the 8th Congress, International Federation of Municipal Engineers, Madrid, 1982.

Table 1

Category of waste	Tonnes/year $\times 10^6$
Household and Commercial	18
General Industrial	23
Building	3
Power Generation	12
Mining	60
Quarrying	50

Source: Waste Management Paper No. 1 (1976). HMSO, London.

Table 2

Method	Percentage processed
Landfill untreated	
Direct delivery	71.5
Bulk transferred	14.5
Landfill after pulverization	3
Incineration	9.5
Other methods	1.5
Total	100

Source: CIPFA Waste Disposal Statistics, 1981/82 Estimates.

However in recent years it has become increasingly difficult to secure landfill capacity for crude household and commercial wastes, particularly near or within large urban areas.

As a consequence the Waste Disposal Authorities, particularly those in England, have had to invest capital in one form of treatment plant or another. The current situation with regard to disposal methods in England and Wales is as shown in Table 2. The figures show that with bulk transfer stations almost 30% of waste is passed through mechanical plant including incinerators. A policy of wholesale incineration has not been pursued, largely on economic grounds, although there are some 40 such plants in operation. Recent studies have been aimed at developing processes which will contain the cost of collection and minimize disposal costs.

Four recently commissioned plants are therefore described as representing evidence of real progress in the fields of waste transport and re-use.

THE BRADFORD BALING PLANT

West Yorkshire Metropolitan County Council are pursuing a policy of controlled landfill for waste, and determined this as a strategy in 1974. In order to rationalize the relationship between collection areas and landfill sites the County Council decided to introduce transfer loading facilities. The pretreatment of some suitable waste by baling was included as a means of achieving further transport economies and improving conditions on landfill sites and, in consequence, the County Council decided to construct two baling plants to serve the major conurbations. Both are now in operation and the second one at Bradford was commissioned in September 1980. The plant serves the whole of the area of the Bradford Metropolitan District Council, which contains a population of 462,000.

Plant operation

The plant is single stream utilizing a three-stage compression baling press. Incoming refuse is discharged onto the floor within the covered reception area. It is deposited by wheeled loading shovels onto a horizontal conveyor below floor level, mixing the refuse in the process. Refuse is passed along a series of plate conveyors to the weigh hopper which automatically stops the input when the predetermined bale weight is attained.

The baler produces bales that are self-sustaining and are 1 metre square and 1.5 metres long. The three compression rams within the press are arranged at right angles to one another and exert progressively increasing pressures up to a maximum of 19,000 kN/m^2 (27,000 p.s.i.). Individual bales have an average weight of 1.33 tonnes and are produced at 90 second intervals. This gives a capacity of 50 tonnes/hour which has been achieved since commissioning.

Bales are automatically loaded in pairs onto specially designed trailers. These trailers are of particular interest, being equipped with hydraulically operated off-side hinged bodies which assist bale removal at the landfill site. Each trailer has a capacity for 14 bales giving a payload of 18.5 tonnes.

Staffing and mobile plant

The plant is operated on a two-shift basis to process 115,000 tonnes of waste during 1981/82. This leaves a margin for growth up to the maximum possible throughput of 200,000 tonnes. The total staff complement at the plant is 38, made up of a manager, 2 shift engineers, a weighbridge attendant, 4 watchmen, 22 manual workers, and 8 drivers. It is equipped with 3 loading shovels, 7 tractor units, and 25 trailers.

Table 3

Description	Cost (£)
Land	156,000
Buildings and civil works	2,131,000
Fixed plant	1,425,000
Mobile plant	668,000
Total	4,380,000

Table 4

Description	Unit cost (£/tonne)
Capital costs amortized	6
Plant operation	4
Transport	2
Total cost	12
Net operational cost	6

Costs

Capital costs are listed in Table 3. Running costs, assuming a throughput of 115,000 tonnes/annum, are as shown in Table 4.

BYKER RECLAMATION PLANT

Tyne & Wear County Council were experiencing an acute shortage of landfill capacity and had a transfer loading station serving part of the City of Newcastle in need of replacement. The options facing the County Council were the construction of either a new transfer facility or an incinerator. However the County Council wished as a policy to reclaim wastes, and Central Government through the Department of the Environment offered financial assistance towards the provision of a pilot plant. In consequence the Byker Reclamation Plant was constructed, opening in late 1979 to serve approximately 150,000 of the City of Newcastle's population.

Design parameters

The plant was designed to process 30 tonnes/hour of household waste and produce a waste-drived fuel (WDF) for use as a fuel supplement in coal-fired boiler installations. Ferrous metal extraction was also required and the plant layout was designed to permit the addition of further process lines and increases in throughput. The flow diagram shows the process (Figure 1).

Input refuse

Plant design was based on the approximate analysis shown in Table 5.

Table 5

Classification	Design range (% by weight)	Tonnage based on 30 tonnes/hour
Screenings below 12 mm	10–15	3.66
Paper and cardboard	30–40	10.25
Vegetables and putrescibles	15–25	5.87
Textiles, rags, etc.	3– 5	1.17
Plastics, rubber—all types	4– 8	1.76
Wood	1– 2	0.43
Ferrous metals	8–10	2.63
Non-ferrous metals	1– 2	0.43
Glass	8–10	2.63
Unclassified	3– 5	1.17
Total		30.00

Moisture Content 20–30%
Average Density 6 m³/tonne
Up to 15% will be in plastic sacks
Up to 5% will be bulky items (i.e. too large for the average dustbin (0.1 m³)).

The waste-derived fuel

The parameters laid down initially for the fuel were as follows:

(1) Calorific value 14,000 kJ/kg gross minimum.
(2) Moisture content (by weight) 20% maximum. (This would obviously depend upon the moisture content of input refuse but 25% is the maximum to make suitable pellets.)

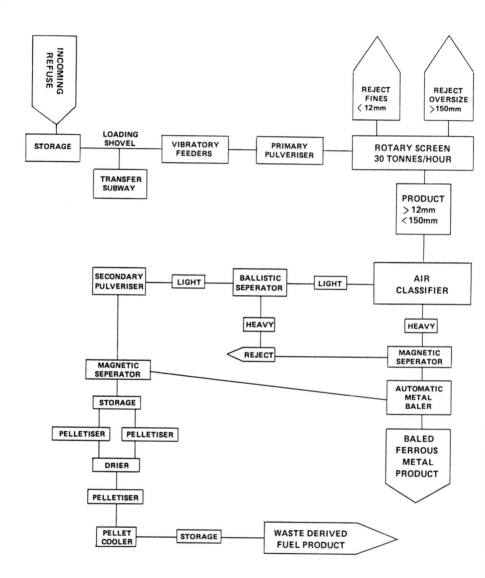

Figure 1   Flow diagram for the Byker project

(3) Non-combustible content—5% maximum.
(4) Fuel to be produced in two forms:
   (a) as a 'fluff'—a non-compacted fuel obtained directly from the secondary pulverizer with 80% less than 25 mm maximum linear dimension;
   (b) as pellets characteristically 25 mm in diameter and 50 mm in length—desired bulk density to be 487 kg/m^3 (30 lb/ft^3).

Early pellets were found to have a lower calorific value than desired, being in the range of 9,.000–12,000 kJ/kg. This was found to be due to the moisture content which, although within the specified range, had an adverse effect. This was caused by the high percentage of fines within the product which should have been eliminated by the rotary screen. Pellets of the desired quality could be produced by selecting only drier, paper-rich material for processing, but in order to process the generality of refuse modifications were made which are shown on the flow diagram.

Modifications have been the addition of the ballistic separator following the air classifier and the addition of a 2.4 metre diameter rotary cascade drier which is used only when the pellets produced are of high moisture content when the product is passed through the drier and re-pelletized before cooling.

Marketing

The plant is close to the Byker Wall housing estate, which has a district heating system. The Newcastle City Council have constructed a new boilerhouse to supply the estate with heat, using pellets from the plant mixed with coal. The reclamation plant can produce more pellets than are used by the boilerhouse, and pellets are sold to industrial and retail customers on a small scale.

Potential large-scale customers have been inhibited by the relatively small volumes of good pellets because of the moisture content problems before the introduction of the drier. Also there has been a reluctance from large-scale users to the introduction of new systems. However, there is interest and the potential market exceeds the supply. Alternative use of the fuel fraction as a core material in the manufacture of chipboard or fibreboard is being explored and a contract has been entered into with a paper mill to provide pellets as fuel and as raw material.

Transfer facility

A transfer subway between the storage and process areas serviced through slots covered by steel plates is incorporated. With this facility available the plant has never needed to divert deliveries elsewhere even when major modifications were taking place to the mechanical plant.

Staffing and mobile plant

The plant is staffed by 16 people including a manager, supervisors, five drivers, four attendants, and office staff. The plant is operated for 60 hours/week and has processed 1,000 tonnes/week for over 2 years. Refuse is received 5 days a week and the plant operates on Saturday mornings on transfer to clear any bulky items or backlog. This means that mechanical maintenance can be carried out, on a contract basis, at weekends. The plant is also equipped with six 30m³ open-bodied tipping vehicles and 2 No. wheeled loading shovels.

Costs

Capital costs are shown in Table 6. At an input of 52,000 tonnes/year the unit costs are as shown in Table 7.

Table 6

Description	Cost (£)
Land	leased
Buildings and civil works	1,500,000
Fixed plant	2,300,000
Mobile plant	leased
Total	3,800,000

Table 7

Description	Unit cost (£/tonne)
Capital costs amortized	4.6
Running costs	
plant operation	8.3
vehicles	3.4
Gross cost	16.3
Income, say £100,000	2.0
Net cost	14.3
Net operational costs (running costs less income)	9.7

Table 8

Description	Energy kW/hour
Energy consumed:	
plant operation	2,000
drier	2,800
total including drying	4,800
Energy potential of pellets	38,900
Thus ratio of production to consumption:	
with drier	8
without drier	19

Energy balance

In calculating an energy balance for the production of fuel pellets it is assumed that the plant processes 30 tonnes/hour of crude refuse and produces 10 tonnes/hour of pellets with an energy content of 14,000 kJ/kg (NB: 1 kWh = 3,600 kJ).

THE DONCASTER PROJECT (Figure 2)

South Yorkshire County Council had a need to provide Doncaster with refuse transfer facilities. Doncaster is a typical medium-sized conurbation in the United Kingdom with a population of some 286,000 in the Metropolitan District Council's area. Central Government offered sponsorship to the County Council to construct a pilot plant based on the resource recovery development work carried out by the Warren Spring Laboratory rather than just a simple transfer loading station. The plant that has been constructed will process refuse from some 250,000 people amounting to approximately 47,000 tonnes/year.

The plant commenced operation as refuse transfer station in September 1979 working a single shift only. The first production of WDF pellets was in June 1980, but it was not until May 1981 that two-shift operation with primary responsibility for processing refuse commenced. The plant currently operates to produce WDF with development work continuing on the glass circuitry and production. The plant has a rated capacity of 10 tonnes/hour whilst the three compactors are rated at 250 tonnes/day in aggregate. The flow diagram (Figure 2) shows the process.

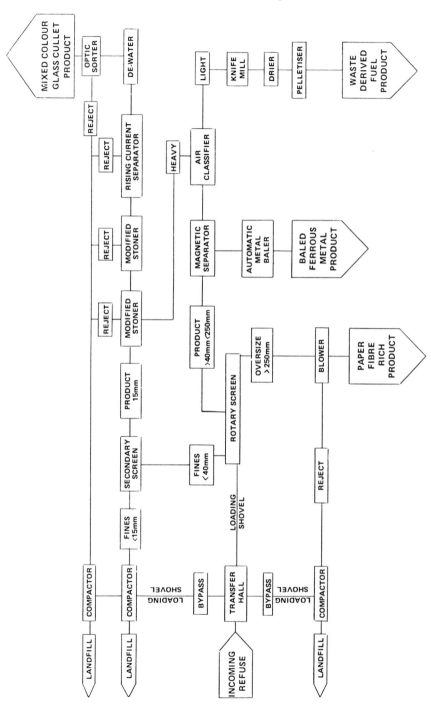

Figure 2   Flow diagram for the Doncaster project

Input refuse

A typical analysis of the input refuse is shown in Table 9.

Table 9

Material	Percentage weight	Percentage volume
Paper	24	49
Plastics	5	9
Textiles	4	5
Unclassified	2	2
Vegetables and putrescibles	28	15
Fines (below 15 mm)	15	5
Ferrous metals	11	13
Glass	11	2
Non-ferrous metals	Trace	—
Total	100	100

Outputs

The potential products from the input are shown in Table 10.

Table 10

Product	Percentage recoverable from total input
Glass	4–6
Fuel	25
Ferrous metals	7
Paper fibre	4
Fines	10–18
Putrescibles	20–25
Textiles	5

Specification of recoverable products

The specifications of the recovered products are:

(1) glass (see Table 11); this product has a value in the range of £7–10/tonne;
(2) waste-derived fuel (see Table 12); this product has a value in the range of £18–22/tonne;
(3) ferrous metals (see Table 13); this product has a value in the range of £7–10/tonne.

Table 11

Contraries		Colours	
Water	Nil	Mixed	85%
Ferrous metals	0.01% (max)	Clear	5–10%
Organics	0.05%	Green	5–10%
Solids	0.05%	Size	
Non-magnetic metals	0.01%		15 mm–40 mm

Table 12

Constituent	Quantity
Energy	18,000 kJ/kg
Moisture	17%
Ash	14%
Sulphur	0.3%
Chlorine	0.9%

Table 13

Material	Percentage by weight
Steel and Tin Plate	90
Aluminium	5
Contraries	5

Other products

There exists the potential to recover the following products:

Paper fibres
 Grade 10 waste paper, UK schedule. Capable of upgrading by merchants for fibre recovery. Value £8–10/tonne.
Fines
 Use as a 'cover material' on landfill sites. Value £4–6/tonne.

Putrescibles and organics
Use as soil conditioner on surfaces of industrial spoil heaps to aid plant growth. Value £2–4/tonne nominal.

Marketing of fuel pellets

The fuel pellets are intended to be marketed as a supplementary fuel replacing industrial coal where this is used to produce energy for process or heating purposes. To date no large-scale users have been found because of a natural resistance to trying unproven products. However South Yorkshire County Council have been able to sell most of their output to the Yorkshire Water Authority who are able to use the pellets to replace oil as the supplementary fuel in the incineration of sewage sludge.

Staffing and mobile plant

The plant has a total staff complement of 22 including a manager, shift engineers, plant attendants, labourers, and drivers. The plant processes 900 tonnes per 5-day week on a two-shift basis. It is also equipped with two 30 tonne GVW and four 24 tonne GVW lorries equipped to handle demountable containers and two wheeled loading shovels.

Costs

Capital costs are shown in Table 14. At an input of 48,000 tonnes/year the unit costs are as shown in Table 15.

Table 14

Description	Cost (£)
Land	100,000
Building services and site development	1,150,000
Process	
special development	570,000
provision and installation	1,440,000
Mobile plant	220,000
Total	3,480,000

Table 15

Description		Unit cost (£/tonne)
Capital costs amortized		12.2
Running costs		
plant operation		6.2
vehicles		1.8
Gross cost		20.2
Income (potential)		6.2
Net cost	14.0	
Net operational costs (Running costs less potential income)		1.8

Energy balance (Table 16)

In calculating an energy balance for the production of fuel pellets it is assumed that the plant processes 10 tonnes/hour of crude refuse and produces 2.5 tonnes/hour of pellets with an energy content of 18,000 kJ/kg (*NB*: 1 kWh = 3,600 kJ).

Table 16

Description	Energy kW/hour
Energy consumed:	
plant operation	405
drier	1,000
total including drying	1,405
Energy potential of pellets	12,500
Thus ratio of production to consumption:	
with drier	9
without drier	21

Probable future developments

Basic work has been done for the recovery of copper- and aluminium-rich ores within the glass circuit. The value of the material obtained would be nominal but should cover marginal costs. Also research shows that the anaerobic digestion of the putrescible and organic materials highly contaminated with paper would produce a marketable methane-rich gas and the resultant sludge will be of value as a soil conditioner. It appears that a substantial economic return could be obtained if the material were retained for 10–15 days yielding 0.4 m³/kg at 50% total solids. The resultant gas could be used by adjacent industries currently utilizing natural gas.

THE WESTBURY PROCESS

Wiltshire County Council were experiencing severe problems in securing landfill capacity within economic distance of West Wiltshire and in consequence had decided to construct a refuse incineration plant. Blue Circle Industries, being aware of this problem, developed the Westbury Process to process household refuse into a form which rendered it capable of burning in their cement kilns at Westbury as a supplementary fuel.

The process is shown on the flow diagram (Figure 3) and consists of double pulverization of the household refuse with magnetic extraction of ferrous metal following each pulverization. The resultant product is then pneumatically fed to the cement kilns as a supplementary fuel. The primary fuel used is pulverized coal.

General information

The plant is twin stream and each stream can handle 17–20 tonnes/hour of refuse. The plant as it now exists came on stream in 1979 and it is believed it currently accepts refuse from a population of 125,000 people. The plant has the capacity to process and burn 60,000–80,000 tonnes/year and each of the two kilns can accept processed refuse at a rate of 4.5 tonnes/hour. In the event of breakdown of short duration the Company have emergency landfill capacity available adjacent to the works.

Plant operation

The plant accepts refuse 5 days a week and processes all refuse accepted each day, operating on a two-shift basis. The feeding of refuse to the kiln is a continuous operation and is carried out on a three-shift basis 365 days a year. The total labour complement is 14 and they are answerable to the refuse processing plant manager. Throughput to end August 1981 is as shown in

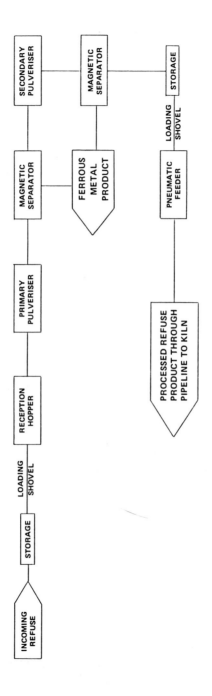

Figure 3   Flow diagram for the Westbury process

Table 17

Description	Tonnes	Percentage of input
Total refuse pulverized	67,669	100
Ferrous metals recovered	5,058	7.5
Rejects sent to landfill	318	0.5
Processed refuse brunt in kilns	62,293	92.0
Coal equivalent	14,562	

Table 17. *NB*: Quantity processed initially limited by availability of refuse and latterly by restricted kiln operation due to depressed cement sales.

Processed refuse specification
   Size: 100% < 50 mm
         95% < 25 mm
   Energy content: 8800 kJ/kg
   Value as a fuel in this usage: £12.00/tonne

Costs

Capital costs are shown in Table 18. Unit costs for the operation of the plant are unfortunately not available.

Table 18

Description	£
Land	
Building and civil works	3,500,000
Fixed plant	
Mobile plant	

Energy balance (Table 19)

In calculating an energy balance it is assumed that the plant processes 20 tonnes/hour of crude refuse and produces 18.4 tonnes/hour of processed refuse with an energy content of 8,800 kJ/kg (*NB*: 1 kWh = 3,600 kJ).

Table 19

Description	Energy (kW/hour)
Energy consumed in plant operation	880
Energy potential of processed refuse	45,000
Thus ratio of production to consumption	51

Development potential

Significant wear has been experienced on the hammers within both the primary and secondary pulverizers and also on the bends in the pneumatic pipelines leading the processed wastes to the cement kilns. The Company are researching methods to partially remove ash and grit from the incoming refuse which, as well as being of benefit in reducing the wear referred to above, will also increase the energy content of the product.

CONCLUSIONS

With the increased awareness of the need to protect the environment and conserve finite resources, the opportunity is being provided for municipal engineers and others involved in collection and disposal of household and commercial waste to make a major contribution in this important field of engineering. At the same time it is essential that any methods that are adopted which appear to result in reduced energy consumption or give an attractive re-use of waste must be justified economically. The marketing of reclaimed materials is as important as their production, and must reliably meet the needs of consumers.

Discounting the climate of opinion which is encouraging recycling and any other energy-saving processes, the continuing increase in population and migration into conurbations and city regions are real problems which in themselves will create, through lack of accessible landfill sites, the pressures and needs to introduce processes which will minimize disposal costs accepting that use of landfill will remain central to waste disposal policies in the United Kingdom.

The four plants in the United Kingdom described above represent the variety of experience being gained in the field of waste disposal. They demonstrate the willingness not only of Central Government, but also the local authorities and commercial undertakings, to invest in new methods.

The Bradford baling plant shows a positive approach on the part of West Yorkshire County Council to the problems associated with the provision of transport to landfill sites. The formation of refuse into bales, rather than compaction into containers, permits the haulage of far greater payloads. Also the landfilling of bales has considerable environmental advantages over the landfilling of crude refuse.

On the other hand the Westbury Process shows the development of a virtually total disposal system. Some 99.5% of refuse input is either recycled in the form of ferrous metal or burnt in the cement kilns as a significant energy source. Further the airborne pollutants normally associated with refuse incineration are neutralized into the cement clinker, thus making the process totally acceptable on environmental grounds.

The two Central Government-sponsored plants at Byker and Doncaster are probably the more exciting and interesting developments. Both plants, although of a commercial scale, are still only pilot plants and consequently are subject to continuous review and development. However, they have already demonstrated that it is a practical proposition to extract a marketable fuel from refuse with a process that is energy efficient and this, in an increasingly energy-conscious world, is in itself important.

For the future, development work at Byker and Doncaster needs to be watched with interest, particularly the work at Doncaster on the anaerobic digestion of the putrescible and organic fractions of the waste which will be of relevance in areas where there is less paper in refuse than in the United Kingdom. It is to be anticipated that the development of markets for the WDFs from both the plants will demonstrate the benefit of their use to users of fossil fuels.

There is no doubt that we are at the beginning of a new field of engineering technology in the production of secondary materials and it is hoped that the United Kingdom's experience will provide, together with the contribution of other countries, a basis for a wider understanding of this important subject.

ACKNOWLEDGEMENTS

Acknowledgements are given to the following for assistance in preparing the paper: C. A. C. Haley, Commercial and Planning Manager, Blue Circle Technical, Blue Circle Industries Ltd; J. M. Hewitt, Executive Director, Department of Waste Disposal, Tyne & Wear County Council; J. R. Holmes, Technical Director, Powell Duffryn Pollution Control Ltd; N. W. Lee, County Engineer and Surveyor, Avon County Council; F. A. Sims, Executive Director of Engineering, West Yorkshire Metropolitan County Council; G. A. Thomas, Chief Environment Officer, South Yorkshire County Council; C. V. Underwood, County Surveyor, Derbyshire County Council.

BIBLIOGRAPHY

Waste Management Paper No. 1 (1976); HMSO, London.

Waste Disposal Statistics 1981/1982 Estimates. The Chartered Institute of Public Finance and Accountancy, London.

Practical Waste Management
Edited by J.R. Holmes
©1983 John Wiley & Sons Ltd

17

Developments in the composting of refuse

J. A. AMBROSE, MIWM
Marketing Manager, Peabody Holmes Limited

ABSTRACT

An important disposal solution in the Netherlands and Switzerland—composting—is in strong favour in the more advanced developing countries, particularly in the Middle East. This chapter reviews the latest technical progress and considers the properties of refuse-derived compost. Nitrogen : carbon ratios, accelerated digestion systems, and a description of a major plant installed in the Libyan Arab Republic complete the work. Illustrations are included of processes by some other principal manufacturers of these systems.

TREATMENT AND DISPOSAL—OPTIONS AVAILABLE

Over 90% of the world's solid wastes are disposed of in landfills. Sanitary landfilling is the main method used in the West: crude dumping is very common in the developing countries.

There is no form of treatment that can entirely avoid the need for land for final deposit. Treatment often enables a proportion of the wastes to be utilized in some way, but there are residues from all forms of treatment, thus sanitary landfilling is usually necessary, although on a reduced scale, whatever form of treatment may be adopted. The most common forms of treatment are:

(1) size reduction of the wastes by shredding or pulverization, in order to improve the landfilling qualities of the wastes;
(2) composting—a system for controlling the natural decomposition process to produce a humus;
(3) incineration—the primary purpose of which is to render the wastes inert, and may sometimes provide a source of energy.

All these forms of treatment provide opportunities for recycling, because facilities for the extraction of saleable materials can be incorporated in the plants.

Pulverization/shredding—landfill

This treatment is used to obtain a fairly homogeneous mixture of wastes, of reduced particle size, which occupy less space than crude refuse at the time of landfill and ultimately decompose to form a consolidated fill without voids.

Composting

The wastes of Middle Eastern countries are often ideal for conversion into organic fertilizer because of their high vegetable putrescible content. Economic forces also favour composting in those countries where high food production is of great importance, and fertilizer imports are limited by foreign exchange constraints.

There are a number of preconditions for successful composting:

(1) suitability of the wastes;
(2) a market for the product;
(3) a price for the product which is acceptable to farmers;
(4) a net disposal cost (plant costs minus income from sales) which can be sustained by the local authority.

When these conditions can be met, a developing country should closely study the possibility of composting because town wastes are a significant potential source of nitrogen, phosphate, and potash as well as an organic soil supplement.

Incineration

For most developing countries incineration can be dismissed firmly as a rational solution to the problems of wastes disposal on the following grounds:

(1) wastes are too low in calorific value;
(2) they are probably too high in moisture content.

EVALUATION

Cost

Given satisfactory standards for the protection of health and the environment, cost will always be the criterion of choice of a waste disposal method. It is necessary, therefore, to consider the probable comparative costs of the main systems. There are dangers in suggesting comparative costs except for a specific city because of wide variations in labour cost and other influences such as site conditions, economy of scale, and the standards of buildings used to house treatment plant.

However, an analysis of recent studies gives the following comparisons as gross capital and running costs, expressed as units of cost. Allowance has, of course, to be made for any revenue earned from the plant.

Sanitary landfill	1
Pulverization/landfill	4
Incineration	15
Composting:	
accelerated windrowing	6
totally enclosed digestion	10

ENVIRONMENTAL ASPECTS

Pulverization/shredding–landfill.

Correctly managed sanitary landfills remove or at least control the environmental problems associated with crude dumping—vermin, flies, and groundwater pollution. The use of shredded or pulverized waste within the landfill is aesthetically more acceptable than crude refuse, can in many instances obviate the need for top covering material, and provides faster consolidation with the consequential earlier return of the landfill to good use.

The use of shredded or pulverized wastes to reclaim scarred landscape which has, say, resulted from mineral extraction is finding increasing interest in the West and its deposit is in this instance seen to have social benefit.

Composting

The possible economic importance of composting to certain developing countries has already been stressed. There may also be significant advantages to public health because it is often the custom for farmers to collect crude wastes and to use them as fertilizers without any proper treatment or control, thus causing risks which would be avoided if the wastes were processed into a hygienic product by the local authority.

Incineration

Incineration has few environmental advantages and invariably presents environmental problems. No doubt as a direct disposal system incineration offers the highest efficiency with volumetric reductions exceeding 90%.

However the financial costs, both capital and operational, are considerably higher than any other disposal method. Furthermore social and environmental costs are high, the atmospheric pollution problems being typical.

CONCLUSION

In terms of both cost and environmental protection, sanitary landfill and composting emerge as the most suitable methods of solid wastes disposal for developing countries. In the case of sanitary landfill, the conclusion is the same as that reached by the great majority of cities in the industrialized countries. Composting, however, has been rejected in most of Europe and USA because of high production cost, and because of the ready availability, at acceptable cost, of artificial fertilizers of guaranteed analysis. None of these factors apply at present in many of the developing countries. Thus in most cases both sanitary landfill and composting may be equally worthy of consideration.

COMPOSTING

General

Converting urban wastes, or a substantial proportion of them into organic manures, has attracted interest over many years. Such a process is known as composting and has two essential features:

(1) the use of methods and equipment which facilitate decomposition of the organic content under controlled conditions, so as to avoid risks to health or the environment; and
(2) the extraction of constituents of the wastes which would be undesirable in the compost; some of these may be saleable.

Thus a composting process usually has three products:

(1) compost for use as an organic fertilizer;
(2) salvaged materials which can be sold for recycling;
(3) 'contraries', which are of no value and must be disposed of by landfill, but which rarely exceed 20% of the original weight.

If composting can be operated successfully, it achieves the following highly desirable results:

(1) conservation of resources by recycling;
(2) support for nature's cycle by returning to the earth organic material.

Compost properties

The aim of composting is to convert a major proportion of solid wastes into a marketable product. It is necessary to begin, therefore, with some understanding of the properties, and the limitations, of compost. Compost is a brown, peaty material the main constituent of which is humus. It has the following physical properties when applied to the soil:

(1) the lightening of heavy soil;
(2) improvement of the texture of light sandy soil;
(3) increased water retention;
(4) enlarged root systems of plants.

Compost also makes available additional plant nutrients in three ways:

(1) it contains N, P, and K, typical percentages being N, 1.2%; P, 0.7%; K, 1.2% but with fairly wide variations;
(2) when used in conjunction with artificial fertilizers it makes the phosphorous more readily available and prolongs the period over which the nitrogen is available, thus improving nutrient take-up by plants;
(3) all trace elements (micronutrients) required by plants are available in compost.

Compost application

Compost is applied to land at a rate of between 20 and 100 tonnes/ha./year; it is commonly used at 40 tonnes/ha. Usually it is applied between harvesting one crop and sowing the next, and ploughed in it can also be used as a mulch to assist moisture retention and inhibit weeds, in which case it is not ploughed in until the harvest.

Because the time of applying compost is determined by the cropping cycle, demand is usually seasonal, thus a compost plant may require storage capacity for its product for several months.

There are two situations in which the production and use of compost may be of great importance to the agriculture of an area:

(1) for bringing into cultivation marginal land suffering from organic deficiency—such areas are most common in tropical climates where hot sunshine tends to destroy organic matter;
(2) in areas where artificial fertilizers are in short supply or are very expensive.

Suitability of refuse for composting

The characters of the constituents of solid wastes have to be analysed to determine how suitable they are for composting.

Before considering a composting project it is necessary to carry out a physical analysis of the wastes, using reliable sampling methods. Although similar constituents occur in solid wastes throughout the world, there are wide variations in relative proportions, not only as between countries, but even between regions within a country. Table 1, which compares Middle Eastern wastes with those of Europe and India, illustrates the importance of adapting composting systems to match waste characteristics.

Table 1

Constituents	Percentage by weight		
	India	Middle East	Britain
Essential to compost:			
vegetable putrescible	75	50	28
Acceptable for composting:			
paper	2	20	37
inert below 10 mm	12	14	9
Compostable total	89	70	74
Salvageable constituents:			
paper (also included above)	2	20	37
metals	0	9	9
glass	0	4	9
textiles	3	4	3
plastics	1	3	3
Total of potential salvage	6	40	61

Carbon : nitrogen ratio

Bacteria use carbon as an energy source and nitrogen for cell building, thus the process of decomposition involves the reduction of the relative proportion of these elements, known as the C/N ratio, from an original level which may range from 20 : 1 to 70 : 1 to a point where the available carbon has been consumed and activity ceases. The final C/N ratio usually lies between 15 : 1 and 20 : 1 but may be higher if the initial ratio associated with the raw waste was near the top of the range.

The initial C/N ratio is a deciding factor in the speed at which decomposition takes place. The ideal initial ratio is between 30 : 1 and 35 : 1; if it exceeds 40 the time required increases considerably. Ratios below 30 : 1 are undesirable for a different reason: there may be excessive nitrogen losses.

In the solid wastes analyses above, the main source of nitrogen is the vegetable/putrescible matter which has a C/N ratio of about 24 : 1, and paper is the main source of carbon. Thus the higher the ratio of paper to vegetable/putrescible matter the higher the C/N ratio.

PRINCIPLES OF COMPOSTING

A composting process seeks to harness the natural forces of decomposition to secure the conversion of organic wastes into organic manure. The purposes of controlling the process are:

(1) to make it aesthetically acceptable;
(2) to minimize the production of offensive odours;
(3) to avoid the propagation of insects or odours;
(4) to destroy pathogenic organisms present in the original wastes;
(5) to destroy weed seeds;
(6) to retain the maximum nutrient content, N, P, and K;
(7) to minimize the time required to complete the process;
(8) to minimize the land area required for the process.

There are two main groups of organisms which decompose organic matter:

(1) anaerobic bacteria which perform their work in the absence of oxygen;
(2) aerobic bacteria which require oxygen.

Anaerobic composting

The main use of anaerobic composting has been in India where for many years it has provided, usually on a small scale, a cheap solution to the combined disposal of solid wastes and nightsoil. These materials are placed in alternate layers in small trenches which are sealed and left undisturbed for many months; the contents are then dug out and used as compost. This, the Bangalore system, is now being abandoned in favour of aerobic methods because of the very large land area required owing to the long retention period. In other respects, however, it is a low-cost system as the amount of materials handling required is much less than for aerobic methods.

Aerobic composting

Aerobic composting is characterized by:

(1) rapid decomposition, normally completed within 1–10 weeks;
(2) during this period high temperatures are attained which achieve speedy destruction of pathogens, insect eggs, and weed seeds;
(3) so long as aerobic conditions are maintained, no offensive odours are produced.

All current composting systems aim to maintain aerobic conditions through-out the process. Many types of organism assist: bacteria, which predominate at all stages; fungi, which often appear after the first week; and actinomy-cetes, which assist during the final stages. The process begins at ambient temperature by the activity of mesophilic bacteria which oxidize carbon to CO_2 thus liberating large amounts of heat. Usually the temperature of the wastes reaches 45 °C within 2 days, and this represents the limit of temperature tolerance of the mesophilic organisms. At this point the process is taken over by the thermophilic phase, which lasts about 2 weeks, in the temperature range 55–70 °C; should the temperature increase beyond 70 °C activity temporarily declines.

The process is dependent, of course, on the provision of a suitable environment for the bacteria. In addition to the nutrients provided by the wastes the main requirements are adequate supplies of air and moisture.

It is important to stress that urban solid wastes, of the character described above, already contain at the time of collection all the organisms required for every phase of aerobic composting.

Moisture content

Moisture content is a critical factor in aerobic composting, for the type of wastes now being considered the following are important:

(1) if moisture content falls below about 20% decomposition ceases;
(2) if it exceeds 55% water begins to fill the interstices between the particles of wastes, reducing interstitial oxygen and causing anaerobic conditions; this results in a rapid fall in temperature and the production of offensive odours.

Middle Eastern wastes of the analysis given earlier probably fall within the optimum range of initial moisture content and are unlikely to require the addition of moisture during the first few days. During the thermophilic stage, however, the high temperature causes rapid loss of water and this must be replaced from time to time until the final fall in temperature.

pH control

A final parameter which is important in evaluating the microbial environment is the pH of the refuse. As in the case of temperature, the pH of the compost varies with time during the composting process and is a good indicator of the extent of decomposition within the compost mass. The initial pH of solid waste is between 5.0 and 7.0 for refuse which is about 3 days old. In the first 2 or 3 days of composting the pH drops to 5.0 or less and then begins to rise to about 8.5 for the remainder of the aerobic process. If the digestion is allowed to become anaerobic, the pH will drop to about 4.5.

Completion and testing

An important facet in the composting process is the determination of the point at which digestion of the solid waste has been completed. Generally satisfactory stabilization is attained when the compost has the characteristics of humus, has no unpleasant odour, high temperatures are not maintained even though aerobic conditions and desirable moisture content exists, and the C/N ratio is such that the humus can be applied to the soil (if the C/N ratio is too high, the compost will remove nitrogen from the soil). To date, most methods used determine chemically when the digestion period has been completed by measuring the reduction in total carbon, cellulose, and lipids and the increase in percentage ash. These methods are adequate for determining the completeness of composting on the assumption that a representative sample has been taken. All have, however, obvious disadvantages such as complicated sample preparation, considerable time, and expensive equipment.

A practical alternative to these methods is the starch–iodine method suggested by Richard D. Lossin of the Bureau of Solid Waste Management. The maximum rate of starch degradation in compost occurs when optimal compost stabilization parameters (temperature, moisture content, pH, etc.) are maintained. Therefore, the amount of starch found in compost decreases with increasing compost age under proper operating procedures. This phenomenon, coupled with the fact that starches form characteristic colour complexes when combined with molecular iodine, forms the basis for the starch–iodine method.

Composting process

A typical process block diagram for a composting plant is shown in Figure 1. Most of these stages incorporate equipment and technology which is already well proven. Primary pulverizing/shredding, for example, is well proven and experience is growing rapidly in preferential separation techniques using magnetic, drum, rotadisc, screening, and air classification techniques.

Worldwide, there were 37 plants identified in 1973. Many of these involved the addition of sewage sludge to the refuse and also required the compost to be matured by windrowing. The use of open windrows is generally regarded as highly inefficient (see the section on Composting) and in particularly hot climates where the rapid loss of moisture greatly retards degradation. Undoubtedly, however, the last few years have seen much greater determination in process development to meet the needs of microbial composting rather

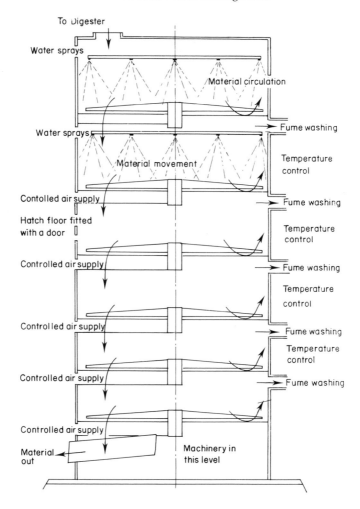

Figure 1   Peabody-Holmes accelerated digester for the composting of refuse.
(Courtesy of Peabody-Holmes Ltd)

than to consider the method purely from a materials handling standpoint. The Peabody–Holmes accelerated windrow and digester systems compare very favourably with its main competitors in this way.

PEABODY–HOLMES ACCELERATION DIGESTION SYSTEM

The heart of the Peabody system is the digester (Figure 2), each digester 7 m diameter × 17 m high is a vertical tower divided into six stages. Shredded

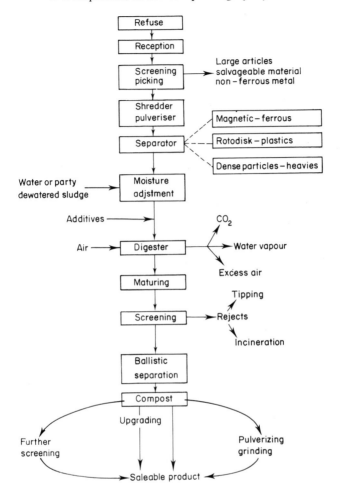

Figure 2 Typical process flow diagram for composting plant—the flow diagram
illustrates various treatment options. (Courtesy of Peabody-Holmes Ltd)

refuse enters the top stage with a load of up to 50 tonnes, and resides there for
1 day. It is then transferred downward to the second stage, where it is held for
a further day, and so on, until it emerges from the base of the digester after 6
days as fully composted humus. Each stage of the digester is fitted with slowly
rotating arms which both aerate the compost and provide periodic agitation
and spreading.

The arm controls can be operated independently to a pre-programmed
requirement for air flow and turning. Water, or sewage sludge, can be added

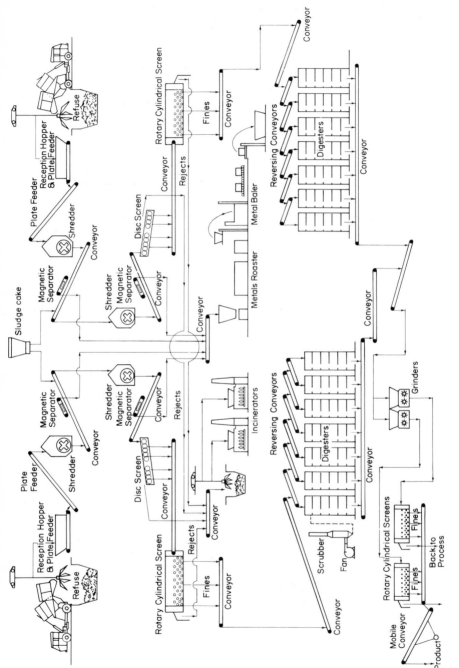

Figure 3 Process diagram for a 500 tonnes/day composting plant for Tripoli, Libya. (Courtesy of Peabody-Holmes Ltd)

to the upper stages of the digester and there is also a separate vent from each stage which is connected to a fume washing plant, or other detoxification system.

The digester can thus vary the operating conditions in each of the stages to provide just the right combination of air flow, temperature, and agitation for the various composting stages. In single-drum or windrow systems it is not possible to achieve such a fine degree of control because, for example, the degree of agitation cannot be varied during the composting period. A further advantage of the compartmentalized digester is that if any toxic substance enters the digester, it can be kept within a single 50-tonne batch without affecting the batches on the other stages. Similarly in the event of any mechanical fault, maintenance can be performed on individual floors whilst the rest of the digester continues to operate.

In normal operation, the progress of the various biochemical steps can be followed by charting the outlet gas temperature from each of the digester stages. This gives the operators a reliable guide to the completion of microbiological action so that when the compost is discharged there can be reasonable certainty that it is pathogenically safe.

This type of digestion system, because of the ease and high degree of process control, lends itself to computer technology to optimization of the system design. This is in fact a feature of the Peabody design and the computer programming has now been extended to include the total refuse processing equipment including handling, shredding, and classification. The computer programme is able to accommodate variations in refuse analyses and moisture contents and offer the optimum process design.

TYPICAL PEABODY–HOLMES COMPOST PLANT

Libyan Arab Republic

Figure 3 illustrates one of three Peabody accelerated composting plants under construction in the Libyan Arab Republic. Plants with in-feed capacities of 600, 500, and 60 tonnes/day are being constructed respectively in Tripoli, Benghazi, and Beida. The plant as shown is a total recovery system as no rejects emanate from the plant, the only products of the composting process are bagged or bulk compost and baled ferrous scrap. Even the heat associated with the combustion of process rejects (plastics, textiles, etc.) is recovered for in-plant use.

The flowsheets and mass balance are shown in Figures 4 and 5 and are typically for Tripoli as follows (tonnes/day):

In-feed:	250
Ferrous metal separation	20.5

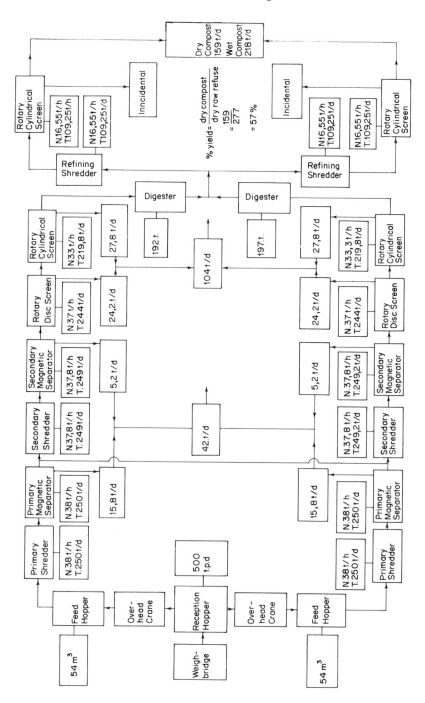

Figure 4   Material balance diagram for a 500 tonnes/day composting plant, Tripoli, Libya. (Courtesy of Peabody-Holmes Ltd)

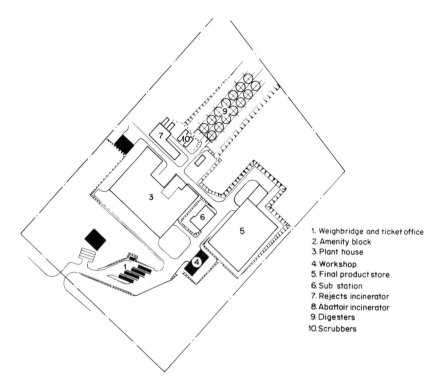

1. Weighbridge and ticket office
2. Amenity block
3. Plant house
4. Workshop
5. Final product store
6. Sub station
7. Rejects incinerator
8. Abattoir incinerator
9. Digesters
10. Scrubbers

Figure 5   The site of a composting plant for Tripoli, Libya

Process rejects (plastics/textiles etc.)	37.5
Feed to digesters	192

The 'normal' flowrate is based on an actual working day of 6.6 hours, all equipment is designed for a 'peak' flowrate which is 10% in excess of the 'normal' rate. The above are 'per stream', as both the Tripoli and Benghazi plants are complete twin-stream installations. The compost will have the following general properties:

Moisture content	25–35% (by weight)
Organic content	not less than 50% by weight
Particle size	less than 15 mm
Copper	150 ppm (max)
Chrome	200 ppm (max)
C/N ratio	23 : 1
pH range	5.5–7.5

ACKNOWLEDGEMENTS

The author acknowledges the following publication from which he has used numerous extracts: *World Health Organisation Report; Management of Solid Wastes in Developing Countries*, by Frank Flintoff.

Practical Waste Management
Edited by J.R. Holmes
©1983 John Wiley & Sons Ltd

18

Selection of refuse collection vehicles

J. M. COTTERIDGE

Deputy Controller of Transport and Works, London Borough of Waltham Forest

ABSTRACT

The factors involved in the proper selection of refuse collection vehicles are considered in this chapter. The definition of perform-ance, reliability, safety, and crew comfort are mentioned. The importance of a professional specification and evaluation procedure is mentioned, and a detailed checklist is included of engineering and operational facets of refuse vehicles. Illustrations show some of the key issues, and mention is made of the behaviour of refuse under compaction conditions.

SUMMARY

A chapter of this length, on this subject, can do no more than offer an appraisal of the topic. Its object is not to provide technical detail but to stimulate some thought on refuse vehicle selection. As such the chapter attempts to concentrate attention on the main issues, to act as a refresher and up-date knowledge of the factors to be considered prior to purchasing. It contains an Appendix which gives a list of factors that may require assessment and recording when vehicles are demonstrated.

INTRODUCTION

The refuse vehicle is such a fundamental part of a solid waste manager's work that after many years a tendency can develop which traps one into assuming that there is not much that can be learnt by an experienced person on this subject. To suggest that the vehicle selected by a trained and experienced officer is wrong, or that the judgement that produced that selection was based on thoughts and feelings—rather than an objective approach to the facts—can lead one on to dangerous ground. Nevertheless most people in the solid waste

415

field, one suspects, have lived to regret in part, if not in whole, the purchase of certain vehicles. To those who were responsible for such decisions, and to those who may be in the future, this chapter is dedicated.

The modern refuse collection vehicle is a complex piece of mobile machinery. The growth in its complexity has developed largely over the past two decades to the point where its purchasing and running costs are amongst the highest, if not the highest, in a solid waste manager's transport fleet. This complexity has resulted from the need to increase operational efficiency but it has had to be balanced against the vitally important need for reliability. The combination of complex engineering and durability faces the manufacturer with tremendous problems and as a result only the best manufacturers and suppliers may be expected to survive in the market.

For the innovative manufacturer a time is reached after all the research, design, construction, and product testing for a new or modified vehicle to be offered to the customer. Since the true quality of a refuse vehicle design is never really apparent until in operation for a year or two, at which time the operator is possibly committed for another 6 or 7 years, that initial purchasing decision must be carefully made.

What should one look for? The list of factors in the Appendix provides a reasonably comprehensive starting point; however, the relative weighting or importance of many vary according to the local conditions or tasks to be carried out by the vehicle.

PERFORMANCE

To define the necessary performance specification of a vehicle is beyond the scope of this chapter and indeed it would hardly be possible as the needs of performance vary according to local conditions of productivity, collection, and disposal. The assessment of the desired performance must always rest with the user.

An inspection of statistics relating to refuse collection may lead the uninitiated to ask if it is necessary, in all cases, to have such large vehicles with the apparently low average payloads being carried. The pat answer, to the uninitiated, might be that since some crews require a large vehicle it is sensible to maintain the whole fleet with large vehicles to obtain benefits of standardization and interchangeability. How true and how valid is this? To some extent it is correct but careful scrutiny of payload requirements in one's own area may cause some further thought and a different conclusion.

It is recognized that a shorter body and wheelbase does not save a great deal on the original purchase price, but even that slight saving could enable extra money to be spent on crew comfort and the gain in manoeuvrability is increasingly important in boroughs where pedestrianization of residential areas is becoming the vogue.

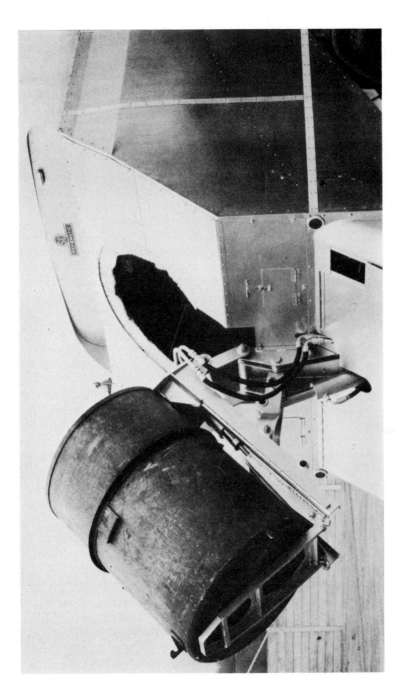

Figure 1   The Shelvoke and Drewry bin lift—this equipment includes many good design features and robust engineering; safety screening, damped hydraulics, and a universal squeeze clamp enabling most types of container to be lifted. (Courtesy of Shelvoke & Drewry Ltd)

There are certain essential requirements which are common to all vehicles and the absence of which should eliminate a vehicle for further consideration if they cannot be rectified by the manufacturer. They are:

(1) the hopper clearance mechanism must be sufficient to cope with the largest expected loading rate without *any* tendency to cause overspill behind the vehicle;
(2) operating dimensions and turning circles must be compatible with collection and disposal sites;
(3) inadequate safety mechanisms or dangerous design features;
(4) maximum loads which overload rear axle.

RELIABILITY

The ability to sustain reliable performance at reasonable cost throughout its planned life, despite indifferent treatment by operatives, must be the most sought-after characteristic of a vehicle. Recognizing and designing for this desired reliability presents a strong test for both purchaser and manufacturer.

One path is to hold to a basically sound design and refine it over a long period. The dilemma in this approach is that new concepts are held back and progress in design may be stunted. The apparent ultimate in design of 10 years ago would look rather ineffective and possibly unsafe by current standards.

This progress has, as in any other field, been achieved by the manufacturers' need to attract custom and keep ahead of the market together with the purchasers' willingness to back their own judgement on new design and innovation. Who, however, is in control of this progressive trend? Are we, the purchasers, inclined to be a little 'fashion-conscious' and be more impressed by the latest compression mechanism rather than the less obvious aspects of potential reliability? Has productivity really increased in the last 6 or 7 years to such an extent that it requires even further redesigning of compression mechanisms? I do not think so. Are the vehicles of today more reliable than those 10 years ago? If not, then I would suggest that little real progress has been made. We have, I believe, reached a plateau in the need for new design to improve loading and payload capability so the emphasis should now be in other areas; the achievement of reliability in particular.

How does one recognize the potential for good operational efficiency with reliability? A day at the factory to see things built is a good start, and it should not simply be regarded as a pleasant day out with the representatives. Searching questions from oneself and an experienced workshop colleague will undoubtedly pay off and probably be welcomed by any manufacturer who is confident of his product. Scrupulous attention should be given to the quality of welding, the quality of the steel, the quality and source of any component

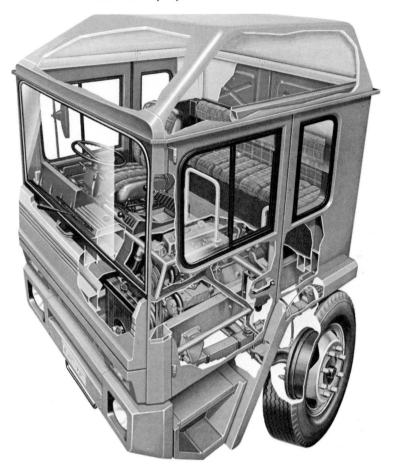

Figure 2   Good cab design—the cab tilts for maintenance, has a flat floor, low steps, and wide doors to ease entry and reduce risks of slipping. (Courtesy of Shelvoke — Drewry Ltd)

parts, the fitting and location of sacrificial wear pieces; the general atmosphere of the production shop, the stockholding in the stores, and after-sales organization. If it is a new design or modification, enquiries about previous product testing are most relevant.

A special mention ought to be given to the 1 m³ (Paladin) bin-lift equipment. A simple calculation will show that a bin-lift attachment may be expected to hoist approximately 120,000–140,000 loaded 1 m³ (Paladin) bins in its working life. Any flaw in design, defect in construction, or inferior material and workmanship will reveal itself well before that total is reached. It may be unreasonable to expect any bin-lift gear, no matter how well engineered, to give total reliability with that degree of workload and it is worthwhile planning for the total replacement of this part of the vehicle after 3 years and costing its replacement into the renewal or capital fund. The need to maintain schedules of 1 m³ (Paladin) container rounds is very important, as much of the work can be either rechargeable or clearing chute feed containers, both of which are sensitive to delays due to breakdown or unavailability. This suggestion of early replacement does not imply that any compromise can be allowed in the quality of engineering of bin-lift gear. Shock loading will produce accelerated wear in the equipment and fatigue in the materials of construction, so it is desirable to have a bin-lift and lowering cycle which has a smooth motion, damped at the emptying and return-to-ground positions. Apart from the improvements in the reliability of the lift equipment a smooth action will also, of course, protect the castors on the containers themselves.

Future reliability is a vital characteristic to be sought when selecting the refuse vehicle. Its assessment, like weather or economic forecasting, is probably more of an art than a science but, like the latter, the case for logical consideration of as many relevant factors as possible has to be stronger than relying on casual judgement.

SAFETY

It seems difficult to deal with any topic in the solid waste field without mentioning health and safety. In this instance it needs no obtuse justification—refuse vehicles can be hazardous. But so is electricity, household bleach, and excessive exercise. There is no need to get hysterical about the safety aspects but there is a need to calmly recognize where a potential hazard exists, what causes it, and take reasonable precautions against it.

The most obvious area of hazard is normally the hopper clearance mechanism and one is continually being confronted by the possibility of someone being severely maimed or even killed by it. However, more accidents and injuries are caused by windblown materials, shattering glass, passing traffic, or lifting heavy objects than by the more obvious dangers of compressing plates or arms. Could it be that the danger in the latter is so obvious that therein lies its safety factor? If frequency of incident is the measure, then this may be so; but the potential for severe injury by compression gear is so high that even a good design of safety gear must be coupled with proper training by operatives if incidents are to be avoided.

The continuous loading or compression mechanism, by definition, can present a greater potential risk than intermittent operation types. Simple tests can quickly reveal that the source of hazard is not always obvious. For

Figure 3   Discharge at height. On an open tip site it may not matter, but checking the discharge height if disposal is carried out inside a building is obviously essential. (Courtesy of Bradley Municipal Vehicles Ltd)

instance the possibility always exists for a man simply to trip and fall against the rear end of the loading hopper and despite the presence of body pressure-activated safety switches it should be impossible for any part of the body to come into contact with a piece of machinery which is likely to crush or maim. During such a test it was found that from the proper ground-level loading position that possibility did not exist with the vehicle under test, but it was fitted with a salvage trailer towing bracket which could have been climbed onto when emptying bins and from this position it was possible for an arm to come into contact with the loading mechanism in the event of a fall. The hazard, therefore, was not the mechanism itself but the towing bracket support about which suitable instructions would have to be given to the operators. If an accidental fall from normal working level could cause such an accident the safety of the vehicle would have had to be severely questioned.

Manufacturers have had to become very conscious of these safety problems and endeavour to guard against these dangers in their design. But the onus is on the person responsible for purchasing to make particular enquiries of these precautions and satisfy himself that they are sufficient. Some of the main points to watch for are:

(1) body pressure and stop devices at the hopper;
(2) location of emergency stop switches;
(3) driver warning systems for operation by loaders;
(4) body cleaning arrangements for ejection discharge systems;
(5) visibility for the driver
(6) windblown dust prevention or restriction;
(7) baffled hydraulics to allow fail-safe or slow-reaction equipment if there is a burst or severing of hydraulic pipes;
(8) wire cables in any lifting system;
(9) control of disposal operations from the cab by the driver;
(10) rear light and other warning systems for other road-users;
(11) automatic tailgate props.

PURCHASE PRICE

How significant is purchase price, and should the lowest tender price be the sole basis for choice? To operate such a policy requires a sure knowledge of the requirements coupled with a precise specification. This policy may not be suitable for refuse vehicles—and the policy of regarding these vehicles as proprietary items, requiring a far wider consideration than price alone, is probably more appropriate. This thinking applies both to the chassis and the body.

This is not to suggest that purchase price is an unimportant consideration but it should be neither the only nor the main one; it should simply be an element within the overall evaluation.

At present costs £1,000 of capital expenditure amortized over 7 years costs about £270 per annum which on a 1,500 hours per annum operating life represents approximately £0.18p per annum hourly rate. As such, price differences of even £2,000–3,000 at purchase are poor value if unreliability, long down-time, and expensive repairs are the true consequences of an unqualified lowest-price purchasing policy. The long-term view of costs is what really matters.

Within the category of the purchase price one ought to consider the price of major components and spare parts, which will obviously influence repair costs at a later date. If, however, the high prices of spares is due to high overheads needed to support an efficient supply and after-sales organization, they can represent good value for money due to consequential savings in down-time and spare vehicle provision.

Only when all the facts, including the long-term costings, have been considered should price influence the final decision.

DRIVING AND CREW CABS

The crew cab is the workplace of the driver and the place for transporting the crew. It should not be overlooked as the degree of comfort provided and the care given to its finish will favourably influence the driver's attitude to the vehicle, although that influence can be rather negative in that it will not necessarily ensure care and consideration in the use of the vehicle. It may, however, help reduce abuse and prevent it being offered up for repair at the slightest pretext in order to obtain one of the preferred vehicles in the fleet—the preferred vehicle being, perhaps, the most comfortable to drive.

At recent conferences and exhibitions it has been encouraging to see more effort being made to introduce engine noise insulation in vehicles. This, with a good vibration-damped and fully adjustable driving seat, together with easy entry and exit from the cab, will go a long way to improving the driver's lot.

Crew seating need not be so well appointed but should preferably have a separate access door on the nearside, grab rails to increase safety when travelling, and suitable storage places for spare clothing.

Attention to working surroundings is so much wrapped up with good labour relations that greater attention to crew cabs is fully justified. It is regrettable that labour relations so often concentrates upon wage rates, bonuses, differentials, etc., and the simpler and cheaper ways of improving employee relations are frequently overlooked.

The cost of providing a high standard of crew and travelling accommodation is fairly small relative to the vehicle cost. For example a vibration-damped and fully adjustable driver's seat, designed to give proper support to the lumbar region and spine, would (on the basis of the calculation given in the notes on Purchase price) add less than £0.20p to the hourly rate and probably be far more worthwhile in the long term than a differential payment of the same amount.

DEMONSTRATIONS

A demonstration of a vehicle within one's own operaton is obviously essential. The value of the exercise will depend upon the amount of effort invested not only during the demonstration but also before and after.

The Appendix to this chapter offers a fairly comprehensive list of the items that should be considered and noted in the demonstration period. The weight or relative importance placed on some items will depend on local conditions. For example, in large cities where night work is a regular and essential factor noise is a most important item justifying investment in a decibel meter and proper training in its use. In rural situations this may have little relevance whereas good dual-purpose manual and bin-lift loading ability may be the primary consideration.

Figure 4   Modular design on the hydraulic and electrical systems is an important aid to maintenance and operations. The hydraulic tray on this Hestair Eagle body can be removed and replaced by a spare unit within minutes to enable repairs to be made on the bench, thereby avoiding expensive down-time. (Courtesy of Hestair Eagle Ltd)

The preparation of an evaluation system and documents is a relatively simple task. However the following points are suggested as being essential to the process. The essential points of assessment for purposes of evaluation and comparison are:

(1) a disciplined and standardized procedure;
(2) a points system of evaluation; this is preferable to qualitative statements of good, fair, or poor, etc.;

(3) the demonstration should be for at least 3 days *in the first instance* and longer thereafter before any final decision;

(4) all vehicles tested should work in selected areas and given a range of tasks kept constant for each series of tests;

(5) an observer would keep an accurate bin check, giving details of:
   (a) total bins per load,
   (b) an estimate of the volume of bulk material loaded,
   (c) the bin positions, kerb site, etc.
   (d) a time check showing loading and travelling time including the inter-round travelling time,
   (e) team size—if the regular driver assists loading it should be noted;

(6) observations and assessments should be sought from a regular group of the department's staff and include the drivers and collectors—facing drivers and collectors with an apparently complicated questionnaire is not generally fruitful and better results are obtained by holding a general discussion with them and completing the assessment document on a consensus basis;

(7) a thorough inspection by workshop staff with a special report on the engineering and maintenance aspects must be prepared;

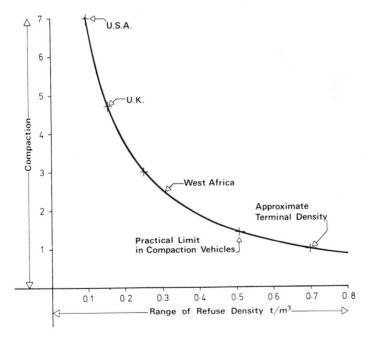

Figure 5   The compaction of refuse

(8) a final report following analysis of the completed assessment documents and reports should be produced and if necessary include any reference from other users.

CONCLUSIONS

The purchase and assessment of refuse collection vehicles is too important and complex to allow decisions to be made from casual inspection, vaguely remembered demonstrations, and past experience alone.

The costs of errors in the choice of vehicles are high, and at a time when the reduction of costs by productivity schemes are increasingly difficult to accomplish reducing transport costs can offer the easiest approach to significant financial savings. Time spent on arriving at the right purchasing decision must, therefore, be cost-effective. To argue that the field of refuse vehicle manufacturing is so competitive that the vehicles produced have got to be sound if the manufacturer is to survive is hardly valid. The manufacturers are attempting to provide a product which will suit all needs—from rural Cumbria to Metropolitan London or Birmingham. To achieve this, certain compromises have to be struck, and their particular compromises may be a vital item in your Authority. In addition the tendency for very similar types of vehicle to be produced by varying manufacturers requires far greater attention to detail.

If I may, I would like to end with a quotation of Oscar Wilde's: 'Experience is the name everyone gives to their mistakes'. I hope this short exposition will be of some assistance to those with 'experience' and reduce the 'pain' for those that will eventually have it.

ACKNOWLEDGEMENTS

For assistance in compiling the Appendix, I acknowledge help from my colleagues on the Research Sub-Committee, Association of London Cleansing Officers.

Appendix: *Evaluation—information and categories*

GENERAL DETAILS

Name of Manufacturer	Chassis..	
	Cab	
	Engine	
	Body	
Chassis	Axle loading	
	Front..	
	Rear...	
	Power steering..	YES/NO
	Type of power take-off	
	Gross vehicle weight	
	Unladen weight...	
Cab	Seating capacity	
	Material of construction	
	Tilt cab	YES/NO
Engine	Manufacturer	
	Optional manufacturers	
Body	Construction	
	Air space	
	Compaction rating (manufacturers)	
	Hopper capacity...	
	Hopper clearance mechanism.........................	
	Rave height unladen	
	Rave width	
	Load discharge type	
Dimensions	Overall length..	
	Overall height..	
	Overall width	
	Wheelbase	
	Rear overhang..	
	Turning circle—swept...................................	
	Turning circle—between kerbs	
	Height when discharging	

428   *Practical waste management*

ITEMS FOR EVALUATION

1. *Safety*
   (a) Hopper clearance mechanism
   (b) Machinery stop systems
   (c) Fail-safe hydraulics
   (d) Tipping/discharge—cab control
   (e) Ejection discharge—body cleaning
   (f) Reversing warning systems
   (g) Crew travelling—adequate seating.
   (h) Driver vision
   (i) Driver warning systems
   (j) Hazard lighting indicators
   (k) Tailgate support system when discharging load

2. *Loading—manual*
   (a) Rave height unladen
   (b) Rave width
   (c) Spillage—hopper clearance
   (d) Hopper clearance rate
   (e) Payload
   (f) Axle loading with achievable payload
   (g) Suitability for bulk material
   (h) Storage of ancillary equipment—sacks, cleaning brooms, etc.
   (i) Trailer towing arrangement
   (j) Compaction overload cut-out
   (k) Intermittent clearance time
   (l) Dust emission

3. *Loading—bin-lift*
   (a) Compatibility of lifting gear with local containers
   (b) Lifting capacity (designed)
   (c) Shock loading protection—lifting gear and containers
   (d) Dust emission
   (e) Clearance of load into body
   (f) Ability to accept potentially damaging material
   (g) Design/quality of lifting gear
   (h) Ease of removal of lifting gear
   (i) Ease of use as manual loader with lifting gear attached

4. *Tipping/discharge*
   (a) Suitability for disposal site—liaison with disposal authority necessary
   (b) Spillage on discharge platform—plant or transfer station
   (c) Full cab-control of tailgate release and discharge

    (d) Time taken for discharge
    (e) Regular ancillary operations required

5. *Driving and crew*
    (a) Ease of entry and exit to driving and crew seating
    (b) Noise insulation in cab
    (c) Driver vision
    (d) Power-assisted steering
    (e) Manoeuvrability
    (f) Instrument layout
    (g) General quality of cab interior
    (h) Driving seat—comfort

6. *Environmental*
    (a) Operating noise—empty/partially loaded
    (b) General appearance
    (c) Dust emission
    (d) Exhaust emission

7. *Engineering/maintenance*
    (a) Quality of construction
    (b) Spare parts availability
    (c) After-sales service
    (d) Warranty
    (e) Access to hydraulics
    (f) Access to main components
    (g) Ease of major repairs to body

8. *Miscellaneous*
    (a) PRICE—basic vehicle
             extras
             options
             spare parts
    (b) Delivery period
    (c) Previous product testing
    (d) References—other users

Practical Waste Management
Edited by J.R. Holmes
©1983 John Wiley & Sons Ltd

19

Developments in the bulk handling of refuse—the transfer station

F. W. STOKES, FINucE, MIM, MISWM
Former Managing Director, Powell Duffryn Engineering Limited

ABSTRACT

This chapter reviews the development of the refuse transfer station. Written by the former Managing Director of one of the country's biggest waste-handling manufacturing companies, it traces the development of the tranfer station and describes where and how it forms part of modern waste disposal strategies. The key aspects of the mechanisms and vehicles involved, their rating, and performance are detailed. Illustrations are given of a range of modern equipment.

INTRODUCTION

The last decades have been periods of immense change in the technology and management of solid wastes in the public and private sectors. The legislative changes of the Local Government Act of 1972, which led to the setting up of the new Waste Disposal Authorities, have had, in the English counties at least, a substantial impact on decision-making and the management of waste disposal. There have been controversies over these measures, but no-one can deny that they have at least generated a new professionalism in the business of waste management.

As far as technology is concerned, the last decade has seen the demise of systems originally thought to be the only future solution. Many technical ideas have had to be discarded, and many hopes for high technology disposal systems remain unfulfilled. Through all this a number of essential truths have remained, and the most important must be the continuing supremacy of sanitary landfill as the overwhelmingly appropriate system of disposal for domestic and commercial waste.

But even here there have been great changes in style, techniques, and the knowledge of the behaviour of wastes in landfills. Coupled to these develop-

431

ments has been a great expansion in the types of mobile and static equipment, enabling waste to be efficiently and economically carried to the new landfills. My company, as one of the largest manufacturers of waste-handling equipment in the United Kingdom, has been at the heart of these changes, serving the public and private sectors of the industry. This chapter is a review of the development of bulk handling systems for waste and it explains how I see these systems playing a part in the future of waste management in the United Kingdom.

THE SIZE OF THE PROBLEM

Recent published papers have given the estimated collection and disposal costs of the 18 million tonnes of domestic and commercial wastes generated annually in the United Kingdom as:

Collection service	£350 million
Disposal service	£150 million
Total	£500 million

The size of these sums must bring home to waste managers the scope that exists for sensible savings, particularly on refuse collection, which absorbs 78% of the financial resources. The introduction of bulk transfer systems to intervene between the refuse vehicle and the landfill site can have a substantial impact on collection costs.

As an industrialist marketing this equipment, I am very concerned that in the English County Waste Disposal Authorities, with their split responsibility for collection and disposal, decisions to invest in bulk transfer, or for that matter any other systems, take into account the complete financial picture of savings to both services. The collection and cleansing costs of a selection of English district councils were given in Mr. J. R. Holmes' paper to the Institution of Municipal Engineers Seminar in November 1978. The following extract from this paper highlights the ratio of refuse collection costs to those of other cleansing activities, and should indicate the potential for studying whether bulk transfer systems have a part to play in using these resources to better effect.

Looking at waste disposal costs, the statistics published by the Society of County Treasurers show the supremacy of sanitary landfill as the premier disposal solution.

WHAT IS A TRANSFER STATION?

A refuse transfer station is a centrally situated complex where local refuse collection vehicles can discharge their contents without having to travel unnecessary distances to a disposal site, thus saving on unproductive crew time and cutting down mileage by the individual vehicles. A transfer station is so designed and equipped that it can accept the loads from whole fleets of collection vehicles and transfer these loads by means of large and powerful hydraulic packers into high-volume steel containers or trailers, which may then be hauled to the disposal site. By efficient compaction each container will accept the loads of many individual collection vehicles and introduce the economies offered by bulk haulage.

Whatever means are used for the collection and disposal of refuse, none can make it disappear. Furthermore, as living standards rise, so the volume of refuse increases, as do costs of disposal. New methods are constantly being sought to improve the environment, and reduce pollution hazards. Whatever means are introduced, at some time the waste has to be hauled to a final disposal site, and it is at this point that a refuse transfer station will show the biggest savings of all. A transfer station is essential, therefore, for civil authorities and big industrial complexes where large volumes of waste arise.

THE GROWTH OF TRANSFER STATIONS

Sanitary landfill is, and will continue to be, the predominant method of waste disposal in the UK for the foreseeable future. Even in the Federal Republics of Germany and Switzerland, countries recognized to have the greatest percentage of incineration techniques, landfill still accounts for 75% and 65% respectively of domestic and commercial waste. In the UK this figure is much higher. But increasingly in the future, landfill will occur at more remote and larger mineral extraction sites, and these will have to be served by transfer stations which will enable the waste to be efficiently compacted, containered, and moved by road, rail, or canal links to the new sites. At present in England only about 14% of gross revenue expenditure on refuse disposal is devoted to bulk transfer systems, but this will increase as local landfills are exhausted and new and more remote sites come into their own.

HISTORICAL DEVELOPMENT OF BULK TRANSFER

My company first entered the bulk waste handling market in 1959 when it negotiated licences with Dempster Brothers of the USA to manufacture their demountable body systems. Manufacturing the Dempster range of vehicle systems, the company soon became an established name in bulk waste handling, not only in the United Kingdom but also in many other parts of Europe. It was during the 1960s that bulk waste handling first became an established part of refuse disposal. Bulk waste-handling stations, where large open-topped container vehicles were gravity-fed with refuse from waste collection vehicles, were set up. The waste was not compacted, except in

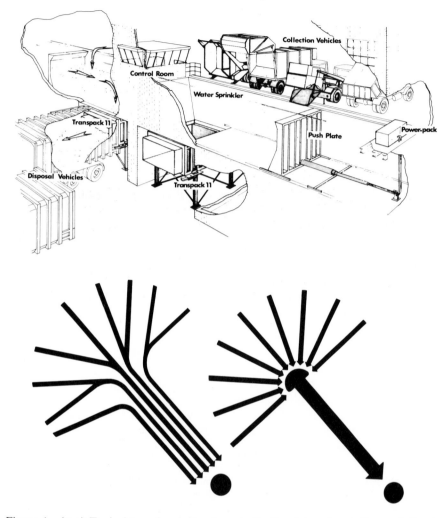

Figure 1   (top) Typical transfer station layout; (bottom) transfer station and direct haul

some cases by simple tamping methods, and once a container was full it was sheeted and taken away. Although a simple and basically effective idea, the economic advantages of bulk handling could not be maximized, as the loose refuse meant that optimum payloads could not be achieved, and the sheeting of refuse was only partly effective in preventing windblown refuse.

Powell Duffryn Engineering Limited became known in the bulk waste-handling market by producing the Dempster Dumpmaster range of compac-

Figure 2   65 yd³ refuse compaction trailer

Figure 3   Fishers Green transfer station

Where no refuse packer is installed at a transfer station, Powell Duffryn offer a range of bulk haulage vehicles which may be fed through the top by hopper, conveyor or grab. These vehicles have an hydraulically operated ejector plate which will effectively compact the refuse loaded into them with the rear door locked; at the disposal site, after unlocking the rear door, the ejector plate is used to discharge the load.

This range of bulk transporters consists of the Dumpmaster RCT65 trailer with a swept volume of 65 cubic yards. It is suitable for coupling to either a 2 or 3 axle tractor unit and the ejector plate is operated by a multi-stage hydraulic cylinder with power derived from a diesel driven pump mounted at its forward end. The range also includes a 52 cubic yard and 42 cubic yard capacity bulk transport vehicle suitable for mounting on to rigid chassis of 30 tons G.V.W. and 26 tons G.V.W. respectively.

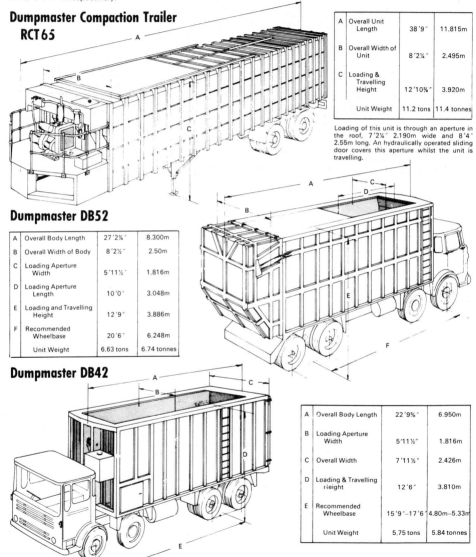

Dumpmaster Compaction Trailer RCT65

A	Overall Unit Length	38'9"	11.815m
B	Overall Width of Unit	8'2¼"	2.495m
C	Loading & Travelling Height	12'10⅜"	3.920m
	Unit Weight	11.2 tons	11.4 tonnes

Loading of this unit is through an aperture in the roof, 7'2¼" 2.190m wide and 8'4" 2.55m long. An hydraulically operated sliding door covers this aperture whilst the unit is travelling.

Dumpmaster DB52

A	Overall Body Length	27'2¾"	8.300m
B	Overall Width of Body	8'2½"	2.50m
C	Loading Aperture Width	5'11½"	1.816m
D	Loading Aperture Length	10'0"	3.048m
E	Loading and Travelling Height	12'9"	3.886m
F	Recommended Wheelbase	20'6"	6.248m
	Unit Weight	6.63 tons	6.74 tonnes

Dumpmaster DB42

A	Overall Body Length	22'9⅝"	6.950m
B	Loading Aperture Width	5'11½"	1.816m
C	Overall Width	7'11½"	2.426m
D	Loading & Travelling Height	12'6"	3.810m
E	Recommended Wheelbase	15'9"–17'6"	4.80m–5.33m
	Unit Weight	5.75 tons	5.84 tonnes

Figure 4   Typical bulk transport self-compaction vehicles—gravity fed

Figure 5   W. Hemmings Ltd transfer station, Bristol

tion trailers and bodies. These bodies had their own built-in compacting ejector blade, and were able to achieve denser loads by compression and thus maximize their carrying capacities. It was the 50 m^3 capacity compaction trailer and the 32 m^3 and 40 m^3 rigid vehicle units which established the company in the bulk waste-handling market and paved the way for the next generation of bulk waste-handling through the development of the stationary compactor-equipped transfer station.

The first examples of refuse compactors were marketed in Europe in the late 1960s when small machines capable of approximately 20 tonnes thrust were introduced into the refuse collection field at supermarkets, hotels, airports, factory estates, and similar locations, where high volumes of waste were generated and transport costs required a more efficient method of achieving a higher payload. These small compaction machines were installed at the most convenient points, e.g. the service entrance of a supermarket or hotel, and closed containers were coupled to the machines. At suitable intervals, or when filled with compacted refuse, the containers were uncoupled and transported to the final disposal site by a suitable container-handling vehicle. Apart from the obvious advantages of higher payload, this method was far more hygienic than the use of open-topped skips, which attracted rodents and were a constant fire hazard.

Figure 6   GLC Brentford transfer station

Many hundreds of these small compactors were installed, and larger models were developed which were capable of handling higher throughput, packing into containers carried on 24 and 30 ton GVW vehicles. This led to the development of the very large transfer station compactor capable of exerting packing forces of up to 60 tons, and Powell Duffryn Engineering introduced the now universally known Transpack 7 and Transpack 11 transfer station compactors, which operated in conjunction with the Dinosaur demountable container-handling vehicle using closed compaction containers of up to 33 m^3 capacity of the 50 m^3 capacity semi-trailer. This allowed greater payloads to be achieved, as the heavy hydraulic compactor units along with the appropriate electrical control equipment could become a fixed installation at the transfer station, and it was no longer necessary for heavy built-in compaction devices to be carried as part of the payload. In 1970 Geesink BV, a Dutch subsidiary of Powell Duffryn Engineering, equipped the first compaction transfer stations in Holland at Almelo, Enschede, and Hengelo. Each station was equipped with two UK-built Transpack 11 transfer station packers, and 50 m^3 compaction semi-trailers were used for bulk transfer to the disposal site. These trailer units were able to operate at the gross train weight of 40 tonnes which is permitted in the Netherlands, thereby achieving very high payloads in the order of 22 tonnes.

Figure 7   Transpack 11 and 65 yd³ transfer trailer

		Loading		
	Length (m)	Width (m)	Height (m)	Capacity (m³/h)
TSP 11	8.6	3.08	3.35	511
TSP 7	6.3	2.20	1.80	286
Unipak 4/35	4.8	1.60	1.30	220
Unipak 2/25	3.04	1.60	1.30	187

In 1971 my company designed and built the first transfer station in the UK, utilizing these high-volume compaction principles. The station was completely designed, built, and commissioned by Powell Duffryn Engineering for Hales Containers Limited at Fishers Green in North London, and it included the Transpack 11 compactor and the 50 m³ capacity closed semi-trailer compaction bodies. This first UK installation is still operating very successfully today, as are the early Dutch installations, and all have been extended to provide greater capacity. The following year further installations were opened at Burnley, Birmingham, Greenford, Twyford, and Barnes in West London, and by 1974 the company had either built or supplied and installed the compaction equipment and vehicle fleets for 4 transfer stations in the UK and Europe. This wealth of experience, far greater than any other company in Europe, led to the nomination of Powell Duffryn Engineering as equipment sub-contractors for the first of the Greater London Council's programme of transfer stations at Newham in North London. This GLC station was, at the time of its opening in 1976, the largest in Europe, and all the mechanical and electrical equipment, including five Transpack 11 compactors, fully air-conditioned control cabins, and the dust-extraction equipment, was supplied, installed and commissioned by my company. This was quickly followed by the GLC's transfer station at Brentford in West London, which is a combined road/rail installation, where refuse is collected and delivered to Brentford by normal collection vehicles, compacted into compaction containers designed to ISO standard dimensions, and transported in 800 tonne train loads to the final disposal site in Oxfordshire. The station is the first of its type in the world, as well as being the largest, with ten Transpack 11 compactors and all the associated electrical equipment, air-conditioning and dust-extraction equipment, together with full service facilities including computerized weighing procedures. Powell Duffryn Engineering were again nominated by the GLC as sub-contractors for the supply and installation of the mechanical, electrical, air-conditioned control, and dust extraction equipment for this, the most up-to-date transfer station in the world.

During the period since 1970 my company has continued to supply and install equipment for, and has in many cases been responsible for the complete design of, transfer stations throughout Europe, including Oslo, Copenhagen, Utrecht, Osnabruck, Slough, and Kilmarnock and, most recently, for one of the largest privately owned transfer stations in Europe for W. Hemmings of Bristol. Some 70 stations have been equipped by the Company to date.

MODERN TRANSFER STATION COMPACTORS

Powell Duffryn Engineering produce a range of waste compactors which includes the two transfer station packers, Transpack 7 and Transpack 11, and the Unipak 4/35 which is used at some smaller stations.

Both the Transpack 7 and the Transpack 11 are large units designed for large quantities of waste, being able to compact 286 and 511 m^3 per hour respectively, and they are designed for transfer station use. However, the smaller Unipak 4/35 is being used increasingly on industrial estates as a small transfer station serving a very localized need, and although it is only about half the size of the Transpack 7 it is still able to compress 220 m^3 of refuse per hour.

Table 1 Estimated transfer station operational costs (£)—a simple bulk
transfer station of 50 tonnes/h single shift

	Distance to landfill site		
	16 km	32 km	48 km
Capital charges	0.91	0.91	0.91
Operating costs	0.62	0.62	0.62
Bulk transport	1.28	2.18	2.87
Landfill	1.78	1.78	1.78
Total cost/tonne	4.59	5.49	6.18

Notes:

Tonnage processed	90,000 tonnes/year
Capital	13% per annum
Payload/vehicle	15 tonnes
Vehicle running costs	25p/mile
Vehicle fixed costs	£283/week
Vehicle capital	5 years @ 13%

MODERN VEHICLES AND CONTAINERS

In addition to normal refuse collection vehicles for primary collection, an
extensive range of vehicles and container combinations is now available to the
waste disposal officer or private contractor. Some examples are as follows:

(a) The Meiller skip-handling vehicle unit. This can be used as a collection
unit feeding waste to the transfer station from small compactor installations. It can also handle self-compacting containers.
(b) The Dumpster container handling vehicle. This fulfils a similar role to the
skip-handling vehicle whilst providing discharge without tipping, via
drop-bottom containers. It can also be fitted to short wheelbase chassis
and is consequently highly manoeuvrable with a low headroom requirement for us in confined areas.
(c) The Dumpmaster front loading self-compacting vehicle. This unit has a
useful and valuable role as a primary collection unit in areas where a large
number of four or 4.5 m^3 capacity containers can be located, such as
airports and trading estates.
(d) The Rolonof and Dinosaur bulk container-handling vehicle units. This
type of large demountable container unit is the most popular in current
use in the UK as a means of transporting large volumes of compacted
refuse from transfer stations to the landfill site. Both are vehicles of great

versatility with the capacity to handle many types and sizes of container, and they are operated by one man, all operations being completely controlled from within the cab.

(e) The 50 m^3 and 57 m^3 capacity transfer trailer. This is a high-volume self-trailer used extensively in the USA and continental Europe. The gross train weight limit of 32 tonnes in the UK does not allow its full capacity to be utilized when used in conjunction with powerful compactors, so consequently this vehicle is not so extensively used in the United Kingdom.

THE ECONOMIC ARGUMENTS

No-one will dispute that there is a valid economic balance between the capital and running costs of a transfer station and the extension of refuse collection routes to reach the more remote landfill sites. The figures must be calculated carefully, and there will be many cases when the transfer station will be ruled out in favour of extended refuse collection routes and a re-specification of vehicle sizes. Possible changes in EEC regulations on gross vehicle weight will also affect the issue. The cardinal point is that these assessments require careful thought. Many papers have been written on the subject, and the analytical work of the Local Government Operational Research Unit is

Table 2   Some examples of the cost of moving waste by road to remote landfill sites or plants

Radial Distance to landfill km	Average speed of journey (m.p.h.)	Tonnes per vehicle per week	Cost/vehicle per week (£)	Cost per tonne carried (£)	Cost/vehicle journey per km (£)
16	16	300	383	1.28	8.0
32	20	188	408	2.18	6.75
48	24	150	433	2.87	5.62
64	32	150	483	3.22	5.06
80	40	150	533	3.55	4.44
96	48	150	583	3.89	4.06
112	48	129	583	4.52	4.06
128	48	113	583	5.16	4.06
144	48	100	583	5.83	4.06
160	48	90	583	6.48	4.06

(Courtesy of Environmental Resources Ltd.)

Notes:

Payload per vehicle	15 tonnes
Vehicle running costs	25p/mile
Vehicle fixed costs	£283/week
Capital costs	5 years @ 13%

available for use by local authorities when they are considering the planning of collection and disposal systems.

BULK TRANSFER AND WASTE RECLAMATION PROCESSES

Bulk transfer systems have a part to play in the steps now being taken to develop waste reclamation processes. Several studies have shown that to make economic sense many of these processes, whether WDF or fibre recovery, need to be positioned at the centres of gravity of the areas they serve, and in effect act as reclamation transfer stations. The reclamation process removes the useful materials from the waste, and the compaction equipment and plant effect the economic removal of the useless residues to landfill. For operational safety, and to secure the protection of the collection service, most new installations have bypass compactors at the front end, so that breakdowns in the reclamation system do not jeopardize day-to-day operations. Other studies indicate that waste reclamation processes are extremely sensitive in their costings to the tonnages processed. Given the time when the technological and marketing problems have been solved, it may be that satellite transfer stations can be installed to arrange economic concentrations of waste at central processing plants.

CONCLUSION

The growth of bulk handling of waste is inextricably linked to the changing style of waste disposal, the higher aesthetic standards expected by the public, and the demise of the local landfill site. Increasingly, waste will be disposed of at larger and more remote landfills backfilling mineral extraction, and access to these must be by an intermediate handling system compacting and conveying the waste from the urban centre to these remote sites. Once the refuse has been compacted and bulk containered, the waste disposal manager and the cleansing officer have at their disposal a tool of immense flexibility, allowing them to reach sites and plants far out of sensible reach of refuse collection vehicles. In this chapter I have tried to sketch out an industrialist's view of the changing problem of waste collection and disposal, and to illustrate some of the new vehicles and other pieces of equipment capable of facing these new challenges.

ACKNOWLEDGEMENTS

The author wishes to thank the Board of Powell Duffryn Limited for permission to publish this chapter and makes it clear that the opinions expressed are those of the author and do not necessarily represent the views of Powell Duffryn Limited or its subsidiary companies.

REFERENCES

Department of the Environment, *Working Party on Refuse Disposal,* 1978.
J. R. Holmes, *The Challenge Facing Waste Derived Fuels.* 1978.

Practical Waste Management
Edited by J.R. Holmes
©1983 John Wiley & Sons Ltd

20

High-density refuse baling as a solution to solid waste disposal

Alan Sowerby
Murphy Solid Waste Systems Limited

ABSTRACT

The compression of refuse into high-density bales so as to effect volume economy and allow a 'building block' system of landfill is increasing in popularity in the United States and Europe. This chapter describes the work of one of the principal manufacturers in this field. Practical experience, capital and revenue costs, the behaviour of refuse bales *in situ* are considered. Examples are given of some of the latest installations in the United Kingdom.

INTRODUCTION

A considerable amount of research has been undertaken in the last decade aimed at developing new methods of refuse treatment with the objectives of reducing capital and running costs, avoiding air pollution, lessening the demands for energy, and increasing the possibilities for recycling. A number of new methods have been developed and offered as commercially viable systems, but the majority have not yet advanced to fully commercial operation or have encountered serious problems when changing from pilot to full-scale operation.

A method which has advanced to full commercial operation and has been running for sufficient time to enable an objective assessment to be made is high-density baling. This appears to offer the best possible solution within the limits of the objectives previously mentioned. High-density baling is basically the processing of solid waste into large, stable cuboids of high density, e.g. 50–60 lb/ft^3. The bales must be sufficiently strong and stable to withstand handling at the plant and transport to the tipping site. It is sensible to produce bales of a suitable size and weight to make the transporting and tipping as economical as possible.

The chapter will explain the operation of a high-density baling plant, evaluate the principles and advantages, and describe the plants at Glasgow, Leeds, Bradford, East Lothian, and Hull. It should be noted that the plants described use Harris economy baling presses as manufactured by American Hoist (Europe) Ltd, Bridgend, South Wales.

PLANT OPERATION

The plants at Glasgow, Leeds, and Bradford use the Harris SWC 2528 baling press which produces self-sustaining bales. The operation at these plants will be described first. The refuse is brought into the plant by collection vehicles and discharged onto the tipping floor. It is then transferred by mechanical shovels onto a horizontal conveyor below floor level, then onto an inclined conveyor which discharges into the loading hopper of the baling press. The loading hopper is mounted on load cells, and when a predetermined load is reached, usually 2,800–3,000 lb the conveyors automatically stop and the baling cycle is initiated. The baler operates on 40 cycles/hour; therefore, approximately 50 tons/hour can be processed.

When the amount of refuse in the loading hopper reaches the predetermined weight the refuse is pushed into the press charging box. The press is a three-ram hydraulic press and the final ram exerts a force of 1,510 tons. The completed bale is ejected from the pressbox and pushed along a diverter

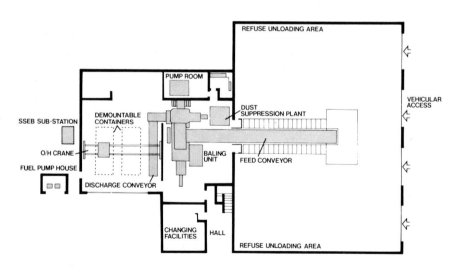

Figure 1   Typical layout of high-density baling plant. (Courtesy of Murphy Solid Waste Systems Ltd)

table. When a second bale is made and pushed next to it, the pair of bales are then pushed onto the transport vehicle by a loader ram.

Each transport vehicle will carry seven pairs of bales to comply with the maximum allowable payload. Each bale is approximately 1 metre square, about 1½ metres long and weighs between 1¼ and $1^1/_3$ tons.

The plants at East Lothian and Hull use the Harris HRB-SWC baling press and the bales are wire-tied. The press is fed in a similar manner to that previously described except the refuse is not weighed before feeding into the hopper. The press is a single-ram hydraulic press and produces a force of 214 tons. Generally three strokes of the ram are required to make a complete bale. The completed bale is ejected from the pressbox and automatically wire-tied at predetermined intervals depending on the stability of the bale. The bale is then either lifted by gravity clamp as at East Lothian, and placed onto the trailer, or pushed along a diverter table as at Humberside, where it is weighed, before being pushed onto the transport vehicle by a loader ram. Each bale is approximately 1½ metres long × 0.8 metre high × 1.1 metres wide (Humberside) or 1.3 metres wide (East Lothian) and weighs about $1^1/_3$ tons.

ADVANTAGES OF HIGH-DENSITY BALING

Some of the advantages of high-density baling are:

(1) Low capital costs and low operating cost due to ease of maintenance and efficient utilization of the energy cost. Furthermore, baling presses are not self-destructive and only require a minimum of replacement parts.

(2) There is no atmospheric pollution (i.e. no chimneys, etc.) and the noise levels are retained to within environmental requirements.

(3) Virtually any kind of solid waste can be processed in the plant; in fact this is one of the great virtues of high-density baling, as practically at no time need the press be stopped due to some object being in the refuse which could jam or damage the machine.

(4) Reduced transport costs, as a maximum payload of bales can be carried per vehicle.

(5) Extends the tipping site's useful life by increasing the in-place density by 60% or more.

(6) Neat and tidy appearance of tip with reduced negative environmental impact, including negligible settlement, reduced litter, dust, odour vectors, fires, traffic, top cover, noise, pollution, and safety hazards problems.

(7) A stable, final landfill area can be provided almost immediately after the bale tip is completed.

The results indicate that bale tipping sites are more aesthetic and environmentally superior to the traditional tipping sites. Because of this planning permission could be granted for sites which would not be allowed for crude or pulverized tipping.

TIPPING SITE CHARACTERISTICS

A bale tipping site, if properly looked after, has a neat and tidy appearance with no 'windblown' paper or dust, as can be seen from Figure 2. The bale transport trailer can be driven fully loaded, over bales already in position on the site with very little risk of punctures from glass or nails, etc. The bales are unloaded by means of a forklift truck. A trailer of 14 bales can be unloaded and the bales positioned in approximately 12 minutes by one experienced operator. The primary advantage in the bale-filling of refuse is the volume reduction. The effective landfill density can be increased by up to 60% compared with a controlled tipping site. For example Glasgow at their Kenmuir tip have an in-place density with refuse bales of 50 lb/ft^3—S.G. 0.8.

Void space within the landfill is reduced and structural stability is increased. Blowing paper, fires, odours, and vectors are minimized and the bale-fill produces a lower-strength leachate than a controlled tipping site. In addition less cover material is required and less equipment needed to manage the site.

ENVIRONMENTAL EVALUATION

Test cells have been built at Glasgow and in the USA in bale tipping sites and/or as separate units filled with bales. These test cells were monitored for settlement, gas, and leachate production and temperature. The following are some of the conclusions drawn from these tests:

(1) The mixing of organic and inorganic constituents in the mix prior to baling results in a more homogeneous refuse mixture, obliterating the high concentration of organic wastes in spots or cells within the landfill, thus reducing concentration of bacterial reaction and pollution potential.

(2) In the course of high-density compression, excessive amounts of both air and water are reduced from the waste mix, thereby limiting bacterial life and biochemical reaction.

(3) The reduction of the permeability of the waste mixture due to the bales' high density limits percolation through the refuse and consequently also affects biochemical reaction time and leachate generation.

(4) The high density of the waste mixture induces excellent insulation properties of the bales, thereby controlling heat generation and heat dissipation which, in turn, results in a cyclical biochemical 'self-cleaning' process.

Figure 2   Aerial view of high-density refuse bales in place on a sanitary landfill site. (Courtesy of Murphy Solid Waste Systems Ltd)

(5) In placing bales in a bale tipping site rather than crude or pulverized refuse in a conventional tip, greater control over some of the factors affecting gas production—such as placement and cover, infiltration, temperature, and aeration—can be exercised, thereby reducing the pollution potential.

It can be concluded in the light of research and studies conducted that the environmental control measures employed are adequate with respect to water pollution and gas control for tipping sites with high-density heterogeneous solid waste mixtures.

The US Environmental Protection Agency conducted some independent tests and the results of these show that methane production is one-tenth of the amount found in conventional tipping sites, leachage generation is practically non-existent, and settlement is zero.

Glasgow Cleansing Department, in conjunction with the Department of the Environment, have built a test cell to monitor the bales produced from their high-density baling plant at Polmadie. Leachate and the related pollution aspects are currently being studied under the direction of a Department of the Environment research programme. The results of these studies will be presented at a later date.

POLMADIE BALING PLANT, GLASGOW

The Polmadie high-density baling plant, the first in Europe, started operating in the summer of 1976. The design throughput is 300 tons per shift and to the end of 1980 it has processed 360,000 tons of refuse. The plant was built at the Polmadie Disposal Works, utilizing the part of the building where two units of the existing incinerator had previously been completely destroyed by fire. The estimated cost of the plant, which in fact is close to the final cost is:

Reinstatement of fire-damaged roof	£75,000
Civil costs	£560,000
Mechanical and electrical plant costs	£1,065,000
Total	£1,700,000

The consultants for the project were Merz and McLellan.

General description of the plant

The refuse collection vehicles are weighed and checked and enter the reception loft where there are eight tipping openings provided with safety barriers and roller shutter doors to seal off the refuse storage area. The collection vehicles tip onto the storage compound, which has an area of 6,800 ft^2 and is approximately 12 ft below the reception loft. The refuse compound

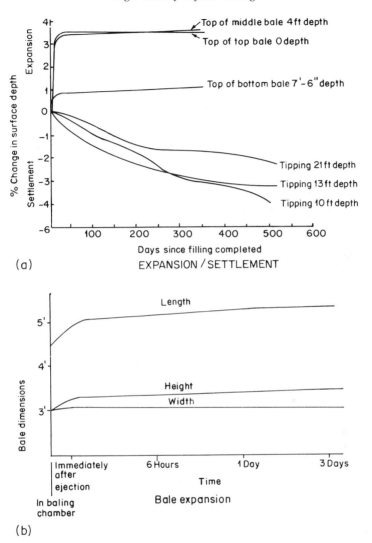

Figure 3 The behaviour of high-density bales

floor is capable of holding 300 tons of refuse. A Volvo BM846 Bucket (4 yd³) loader mixes and transfers the refuse into the feed hopper of No. 1 conveyor. The No. 1 or primary conveyor, which is an apron plate conveyor, is 8 ft wide and inclined at 23°. Variable-speed drive is incorporated giving infinitely variable speeds from 7 to 20 ft/min. The delivery of refuse from the primary

Table 1

	Annual expenditure (£)	Per ton (£)
Baler		
Wages—disposal (including shovel)	49,026.16	
Wages—maintenance	48,558.64	
Wages—miscellaneous	1,487.09	
Superannuation	2,789.89	
National Insurance	8,877.13	
Local rates	82,577.42	
Electricity	70,113.11	
Water supply	2,013.93	
Plant and property repairs	17,283.47	
Machine rental and maintenance	1,370.89	
Vehicle hires	1,894.00	
Telephones	456.50	
Security and fire	11,939.47	
Costs—stores and workshops	21,982.34	
Subtotal	320,370.04	3.98
Transport		
Wages	45,577.42	
Repairs, including tyres	44,289.57	
Fuel and oil	12,354.49	
Licences, insurance, garaging	6,922.38	
General Expenses	4,557.74	
Subtotal	113,701.60	1.41
Disposal sites		
Wages, rates, materials, etc.	71,507.14	0.89
Capital charges		
Estimated expenditure	193,000.00	2.40
Total 1977–78 costs =	698,578.78	8.68

conveyor passes over a Unimag 6 ft diameter × 10 ft long drum-type magnetic separator, which extracts most of the tins and ferrous metals and deposits these into the metal compound which is situated below the refuse compound at ground level.

The tins, etc. are transferred from the compound into a Lindermann metal baling press by a Unimag travelling circular lifting magnet, which has a capacity of approximately 350 lb of domestic cans. The cycle time for the lifting magnet is approximately 2 minutes, thereby providing a transfer capability of about 4 tons/hour. The refuse from the primary conveyor, after passing over the magnet drum, falls on to the No. 2 or secondary conveyor which is inclined at 30° and has a single speed of 30 ft/min. The refuse passes from the secondary conveyor into the loading hopper of the Harris economy high-density baling press, the operation of which was described earlier in the chapter. To control dust and fumes arising in the refuse compound and various refuse transfer points, two Rotoclone Hydrostatic Precipitators of American Air Filter Inc., are installed, each with a capacity of 48,000 ft^3/min.

A comprehensive fire detection and fighting scheme was installed by AFA Minerva Limited; the fire fighting is by a deluge system as this was thought to be the most effective method to control a fire on the refuse compound floor.

The baled refuse was at first taken to the Kenmuir tipping site where the Department of the Environment test cell is situated. This site was filled in 1977 and Glasgow have since used the Wilderness tip where 276,000 tons of baled refuse have been placed up to end of 1980. The operational costs for year ending 31 March 1978 are shown in Table 1.

Glasgow officials' comment[1] is that the plant is comparatively straightforward to operate and maintain, and since commissioning has achieved a good plant availability record.

The bales that are produced can be transported and handled at the landfill site to give a satisfactory in-place density. Environmental conditions at the landfill are satisfactory, an important factor being the virtual elimination of windblown material.

The only major problem encountered at the plant during the first 5 years of operation has been the pressbox. During a routine examination of the pressbox in September 1977 cracks were detected. They were thoroughly investigated by American Hoist/Harris, and the results of detailed structural analysis showed that the pressbox was under-designed. After careful consideration American Hoist decided to replace the pressbox with a new redesigned box. This decision was taken in preference to carrying out an on-site modification exercise to the existing pressbox. The new pressbox has been designed for a minimum life of two million cycles.

LEEDS BALING PLANT

The City of Leeds have been interested in high-density baling for a number of years (in fact before Glasgow) but due to a number of reasons—one of which was the County reorganization in April 1974—they were not able to go out for tenders until late 1974. The disposal of wastes is controlled by the West

Yorkshire Metropolitan County Council's Solid Waste Disposal Unit under the Executive Director of Engineering. They awarded the contract in March 1975 and it was completed by 31 December 1976. The plant started operation in the spring of 1977 and to the end of 1980 had processed 390,000 tons of refuse an average of 511 tons every working day.

Like Glasgow, West Yorkshire were interested in high-density baling because of the low capital and operating costs, the high plant availability, the simple and efficient plant operation, and the environmentally acceptable conditions. The plant was built on a prepared site of 5 acres off Kirkstall Road, Leeds. The plant building is 71 m long × 48 m wide × 18 m high at the baling press end and 9 m high over the refuse reception area. The final cost is approx. £2,650,000 which is very close to the original estimate. The civils and buildings, and mechanical and electrical, are each approximately half of the total cost. The land cost of £125,000 is included in the final cost.

General description of the plant

The refuse collection vehicles are weighed and checked on entering the site and then proceed into the refuse compound through one of the seven roller shutter doors in the building. The particular door through which the vehicle is to enter is controlled from the control room, or from the compound floor. The collection vehicles tip onto the refuse compound floor, which has an area of 1,800 m^2, and is capable of holding 600 tonnes of refuse. One or two 5 yd^3 loaders mix and transfer the refuse on to the No. 1 or primary conveyor which is horizontal and positioned at 3.5 ft below floor level. This conveyor operates at a fixed speed of 15 ft/min. The primary conveyor transfers the refuse directly on to the No. 2 or secondary conveyor. This conveyor is 95 ft long and is inclined at 30°. It has a variable speed drive giving infinitely variable speeds from 22 to 37 ft/min.

A dust control plant similar to the system used at Glasgow is installed.

Unlike other high-density baling plants based on the American solid waste systems concept, the refuse on leaving the secondary conveyor falls on to a steel chute which feeds the loading hopper of the high-density baling press. The reason for the chute is that provision has been made in the plant design for rotary drum magnetic separation at some future date. If the drum magnet is installed the chute will be removed and the refuse from the secondary conveyor will pass over the drum magnet and then into the baling press.

The Leeds plant is designed for a throughput of 50 tonnes/hour. To ensure there are no delays in changing trailers, a twin loading system is installed. This system allows two trailers to be positioned side by side in front of the diverter table, and when a trailer is filled with its complement of 14 bales the next bales are automatically placed on the next trailer, while the full trailer is taken away and an empty one placed in position. The baled refuse is taken approximately 6 miles to a new tip site at Middleton Broom which has 2½ million yd^3 of space, giving an anticipated life of 7–10 years.

As the Middleton Broom site can only be used during the day, and the plant will operate on two shifts, i.e. 6 a.m. to 2 p.m. and 2 p.m. to 10 p.m., extra trailers are required to accommodate the bales made during the evening shift. In all there are 25 trailers for the plant, with six cabs and one shunter.

The operational costs, including transport, landfill and capital repayment for year ending 31 March 1979 were approximately £10/tonne. These costs are based on the 1978 throughput of 99,682 tonnes.

As at Glasgow the only major problem encountered has been the pressbox. This has also been replaced with the new 2 million cycle box. During the early operation of the press West Yorks and Glasgow engineers were concerned about the relatively rapid wear of the liners in the pressbox. American Hoist investigated the use of a special hard facing on the liner plates and were successful in producing liner plates which have proved to give a very good life factor, infact the press has processed 145,000 tonnes of refuse since these new liners were fitted and indications are that a further 50,000 tonnes can be processed before the liners need be changed. The liners were also changed at Glasgow and to end-January 1981 approximately 150,000 tonnes of refuse has been processed. They too expect a further 50,000 tonnes before the liners need changing.

A measure of the confidence which West Yorkshire Metropolitan County Council have in high-density baling as a system of waste pretreatment is that following the successful operation of the Leeds plant they decided to build another plant at Bradford.

BRADFORD BALING PLANT

The Bradford high-density baling plant, Figure 4, started operating in the Autumn of 1980. The plant, like the Leeds Baling Plant, was engineered by the West Yorkshire Metropolitan County Council and is of the same design capacity e.g. 600 tonnes of waste per day. The plant was build on a prepared site of approximately 10 acres at Bowling Back Lane, Bradford, of which approximately 21,000 yd^2 are surfaced; the remainder has been landscaped. The plant building is 72 m long × 48 m wide and is on average 10 m high with a central area of 17 m high.

The final costs are:

Land	£0.156 million
Works	£3.591 million
Mobile plant	£0.698 million
Total	£4.445 million

Figure 4   External view of a high-density baling plant. (Courtesy of West Yorkshire Metropolitan County Council)

The description and operation of the plant is very similar to the Leeds plant.

The baled refuse is taken approximately 3½ miles to a new tip site on the outskirts of Bradford City. The vehicle fleet consists of six articulated tractor units and 25 purpose-built trailers. The trailers have hydraulically operated offside hinged bodies which enables the bales to be removed easily at the disposal site and cost £11,400 each (Figure 3).

EAST LOTHIAN BALING PLANT

The East Lothian baling plant at Barbachlaw, Musselburgh, started operating early in 1980 with the official opening on 17 March 1980. The disposal of waste in East Lothian is under the control of Mr J. B. Cunningham, Director of Environmental Health, whose staff and the appointed consultants, National Industrial Fuel Efficiency Services, started in 1975 to study various methods of refuse disposal. After careful consideration they decided to adopt the Harris HRB-SWC type of refuse baler as the unit around which the plant should be designed.

The plant was built on a prepared site adjoining the district sewage works and is to cater for a population of 79,000 from an area of 250 square miles. The throughput of refuse has been increased in stages and the plant now deals with 24,000 tonnes of the annual 28,000–30,000 tonnes of refuse produced in East Lothian. This is well within the plant capability of 30 tonnes/hour. The building (Figure 4), for which special care was taken with the design as the area is within the Edinburgh Green Belt, is 43 m long × 23 m wide and has a refuse reception area of 620 m^2.

The final approximate costs are:

Building, etc.	£290,000
Mechanical and electrical plant, etc.	£222,000
Vehicles	43,000
Total	£555,000

General description of plant

The refuse collection vehicles enter the building through one of the four roller shutter doors and tip on to the reception area floor. A front shovel loader then transfers the waste onto a 1.8 m wide steel belt conveyor which discharges into the hopper of the baling press. The hopper is covered by a dust hood and dust extraction is incorporated in the plant. A compression ram forces the refuse into the baling chamber of the press until a sufficient charge to form a bale is present. This is determined by pressure and generally three strokes of the compression ram are required to form a bale. During the acceptance tests an average bale density of 53 lb/ft^3—S.G. 0.85—was achieved.

The baling press will produce a bale every 130 seconds and on exit from the press the bale is automatically wire-tied at predetermined intervals and then pushed onto the discharge conveyor. Up to six bales are assembled on the discharge conveyor, from where they are lifted by gravity clamp from an overhead crane and placed on the transfer vehicles.

The baled refuse is taken approximately 2 miles to the disposal site which is a rehabilitation area leased from the Scottish Development Agency. The bales are discharged from the transporter vehicle by raising the body and allowing the bales to slide off on to the ground. The bales appear to suffer no damage from this method of unloading. They are stacked in double layers and covered at the end of each day's operation on the top and exposed faces. East Lothian report that the relative ease with which a satisfactory environmental standard is achieved is one of the major factors justifying the baling project.

There are a total of five employees at the baling plant and disposal site. The baling/disposal cost per tonne is presently (March 1981) £8 based on a 24,000 tons/year throughput. This cost will be reduced as more refuse is brought into the plant from other areas.

Initially there were problems with the wire-strapper which were aggravated by the excessive fine ash content in the refuse. The wire-strapper was recently modified and the East Lothian officials now comment that the plant has settled down and is working well.

HUMBERSIDE REFUSE BALING PLANT

The building of the Hull refuse baling plant at Wilmington commenced in June 1980 and is now in operation. The disposal of wastes is controlled by Humberside County Council.

The plant which is dual-stream, will handle up to 130,000 tonnes/year of domestic and bulky household waste arising in the Greater Hull area together with as much baleable commercial and industrial waste as can be attracted to the plant up to an overall waste input of 65 tonnes/hour.

The plant building is a single-multistorey steel portal frame structure 72 m long × 54 m wide × 10 m high to eaves.

The capital cost, based on July 1979 estimated costs, is:

Civils and buildings	£1.5 million
Mechanical and electrical plant	£0.7 million
Movable plant	£0.475 million
Total	£2.675 million

GENERAL DESCRIPTION OF THE PLANT

Waste will arrive at the plant by both collection and private vehicles and will be discharged on to the floor of the reception hall, which has an area of 1,900 m^2 and has been designed to accommodate 1 day's input. A front shovel

Figure 5   Specially designed vehicle to carry high-density bales to the landfill site.
(Courtesy of Murphy Solid Waste Systems Ltd)

loader will then transfer the waste on to either of the two reception conveyors, which are 1.8 m wide × 21.7 m long and are inclined at 40°. The delivery of waste from the conveyor passes over 1.8 m diameter × 3 m long drum-type magnetic separators, which extract most of the tins and ferrous metals and deposit these on to a rubber belt conveyor for feeding into a Harris TGS25 automatic metal baling press. The waste from the conveyors, after passing over the magnetic drum separator, falls into the loading hopper of the Harris HRB-SWC3 refuse baling press, the automatic operation of which was described earlier in the chapter. The bale, when ejected from the pressbox chamber, is automatically wire-tied at predetermined intervals and then moved directly in front of the diverter ram by a moving platen which also weighs the bale and records and prints out the weight.

The bale loading onto the trailers is similar to the operation at Leeds and Bradford, i.e. a twin loading system as described earlier in the chapter. The baled refuse is taken across the city to a landfill site near Hessle, a distance of approximately 8 miles.

The estimated operational costs of the plant, based on 520 tonnes/day (i.e. 135,200 tonnes/year) are £7.50 per tonne.

In addition to the refuse baling, the Wilmington plant has a small incinerator fired by natural gas and capable of burning up to ½ tonne/hour of confidential papers, animal carcasses, and pharmaceutical waste.

CONCLUSION

This chapter sets out to show that high-density baling is a solution to solid waste disposal on environmental, transportation, and cost considerations, and from the number of plants installed in Britain (as well as those world-wide), all of which are operating successfully, the future for refuse baling looks assured.

ACKNOWLEDGEMENT

The author wishes to thank Glasgow Cleansing Department, West Yorkshire Metropolitan County Council Waste Disposal Unit, East Lothian Environmental Health and Solid Wastes Management, and Humberside County Council Waste Disposal Unit for the information and assistance given in the preparation of this paper, and the directors of J. Murphy & Sons Ltd for their support and permission to present the paper.

NOTE

As the subject is based on United States imported equipment to the United Kingdom certain quantities are expressed in Imperial units. Convenient conversions are stated below.

1 lb/ft^3 $= 16.05 \text{ kg/m}^3$
1 lb $= 0.4545 \text{ kg}$
1 ft/min $= 0.3048 \text{ m/min}$
$1 \text{ ft}^3/\text{min}$ $= 0.0283 \text{ m}^3/\text{min}$

REFERENCE

1. Shepherd, I. W. *Solid Waste Disposal by High-Density Baling. An Operational Review*. Glasgow, April 1979.

Practical Waste Management
Edited by J.R. Holmes
©1983 John Wiley & Sons Ltd

21

The case for incineration of municipal wastes

W. A. Clennell, CEng, MIMechE
Director and General Manager, Motherwell Bridge Tacol Limited

ABSTRACT

The case for refuse incineration is ably made by an executive of one of the principal companies in the field. The European experience in energy recovery and power generation is described, together with some similar processes in the United Kingdom. The combustion process, gas cleaning, district heating, and pyrolysis are mentioned. Illustrations show a selection of processes that now operate in this country. While much emphasis is given to sanitary landfill, this chapter reminds us that refuse incineration is still an important solution for the disposal of urban wastes.

INTRODUCTION

Incineration is mainly used as a means of achieving maximum volume reduction of municipal wastes. It is usually a more costly method of disposal than controlled tipping, although not necessarily so, and for this reason is not generally adopted when the Authority has access to long-term tipping facilities. Incineration was becoming increasingly used as a method of disposal by the urban authorities in Britain, up to the time of Local Government reorganization in 1974, because of shortage of tipping facilities within their boundaries; but, in many cases, the reorganization gave access to land in rural areas and this, together with tightness of money and high rates of interest, has resulted in few incinerators being ordered at the present time.

Prior to the middle 1960s, almost all incinerators constructed in Britain were of the multi-cell batch type, facilities being provided within the plants to screen out surplus fines, and to salvage ferrous metals. These installations tended to be labour-intensive, and from 1966 were superseded by incinerators with continuously moving grates. The reducing amount of fines in the refuse allowed these units to be operated without any pre-screening of the refuse and the reduced labour requirements allowed plants to be operated on a

Table 1

Contract	Year	Incinerator contractor	No. of furnaces	Furnace size (tonne/h)	Special features
Middleton	1966	MB	1	8.1	Cyclone gas cleaning
Sutton	1966	RHF	1	10.15	
Dudley	1966	JT	2	6.3	
York	1967	RHF	1	8.25	
GLC—Edmonton	1967	MB	5	14.2	Heat recovery with power generation
Derby	1967	IC	2	7.6	Cyclone gas cleaning
Birmingham—Perry Barr	1967	HW	2	12.2	
Lichfield	1967	SH	1	5.1	
Glasgow—Dawsholm	1968	MB	2	12.2	
Bristol	1968	MB	2	15.2	
Exeter	1968	HW	1	9.5	
Bolton	1968	RHF	1	16.25	
Basingstoke	1968	RHF	1	9.1	
Edinburgh	1969	RHF	2	12.7	
Tynemouth	1969	RHF	2	10.15	
Sunderland	1969	IC	2	10.15	
Gateshead	1969	IC	2	10.15	
South Shields	1969	IC	2	10.15	
Blackburn	1969	RHF	1	11.5	
Mansfield	1969	B&S	1	5.6	District heating scheme
Renfrew	1970	MB	2	8.1	
Coventry	1970	HW	3	12.2	Heat recovered for sale to Chryslers
Nottingham	1970	HW	2	11.8	District heating scheme
Wolverhampton	1970	MB	2	10.15	
Upton	1970	B&S	2	2.0	Cyclone gas cleaning
Winchester	1971	RHF	1	9.1	
Altrincham	1971	RHF	2	4.6	Sewage sludge co-incineration (not now required)
Salford	1971	RHF	2	6.6	
Folkestone	1971	CJB	1	4.9	
Blaby	1971	RHF	2	10.15	
Rhonnda	1971	IC	1	9.1	
Swindon	1971	CJB	1	12.2	Heat recovery
Havant	1971	JT	1	14.2	Sewage sludge co-incineration
Otley and Ilkley	1971	B&S	1	4.1	Heat recovery
Rochdale	1972	RHF	1	8.1	
Huddersfield	1972	CJB	2	6.0	
Dawley	1972	RHF	1	10.15	
New Forest	1972	MB	1	11.2	
Teesside	1972	RHF	2	16.0	Sewage sludge co-incineration
Sheffield	1973	B&W	2	10.15	District heating scheme
Portsmouth	1973	IC	2	10.15	
Belfast	1973	RHF	1	4.1	
Stoke-on-Trent	1973	RHF	2	10.15	
Birmingham—Tyseley	1974	RHF	2	15.0	
Birkenhead	1974	RHF	2	14.0	
Dundee	1975	RHF	2	7.0	

two-shift or three-shift per day basis. This change to incinerators with mechanical grates followed developments on the Continent. Continental practice, however, favoured recovery of waste heat from the incinerators, whereas in Britain energy recovery was the exception rather than the rule.

A list of British continuous operating type incinerators is given in Table 1.[1]

REFUSE COMBUSTION

From the combustion point of view, municipal solid wastes can be considered to be a mixture of ash, moisture, and combustible materials, mainly cellulosic in type but with a certain amount of plastic. A typical average ultimate analysis is as shown in Table 2.[1]

Table 2   Ultimate analysis of sample of refuse (% by weight)

Carbon	21.5
Hydrogen	3.0
Oxygen	16.9
Nitrogen	0.5
Sulphur	0.2
Water	31.0
Ash and inerts	26.9
	100.0

Gross CV as fired: 9,000 kJ/kg

To enable the best possible design for an incinerator to be produced, the designer needs to know as much as possible about the quantity and quality of the refuse to be handled.

The basic design parameter of a municipal refuse incinerator is the 'maximum furnace heat release'. This is derived by combining the maximum calorific value of the refuse which the incinerator is expected to receive, and the specified incineration rate of refuse having that maximum calorific value. Having ascertained the condition of 'maximum furnace heat release' it is then necessary to produce the incinerator thermodynamic design based upon that figure, and this design work is normally accomplished by the use of a computer program. However, for this to be undertaken, specific input data are required. This leads to a need for the following information, which also calls for a physical description of the refuse; a further necessary design tool.

Refuse analysis specification

The nature and composition of municipal solids wastes varies continuously, according to climatic conditions, seasons of the year, and location. To ensure that adequate provision has been made in the design, it is necessary to analyse a number of seasonal samples. The methods by which these analyses can be undertaken are adequately described in the monograph prepared by A. Higginson on behalf of the Institution of Solids Wastes Management.[3]

By common practice, and because of the fixed grate size and other physical dimensions of the furnace, it is normal to specify a fixed refuse throughput for a range of calorific values. However, the furnace will have a capability of burning in excess of the specified throughput at particular values of calorific value between the two specified extremes.

The maximum stipulated gross calorific value of refuse to be consumed by the incinerator can be shown illustratively along the line A1/A2 in Figure 1. This is a line of constant combustible content and of varying inerts and moisture content. Hence, the 'typical' ultimate analysis of Table 2 may be adjusted to show the anticipated ultimate analysis of refuse of the stipulated gross calorific value by inserting into the 'typical' analysis the appropriate refuse percentage inerts and percentage water contents pertinent to the extremes of the line A1/A2.

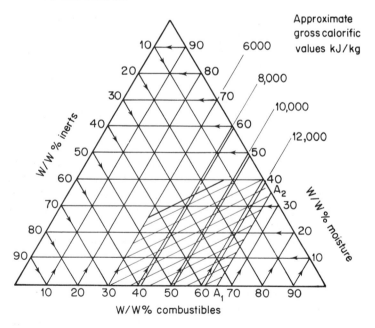

Figure 1   Typical incinerator operating envelope. (The area shown shaded is for illustrative purposes only)

If it is anticipated that industrial waste will be burnt in the incinerator, it is necessary to know the composition and throughput, and the effect of this industrial waste on the refuse composition of Table 2 should be recognized. The incinerator operational area shown in Figure 1 can then be adjusted accordingly.

Bulk density of refuse

The design of the refuse-handling equipment for the incineration plant equipment requires that the bulk density of the refuse be known and this is usually determined at the time of sampling the refuse to ascertain the typical analysis. The average and the maximum and minimum densities should be established.

Throughput

The designer needs to know the philosophy of the plant use and availability as well as the required throughput and variations in throughput.

THE COMBUSTION PROCESS

In general terms, municipal refuse has a GCV of about one-third that of coal and a density of about 200 kg/m^3.

For the above reasons, it is not generally incinerated satisfactorily on conventional coal-burning stokers and a series of specially designed incineration grates have been developed. The bed thickness of the refuse at the feed end of the grate may be in excess of 1.5 m, and agitation is necessary to ensure that oxygen can penetrate to all materials and all surfaces. Most incinerator grates are therefore inclined and have some mechanical means of producing this disturbance in the bed of refuse.

In general terms there are three combustion zones. The front part of the grate is regarded as a drying-out section, the incoming refuse being exposed to the radiant heat of the combustion chamber, resulting in evaporation of moisture. This stage is followed by ignition of the substantially dried refuse and combustion of volatiles in the middle part of the grate. In the third and last zone, burning-out of the residue takes place, including combustion of much of the fixed carbon and, as combustion is completed, the residual clinker cools before discharge at the end of the grate. Conventional practice assumes a burning rate of about 300 kg/m^2 of grate area and a total retention time on the grate of about 1½ hours.

Due to the heterogeneous nature of the refuse feed on to the grate, it is necessary to supply excess combustion air, mainly as primary air through the grate. This excess air may vary from 60% minimum to more than 100%, with the object of obtaining a mixed gas temperature of more than 800 °C (the odour limit) and below 1,100 °C.

GAS COOLING AND CLEANING

The temperature of gases leaving the combustion chamber of an incinerator (usually about 1,000 °C) is much too high for direct discharge to conventional gas-cleaning equipment. Some form of cooling is therefore necessary and this is likely to depend upon:

(1) the size of installation;
(2) the cost of energy;
(3) the requirements of Clean Air Legislation.

Present Clean Air Legislation is not specific in relation to discharge of particulates from incinerators; but best conventional practice leads to use of electrostatic precipitators (ESPs) to achieve a dust burden to the chimney not exceeding about 200 mg/m^3. The maximum temperature tolerated by ESPs is about 300 °C; however, and in the majority of cases in the UK, cooling of the combustion gases to this temperature takes place in vertical spray towers in which water is injected into the gas stream in a finely atomized state. About 3 kg of water is required to cool the gases for every 1 kg of refuse incinerated. The cost of water, and of the power required for atomization, justifies consideration of incinerator gas cooling by integral water-tube boilers. This method of cooling has the following advantages.

(1) replacement by water-tubes of most or all of the combustion chamber refractory lining, which is usually expensive to maintain;
(2) minimum gas quantity passing to the gas cleaning equipment, induced draught fans, and chimney;
(3) opportunity for energy recovery.

ENERGY RECOVERY

As pointed out earlier, the refuse collected and delivered to an installation by a municipality has a calorific value of about one-third that of coal. There is an obvious potential benefit in exploiting this energy in an incineration plant receiving several hundred tons of refuse per day. In Britain, this energy recovery opportunity has generally been neglected and few incinerator installations undertake significant energy recovery. On the Continent, however, energy recovery from incineration is almost always undertaken, as Table 3 shows.

Table 3 Energy recovery from refuse in Western Europe[1]

Plant type	Number of installations	Total refuse burning capacity (tonnes/hour)
Electric power generation	22	786
District heating	78	912
Combined power generation and dis-trict heating	15	465
Other types	10	391

Condensing power station

A typical flow diagram for the first alternative is shown in Figure 2. The GLC refuse incineration plant at Edmonton is of this type having five incineration units with integral boilers of the water tube type each designed to incinerate refuse at the rate of 14 tonnes/hour, and to generate steam at up to 40 tonnes/hour. The plant is equipped with four 12.5 MW steam turbogenerators at 11 kV and two 2.5 MW house sets at 3.3 kV. Electricity is exported to the adjacent Deephams sewage works and to the Eastern Electricity Board, a considerable revenue being obtained to offset the plant operating costs.

As indicated in Table 3, the heat from refuse incineration is utilized for power generation in many installations on the Continent. The incineration plant at Dusseldorf is designed to produce 100 tonnes/hour of steam at 88 bars, and this steam is transported by pipeline to a nearby central power station. Condensate is returned to the incineration plant.

At Stuttgart the central power station is equipped with dual-fuel fired boilers, the first part of each boiler being equipped with a conventional incinerator grate designed to burn 20 tonnes/hour of refuse. The high-temperature gases rise through the radiant shaft of this first section and pass into the main section of the boiler, where most of the total amount of steam required is produced by conventional oil-firing equipment. The total designed output of the boiler can be obtained by oil-burning if the refuse burning section is not available. A large power station at Munich also operates with similar dual fuel firing.

District heating power station

A typical flow diagram for the second alternative is shown in Figure 3. Incineration plants operating in this way are installed at Nottingham, Coventry, and Sheffield. The Nottingham incineration plant consists of two

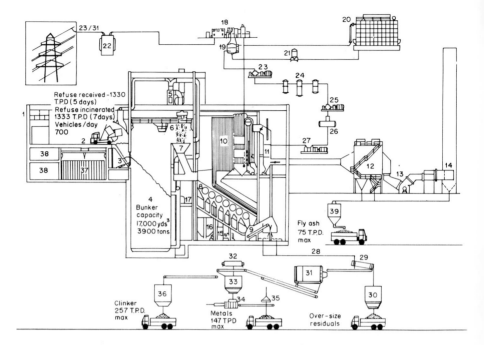

Figure 2   Schematic arrangement of plant

incinerator units each of 12 tonnes/hour capacity with integral boilers, and operating in conjunction with conventional coal burning boilers to provide steam as part of a large district heating system within the City of Nottingham. The Bernard Road, Sheffield, incineration plant consists of two incinerators, rates at 10 tonnes/hour with integral waste heat boilers designed to produce a maximum of 32 tonnes/hour of dry saturated steam at 12.8 bars. The steam is used for district heating purposes in a large block of flats located about 800 m away.

Combined power generation and district heating

This type of plant is very similar to those undertaking power generation, except that the turbogenerators are of the steam pass-out type. The steam is condensed in heat-exchangers serving the district heating plant. There are no installations of this type in Britain, but the incineration plant at Offenbach, West Germany, is typical of several in Europe. At Offenbach three incinerators, each of 10 tonnes of refuse per hour burning rate, are installed. The associated boilers are designed to produce 26 tonnes/hour of steam at 22 bars.

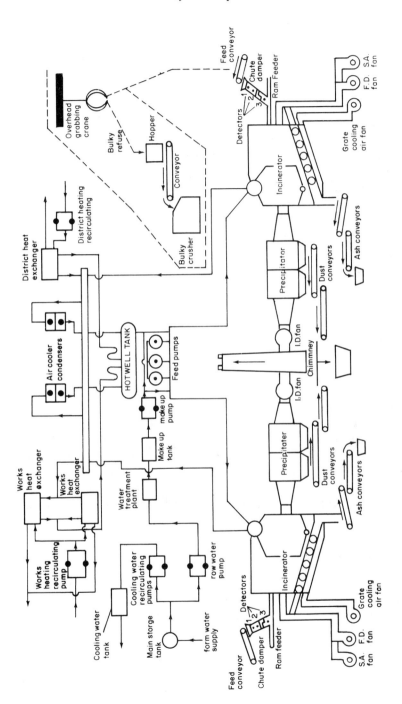

Figure 3 Schematic of the Bernard Road incineration plant and district heating scheme

Figure 4   The 20 tonnes/hour Bernard Road incineration plant in Sheffield. This is a steam-raising, heat-exporting installation. (Courtesy of South Yorkshire County Council)

Other types

The Coventry incineration plant consists of three incineration units with integral boilers, very similar in size and configuration to those installed at Nottingham. This installation has particular interest in that the main auxiliary plant is operated by steam-driven prime movers. The remainder of the steam is taken by a local motor manufacturer.

Some incineration plants in Britain have been designed to burn sewage sludge in addition to municipal refuse. The sludge is injected into the vertical shaft above the incinerator grate at a moisture content of 95% or more, and is flash-dried and burnt. There is a limit, how-ever, to the amount of sludge which can be disposed of in this way, as care must be taken to avoid a fall in gas temperature to below 800 °C—the lower odour limit. A number of refuse incineration plants have been constructed in West Germany recently, however, to burn sludge from filter presses or centrifuges typified by the installation at Krefeld. The sludge, at about 70/75% moisture, is fed into hot gas swept mills and discharged as a semi-dry powder at a temperature of about 200 °C into the common combustion chamber. Most of the sludge burns in suspension and the remainder is well distributed on the grate. Both fuels, refuse, and sludge, are, therefore, incinerated under optimum conditions, and the steam generated in the boiler unit is used for power production.

Another possibility for waste heat usage is to locate the refuse incinerator/boiler unit adjacent to a conventional fuel-fired boiler in an electricity generating station. The incinerator boiler can be used as an economizer or as a stage reheat unit operating in conjunction with the main steam generator. Studies for this application have been made in relation to a 600 MW power station.

USE OF CLINKER

The clinker remaining from the effective combustion of the refuse is generally discharged from the end of the grate into a quenching unit (water bath) where the temperature is reduced to acceptable levels before transfer to conveyors. It is a general requirement that the clinker shall contain not more than about 5% of unburnt carbon, and not more than about 0.3% of organic materials. A large part of the volume of this clinker consists of metal containers, and it is not very suitable for fill material unless this is removed. Most incineration plants, therefore, include means of removal of the ferrous metal from the clinker, this ferrous metal then being recycled as low-grade scrap.

The remaining clinker has a use as fill material. A study on the use of incinerator clinker for this purpose was undertaken by the Transport and Road Research Laboratory[3] and concluded that the clinker is a material having considerable roadmaking potential. At some incinerator plants this

Figure 5 A major public housing complex in Sheffield heated with refuse-derived heat from the adjacent incineration plant. (Courtesy of South Yorkshire County Council)

clinker is already sold to contractors, and there is no doubt that, by closer attention to good combustion, and by more intensive marketing, most of the approximately 1 million tonnes of incinerator ash produced each year in Britain could be effectively used in this way.

PYROLYSIS

In recent years much research has been undertaken into incineration of refuse under oxygen-deficient conditions. This technology has been used for many years in the treatment of coal to produce coal gas and the objective of pyrolysis of refuse has generally been to produce a gas which could be stored, and used when advantageous, whereas it is not practicable to store the energy recovered from normal incineration of refuse. Other possible advantages of treating the wastes by pyrolysis include a reduced volume of waste gases to be handled in the gas cleaning plant, fans, and chimney, with consequent lower cost for these items, and (in slag tap systems) the discharge of the clinker as a dense granular material. A major disadvantage, however, is that pyrolysis designs developed to date have not been able to take in crude refuse as do large incinerators; and the cost to produce a prepared material suitable for pyrolysis is considerable.

Experience in pyrolysis plants constructed in the USA has generally been disappointing. Continuing research, however, may lead to the development in the near future of pyrolysis technology comparing attractively with conventional incineration. In Britain research tends to be concentrated on the pyrolysis of special industrial wastes of homogeneous characteristics.

CONCLUSION

Incineration of municipal refuse has been proved over many years to be a most effective method of disposal, particularly in urban areas. One attraction of this method of refuse treatment is that the reduction in volume (95%) is much larger than by any other process. The return to industry of energy in the form of electricity or of heat, of ferrous metal, and of clinker as a fill material assist in the economics of incineration.

Incinerator installations are capital-intensive, however, and are therefore most applicable to large urban areas, where advantage can be taken of their location in a central area to minimize refuse collection and transportation costs.

REFERENCES

1. *A Symposium on the Supply of Incineration Plant.* Process Plant Association, April 1976.
2. Higginson, A. E., *The Analysis of Domestic Refuse.* The Institution of Solids Waste Management, London.
3. *The use of Waste and Low-Grade Materials in Road Construction.* Incineration Refuse Transport and Road Research Laboratory, London. 1976.

Practical Waste Management
Edited by J.R. Holmes
©1983 John Wiley & Sons Ltd

22

Domestic solid waste as a fuel in cement making

COLIN H. C. HALEY
Blue Circle Industries Limited

ABSTRACT

From the mass of research and development into refuse recycling, the pioneering work of Blue Circle Industries has been chosen for inclusion in this book. A realistic working proposition, with a credible technical and commercial viability, this chapter describes the work of Blue Circle to develop the use of treated refuse as a support fuel in the manufacture of cement. A full-scale plant for the processing and use of domestic refuse as a fuel has been in operation at Blue Circle's cement works at Westbury in the UK since the latter part of 1979. During its first 18 months of operation 47,000 tonnes of refuse were processed in the new plant resulting in a coal saving of 10,000 tonnes and the recovery of over 3,000 tonnes of ferrous metals.

INTRODUCTION

Cement making is an energy-intensive industry. Energy accounts for more than 40% of cement manufacturing costs in the UK, and over 80% of this energy is used directly as fuel in the cement making kilns. There is therefore a great incentive to reduce primary fuel consumption.

The heart of any cement works is the rotary kiln. This is a steel tube mounted on rollers and turned slowly by electric motors. It is inclined to the horizontal to facilitate flow. Raw materials—typically a mixture of clay and chalk—are fed in at the upper end of the kiln in slurry form in the wet process operation as at Westbury. At the lower, discharge, end the fuel firing the kiln is introduced—either pulverized coal, oil or gas—producing a temperature of 1,450 °C in the burning zone. As the slurry travels down the kiln it first dries and then reacts chemically to form cement clinker.

Thus, a characteristic of the cement-making process is that the fuel ash reacts in with the basic raw materials. This has the advantage of eliminating any fuel ash disposal problems but, because of cement quality considerations, places a limitation on the amount of low grade–high ash–heat-containing materials that may be used to supplement, or replace, the primary fuel employed.

DEVELOPMENT OF THE PROCESS

Experiments to employ cement kilns for the disposal of domestic refuse were initiated by Blue Circle as long ago as 1971. The early trials showed that whilst this means of disposal was practicable, the method of refuse addition employed was cumbersome and difficult to control.

The explosion of world fossil fuel prices gave additional impetus to the work, since for the first time the fuel-saving element of the work became really significant. Full-scale trials involving the injection by pneumatic means of some 1,200 tonnes of pulverized refuse into one of the Westbury kilns were carried out in 1974. These were sufficiently encouraging to enable a decision to be taken to install a demonstration plant at Westbury to handle and burn up to 80,000 tonnes/year of locally derived refuse, representing a saving of some 12–13% of the primary fuel normally employed. It was decided to limit operations to this level of refuse input initially, in order to avoid any prospect of an effect on cement product quality, which was, and is, of paramount importance.

The original demonstration plant, which was designed to be mobile, or at least movable, came on-stream in 1977 and operated to the end of 1978. During this period some 30,000 tonnes of household refuse were pulverized and fed to the kilns. The insufflation section of the plant operated well; however, the pulverizing equipment was not so successful. The processed refuse was coarser than specified and the equipment itself lacked robustness.

It was therefore decided that for a permanent operation, the original processing plant would have to be revamped. The new Westbury plant came on-stream in August 1979: acceptance tests were passed in a satisfactory manner, and the plant has been running in a reliable and consistent fashion since that time.

THE NEW WESTBURY PROCESS DESIGN

An outline flowchart of the process is shown in Figure 1. The plant is designed to handle up to 80,000 tonnes/year of crude domestic refuse received off the streets or through a transfer station. Two processing lines are in use in parallel. They are designed to operate 5 days a week with two 8-hour shifts per day. Throughput is 20 tonnes/hour on each line.

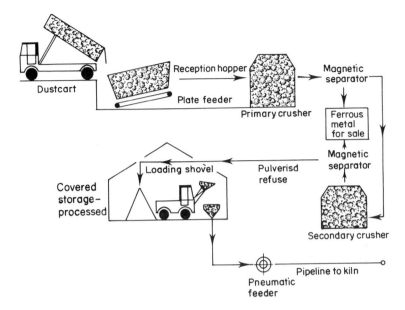

Figure 1   Refuse in cement manufacture

Variable-speed plate feeders transport the crude refuse to two Hazemag Universa 1620 gridded primary crushers. These are of fixed-hammer design. A vibratory conveyor collects the material from the grids and carries it to a secondary crusher. One line uses two Hazemag E.M. mills for the secondary crushing: the other uses a Gondard Civic mill. Both types are gridded swing-hammer mills. Overband magnetic separators remove ferrous metals between the two stages of pulverization and after the second crushing stage. At Westbury the ferrous metal content is around 7%. This is sold for recycling.

An adequate stockpile of finished material is essential for continuous feeding of the two rotary kilns, hence the final product store has a capacity of around 1,000 tonnes, enough for several days' supply to the kilns. Each of the two kilns at Westbury is provided with identical weighing and pneumatic handling facilities which convey the pulverized refuse through pipelines, some 300 metres in length, before it is burnt in the kilns. The design feed-rate to each kiln is 4.5 tonnes/hour of processed refuse, equivalent to around 12% by weight on cement clinker produced.

At Westbury, separate streams of pulverized coal and refuse are blown into the furnace ends of the kilns to provide fuel for the process, although a concentric combined fuel burner has been designed for use elsewhere.

Figure 2   Refuse intake facilities at Blue Circle Westbury plant. (Courtesy Blue
Circle Technical)

The processing plant is all operated from a central control room which is
air-conditioned and sound-proofed. The whole refuse plant, including insufflation, employs 14 men under the control of the plant manager.

PLANT PROCESS DATA

Relevant input and output data for the new Westbury plant are given below,
for the period to end May 1981:

Crude refuse pulverized, tonnes	59,933
Ferrous metals recovered, tonnes	4,442
Undesirables sent to landfill, tonnes	290
Pulverized refuse burnt in kilns, tonnes	55,027
Coal saving, tonnes	12,863

Figure 3   Primary refuse crushers provided by the Hazlemag Company at Blue Circle Westbury Works. (Courtesy Blue Circle Technical)

During most of 1980, input was limited by availability of refuse. Plant availability was high and at no time was refuse taken to landfill because of plant breakdown.

Crusher hammer wear is a major cost factor: hammer life on the secondary crushers is somewhat better than predicted at 3,000 tonnes of refuse for the Hazemag E.Ms and 2,000 tonnes for the Gondard machine. Hammers on the primary Hazemag crushers are rewelded after each day's run *in situ*.

A typical size analysis of the pulverized end-product is given below:

Sieve data: percentage finer than

100 mm	100
50 mm	97–100
25 mm	92–97
12.5 mm	80–85
3 mm	40–45

It is important that no long pieces of rag or plastics—so called 'streamers'—penetrate into the fuel system, and crushers and process line have been chosen so as to eliminate such materials.

CEMENT QUALITY CONSIDERATIONS

Typical key data on the pulverized refuse as burnt at Westbury are given in Table 1. Moisture at these levels is simply an economic debit. The ash consists largely of silica (in the absence of glass removal) but in addition to the main elements listed also contains smaller quantities of zinc, lead, chromium, and so on. It is the capacity of the cement clinker to absorb these materials without detriment to cement quality that limits the amount of refuse that can be employed. In general terms the lower the ash the more refuse can be burnt.

A substantial body of data has been built up to confirm that there is essentially no effect on cement quality under the conditions and with the quantities employed. Typical figures for compressive strengths of concrete

Table 1

Composition, (weight %)	Water	30
	Ash	30
	Sulphur	0.3
	Chloride	0.2–0.4
Ash analysis (weight %)	SiO_2	52
	CaO	16
	$A\ell_2O_3$	11
	Fe_2O_3	7
	Na_2O	7
	K_2O	1.5
Gross calorific value, (kcal/kg)		
as fired		2,300
dry basis		3,300

cubes prepared to BS 12 specification from Westbury cement are shown below:

Refuse input, percentage on cement clinker	0	3.4	4.7	6.5
Compressive strength, N/mm^2				
after 3 days	22.3	21.0	20.7	22.5
after 28 days	43.1	43.2	42.0	43.2

These figures are the averages of production runs over prolonged periods.

ENVIRONMENTAL ASPECTS

Adequate pollution control is a standard feature in most modern cement works. Electrostatic precipitators remove all dust from the exhaust gases. Moreover, the process described has the basic advantage that the acidic gases normally produced during incineration of wastes are neutralized by the alkaline constituents of the raw materials used in cement making. It has previously been shown that the total pollutant emission to atmosphere resulting from combustion of main fuel and domestic waste in a cement kiln, under proper conditions, is significantly lower than that arising from the operation of a cement kiln and a refuse incinerator separately.[1]

FURTHER APPLICATION OF THE PROCESS

In addition to its commercial application on the coal-fired wet process works at Westbury, full-scale plant trials have also been successfully carried out on an oil-fired dry process kiln at Plymstock in England. During these trials some 3,000 tonnes of household refuse were pulverized and burnt in the kiln, showing an immediate and noticeable saving in process fuel of up to 20% under the conditions employed. Details have been reported elsewhere.[1]

It is therefore considered that a system of replacing primary fuel by domestic refuse in cement making has now been successfully demonstrated in both wet and dry process cement kilns, with a consequent saving on primary fuel, whether coal or oil.

Patents have been granted in the UK[2,3] and in a number of other countries. The experience gained, and the lessons learnt over almost 10 years of development, are capable of wide application in many countries of the world. Further developments to refine the process and extend still further its applicability are planned.

ACKNOWLEDGEMENT

I wish to thank my technical colleagues in the Blue Circle Group for their assistance and information. Their contribution to the development of the process described is hereby acknowledged.

REFERENCES

1. Haley, C. A. C. *Use of Domestic Refuse as a Fuel in Cement Manufacture.* International Recycling Congress, Berlin, 1979.
2. British Patent No. 1,405,294, 'Conversion of Refuse', dated 27 July 1973.
3. British Patent No. 1,510,392, 'Portland Cement Manufacture and Utilisation of Waste Matter', dated 19 January 1977.

Practical Waste Management
Edited by J.R. Holmes
©1983 John Wiley & Sons Ltd

23

The solid waste problems of Poor People in Third World cities

JOHN PICKFORD*
Group Leader, Waste and Environment in Developing Countries, University of Technology, Loughborough

ABSTRACT

Turning aside from the high technology solutions that apply in the developed world the author defines the solid waste management problems as seen by poor nations, the bulk of humanity. The paper extends the use of simple and appropriate technology, looks at composting and shows how solid wastes management fits in to the overall picture of sanitation.

The term 'developing country' encompasses a great variety of economic and social conditions. At one end of the range are the oil-rich nations who can afford to improve their public services rapidly. More resources can be put into the collection and disposal of solid waste than can be afforded in the older industrial countries of Europe and North America. In a few small and exceptionally rich countries good public services are available for everyone. However, national prosperity usually fails to benefit the bulk of the people, whose conditions remain fairly squalid.

The majority of Third World countries are poor. Many are very poor. In fact 'developing country' is generally synonomous with 'poor country'. In nearly all developing countries, whether the average GNP is at the top or the bottom of the range, most people are 'underprivileged'. Half the world's population lives in conditions of deprivation. These people lack adequate nutrition, education, or health care. Most of them are without a safe, clean water supply or reasonable sanitation.

In the large cities of the Third World there are areas with all modern conveniences. This particularly applies to national capitals. Often the commercial areas boast multi-storeyed reinforced concrete or steel-framed build-

---

* Past President, Institution of Public Health Engineers.

ings which compare favourably with those of the 'West'. The residential areas occupied by the elite are opulent. The successful businessman and politician lives in a well-paved, well-lit area in a large house surrounded by a large garden. He has electricity, servants, cars, refrigerators, water-carried sanitation, and perhaps coloured television and video-recorders.

Around these favoured areas live the poor—crowded, ill-served by amenities, maybe walking several miles to work or travelling in public transport carrying more passengers outside than inside. These are the people dealt with in this chapter.

In many Third World cities more than half the people live in shanty towns, squatter settlements, bustees, and other 'transitional settlements'. Far from being temporary, these depressed and depressing districts are 'home' for the whole lives of some of their inhabitants. Some squatters live in makeshift hovels built from old packing cases and second-hand corrugated iron sheets. Some have flimsy canvas for shelter. Others build in the way which is traditional in their rural background—mud or bamboo walls and thatched roofs. Surprisingly, there are a few substantial buildings in many of the older squatter areas. A squatter may make good, often by trading, and decide to stay put, gradually improving his accommodation. He remains a squatter because he has no legal right to the land—no security of tenure.

Other 'underprivileged' people in Third World cities are crowded, a family to a room, in derelict buildings abandoned by the better-off, or in tenements reached by dark and stinking staircases. Some live in the old city areas, criss-crossed by alleys too narrow for the smallest car.

For all these people existence is a struggle. Many have to queue for a trickle of water from a public standpost or go to distant pools or streams or wells for polluted water. Their 'lavatory' may be a ditch, the roadside, or a piece of vacant land. Their dirty water is thrown in the unpaved roadway. So is their refuse. The characteristics of these areas include a putrid stink, flies by the million, and pools of fetid liquid between heaps of rotting vegetation. Solid waste is the source of many of these nuisances.

Even where public works have improved conditions refuse is still a problem. Where concrete 'monsoon drains' replace earth ditches, they become choked with rubbish. Where sewerage is provided, manhole covers are removed so that garbage can be dumped there; then the sewers become blocked. Where special dumps or public 'dustbins' are provided, refuse is heaped outside. Wandering dogs, goats, and pigs forage amongst the rubbish, spreading it around. In Hindu areas the ubiquitous protected and sacred cow joins in foraging.

The refuse itself in these areas is quite different in character from that of the wealthy 'residential' and commercial areas. The way of life of the well-heeled may closely resemble that of Europeans. Anyway some of the privileged people come from Europe and North America. The expatriate communities expect a standard of living even higher than at home. But the poor people buy local food packaged in the local way—rice measured by the tinfull and wrapped in a leaf or a little heap of peppers wrapped in a twist of paper from an old exercise book. Refuse is therefore largely vegetable. Any bottles and jars and tins these people get are kept and used again. In some areas the dry vegetable waste is used as fuel for cooking, and then there is a high proportion of ash. In any case the density is high. The calorific value is low. To add to the difficulties, refuse often contains a fair amount of animal and human excreta.

Other problems are climatic. Developing countries straddle the equator with the rich industrial countries away to the north—a few to the south. So temperatures in the Third World tend to be higher. Something over 25 °C is usual. This results in much quicker putrefaction, stronger stinks, more flies. Rainfall when it comes is often intense, and uncovered refuse becomes soggy.

Even if the temperature did not make it desirable to collect refuse more frequently than the usual weekly collection of Europe, living conditions preclude much storage. A family occupying a 10 × 10 ft (3 × 3 m) room has no space for a dustbin. If they had space they could not afford one. If it were provided 'by the Council', it would be used to store water or food or for some other purpose more important than keeping rubbish. The most that can be expected for household refuse storage is an empty tin. Sometimes even that is impossible because religious taboos forbid keeping any waste in the house or in the yard.

At the 'municipal' level the collection of refuse presents special difficulties. As already mentioned, maybe half of the inhabitants of some Third World metropolitan areas are squatters of one kind or another. By definition a squatter has no 'right' to live where he is living. In the past some squatter areas were cleared. Bulldozers moved in, often at dawn, and hardly gave the people time to move their meagre belongings before flattening their homes. Now 'authority' generally turns a blind eye to the illegality of the squatters.

In many cities attempts have been made to provide decent low-cost housing. Unfortunately the urban drift is so great that as soon as one group of shanty-town dwellers have been moved, others take their place. It is also incredibly difficult to provide even the minimum reasonable standard of housing and amenity at really low cost. So what is intended to solve the squatter problem houses those higher up the social and economic ladder. The squatters remain, and every year more people join them.

Squatters are illegal. They should not be where they are, but there is no easy way to move them. If authority turns a blind eye to their illegality it may not unreasonably turn a blind eye to their need for services, including refuse removal. Nevertheless, fear of an epidemic of cholera or growing awareness that waste-associated disease spreads throughout the city may demand that something is done for the poor. International agencies also have some influence. Following the missionary zeal of Robert McNamara, the World Bank President, loans and grants required for water supply and sanitation are sometimes conditional on improvements for poor people as well. In order to improve the water supply of the city and so make sure of enough water to sprinkle their own well-cut lawns, those in authority have been forced to provide standposts for squatters. To obtain loans to import refuse vehicles to take away their own beer cans and Weetabix cartons something has to be done about refuse in the slums.

But what can be done? The temperature and humidity and the lack of household storage space make frequent collection essential. In Europe our high rates and taxes can hardly support a weekly visit by the dustman. So how can these poor people without enough income to pay for adequate food justify a costly more frequent service? Most of them pay no rates or taxes anyway.

Modern refuse-collecting vehicles are out of the question. Streets are too narrow. Most are unpaved so they become muddy streamlets in the rainy season. Anyway the refuse as collected would already be as dense as pulverized and consolidated 'Western' material. Specialist compacting vehicles are fantastically expensive, even if foreign exchange is available. Spares are likely to be difficult to obtain and maintenance of uncommon equipment always presents difficulties.

Unskilled labour is often easily available. It may be very cheap in terms of the rate per man-day. Women and girls are the customary labourers in many countries and may be cheaper still—in terms of the rate per woman-day. It does not follow that labour-intensive work such as refuse collection is cheap in terms, for example, of the rate per tonne moved. As in industrial countries, in the Third World good management, good labour relations, and good supervision make all the difference. Without them productivity may be abysmally low. Hundreds, or thousands, of sweepers are employed. They squat all day forming little heaps of dust and other rubbish with the local brush, a bunch of sticks or reeds. Sometimes other sweepers collect the little heaps in baskets and carry them to bigger heaps. It is not surprising that India spends four times as much on street sweeping as on refuse collection, and in Jakarta there are over 4,000 sweepers but less than 1,800 collectors including drivers.

In Manila good management has resulted in a tremendous improvement in solid waste practice. The status of personnel has been changed by issuing them with attractive uniforms, by providing free insurance, and by incentive payments. Small handcarts with portable bins are used to collect street sweepings.

In some cities private enterprise has taken over refuse collection. Perhaps the best-known of these entrepreneurs are the Coptic Christians of Cairo whose donkey carts are seen all over the city. The conditions of these Cairo refuse collectors are not to be emulated as they live amongst their collected refuse. Pigs live with the people and eat the vegetable waste. Anything capable of being recycled is taken out. Another example of private enterprise is Saigon where self-employed refuse collectors use tricycles.

Cairo's donkey carts are an example of appropriate transportation. They are locally-made and need virtually no maintenance. In parts of the Indian sub-continent bullock carts are used. The advantage of a transport system which does not require petrol or diesel oil are obvious enough. Escalating fuel prices have had a disastrous effect on the non-oil-producing countries of the Third World. In Tanzania, for example, the value of oil imports is 60% of total imports in spite of rigorous control of the use of fuel. Unlike a motor vehicle, donkeys and bullocks do not need a non-working driver. The animals stand still untended while collectors collect. They may even be trained to move from one collection point to the next at a word of command.

Where streets are too narrow for carts, and where animals are not usual for hauling loads, some form of man-powered conveyance may be suitable. Garden-type wheelbarrows are still very common, although their capacity is much too small for efficiency. They also share with the box-cart the disadvantage that their contents must be unloaded either by tipping or by lifting out by hand. A more satisfactory hand-cart carries up to six portable bins or old oil drums which can be lifted from the cart to discharge their contents directly into a larger vehicle. Tricycles, with the load either in front or behind, are used in some towns.

Animal-drawn and man-pushed vehicles have the disadvantage of a very limited range. Their speed is so slow that any lengthy journey to a discharge point takes an inordinate time. This naturally leads to the use of larger vehicles to carry the load to tip or processing yard. Tractors and trailers have many advantages for this further transport of collected household refuse and street sweepings. Trailers can be locally made. Rear axles of worn-out lorries and cars may be suitable if money is too scarce for completely new trailers. A refuse trailer should have a low loading height—1.6 metres is a recommended maximum. The capacity can be 8 m^3 or more. If spare land is available for transfer stations one tractor can operate with three or four trailers. A tractor only costs half as much as a lorry and there are likely to be spares and maintenance expertise available.

In one Third World city open sheds have been constructed for transfer stations. Each is large enough for two trailers for easy handling. A tractor simply leaves an empty trailer and takes a full one away with minimum delay. These trailers are available for refuse dumping by nearby householders as well as the transfer of refuse collected in hand-carts. Ideally a transfer station should be at two levels and this can sometimes be easily arranged on sloping ground. The trailers are brought in from a roadway at a lower level. The collection carts or barrows enter from a higher roadway and their contents can be tipped directly into the open-topped trailers. However, in the majority of poor areas of metropolitan cities land is precious and only the minimum of space can be spared. A trailer may be left at the roadside. Under cramped conditions there is obvious advantage in using portable bins. With hand-cart systems operated by one man or woman it is desirable to have a labourer on duty at the transfer point. He can help the collectors lift full bins and be responsible for the cleanliness of the area. Householders' refuse is likely to be spilled, especially if it is brought by children.

In several Third World cities more elaborate trailers or skips have been provided. These may be used for transfer of collectors' loads or dumping of refuse by householders. Some of these containers have sliding or hinged access covers to keep the refuse dry during rainfall and to prevent ingress and egress of flies. However, my observation, confirmed by others, is that the covers are rarely used on site and that they soon become broken. There are obvious advantages in covering refuse while it is taken from a transfer point whether a specially made skip or any other vehicle is used. In wet weather a cover of some sort prevents refuse becoming too soggy. In windy weather a cover prevents the often-seen trail of refuse behind a moving vehicle. It reduces smell and complaints from the public. The simplest cover is a tarpaulin or plastic sheet. However, experience indicates that unless supervision is exceptionally good such covers are not used. With tractor-hauled trailers the maximum speed is less than with higher-speed lorries, and spillage is not so great. Old-fashioned side-loaders with curved sliding covers are suitable for some areas, but cannot be used economically as householders' dumps.

In the long term a well-run tractor–trailer system can be most satisfactory and cheapest. But in the majority of existing situations, where there is any motorized system at all 'ordinary' tipper lorries are most likely. There are several reasons put forward for their use. They are readily available. Their maintenance and repair can be carried out easily with that for other vehicles, such as those of the roads and water departments. When not required for refuse collection they can be used by these other departments for carrying sand, pipes, and the like. Often refuse collection starts before dawn and is completed by mid-morning, permitting this 'other use' in the afternoon.

These arrangements are fallacious in relation to a properly organized refuse collection system. Tipper lorries, as already mentioned, are likely to be twice as expensive as tractors. The type commonly employed has a high loading level and is designed for heavy material. When carrying refuse, even the comparatively high-density refuse of the tropics, the load–power ratio is wastefully low. 'Shared use' is often a delusion. Either the other departments want the lorries in the morning, making them unavailable for refuse collection, or the vehicles stand idle in the afternoon. The most efficient use of transport is undoubtedly to have special refuse vehicles with two or more shifts of collectors. Such a system requires very good management and supervision—qualities which are in particularly short supply. A shift system also involves a change in habitual working practice, which is itself difficult to achieve.

Any kind of efficient collection system almost certainly requires a change of householders' habits. This is likely to be more difficult than organizational improvements. Where it is customary to throw rubbish 'outside the front door' (outside the only door) or at the corner of the street people usually accept the resulting mess as part of their inevitable environment. In Nepal cooking is usually carried out at the top of buildings in the congested urban and semi-urban areas. The waste (vegetable peelings and the like) is thrown out from the fourth or fifth floor and is well dispersed by the time it reaches the ground.

Changes can occur. Complete transformations of areas of squatter and slum housing have taken place in many cities. Removal of rubbish can be the focal point of transformation. But introduction of a refuse collection programme, however well-managed, however efficiently planned, however well-equipped with suitable vehicles, cannot by itself affect the change. The people's co-operation is essential. Such co-operation cannot be obtained easily. True collaboration certainly cannot be aroused by a paternalistic attitude by those in authority. It is easy to fall into a trap. Whether expatriate or local, experts can be tempted to say 'we are providing you with a wonderful collection system with fine new tractors, trailers, and hand-carts; all you have to do is carry your rubbish to the nice new concreted collection place two blocks away. Then the streets will be tidy and you will be free from flies, smell and disease'. If the words are not spoken, the thought may be there. The zeal of the reformer telling people what they *ought* to do is seldom effective.

An efficient refuse collection system is not enough. It must be part of an overall improvement actuated by a wish to change the environment. Other physical components of such schemes may be the paving and lighting of streets, drains to remove wastewater and rainwater, and provision of household or communal latrines. Occasionally upgrading has been successfully introduced by governments—local, regional, or national. Examples are the bustee improvement schemes of the metropolitan development authorities in India. In other places an awareness of the health dangers of ubiquitous human waste has led to improvements. A few successful upgrading schemes have resulted from what may best be described as 'community spirit'. A group of citizens decides to take things into its own hands and improve its locality. It may be only one alley, or a group of houses or shacks occupied by a particular ethnic or occupational community. Sometimes this community spirit spreads. Other groups follow the example of the first. Competition for the greatest improvement may follow. Even in the slums there may be the equivalent of 'keeping up with the Jones'. It must be admitted that such spontaneous activity for environmental improvement is rare. Most of the Third World poor are so concerned with the effort of keeping body and soul together that nothing further is possible. Nevertheless these communal improvement schemes do occur; sometimes where least expected.

Wherever there are improvements to the total environment—paving, lights, drains, and latrines—satisfactory solid waste collection is absolutely essential if conditions are not to revert to their previous anarchy. Ongoing maintenance of all public works is of course important, but in no other sphere is regular and continuous activity so important as in the removal of refuse. Herein lies its tremendous importance. Good solid waste management is crucial to environmental upgrading.

The final disposal of solid waste from poor areas is usually in association with the waste from the well-off districts. Crude dumping, often at quite unsuitable sites such as the banks of rivers, is still by far the most common method. For example, 90% of Asia's refuse is dumped. Exceptionally, controlled tipping is carried out to some extent in Korea, Taiwan, Singapore, and Hong Kong. In New Delhi a reclamation scheme with its own nursery for trees and flowering shrubs has converted a derelict area into parkland.

The very high vegetable content of refuse from the poorer areas points to the use of composting, particularly as there is pre-collection separation in many 'countries'. Bottles, tins, paper, and rags are taken out for re-use. Erosion of soil by tropical rainfall, the demand of growing populations for greater food production, and the multiple cropping associated with the 'green revolution' all suggest that compost is a much-needed commodity. In a few places farmers take refuse to produce their own compost or spread raw refuse on fields. Some towns have simple manually-operated windrow composting yards. But composting of refuse at the municipal level is no more successful in the Third World than in the United Kingdom. Several expensive plants have been built with expectations of good sales. In nearly all cases marketing has not been successful and most plants have closed down or run at well under capacity. In spite of these failures efforts to find cheap and effective means of recycling vegetable matter and returning nutrients and humus to the soil must surely continue.

Providing for the needs of the poor of the world is a tremendous challenge. Much more than good engineering is required; more than good management. There has to be a combination of these with an understanding of the people and their way of life. Yet their welfare is tremendously important. The satisfactory removal of solid waste and the consequent reduction in flies and disease are vital elements in improving the miserable existence of half the world's population.

FURTHER READING

Flintoff, Frank. *Management of Solid Wastes in Developing Countries.* WHO, New Delhi, 1976.

Pickford, John. Solid waste in hot climates. in *Water, wastes and health in hot climates* (Ed. Feachem, McGarry and Mara). Wiley, London, 1977.

World Health Organization, Expert Committee, *Solid Waste Disposal and Control.* Technical Report Series No. 484. WHO, Geneva, 1971.

World Health Organization, *Solid Wastes Management.* Report of Regional Seminar, Bangkok. SEA/Env. San/158 WHO, New Delhi, 1975.

Practical Waste Management
Edited by J.R. Holmes
©1983 John Wiley & Sons Ltd

24

Metropolitan waste management planning in developing countries— A case study in the Istanbul metropolis

P. K. PATRICK*, MBE, DFC, MIPHE
Waste Management Consultant, D. Balfour & Sons, Consulting Engineers

ABSTRACT

This chapter looks at the principles of solid wastes management in the large metropolis of the developing world. The author considers the correct ordering of cleansing priorities and the proper consideration of economic and sociological constraints. The necessity for master plans, waste analyses, organization and methods are debated together with the financial implications of the solutions chosen.

INTRODUCTION

The creation and growth of metropolitan areas in any country give rise to problems in waste management. In developing countries, metropolitan authorities exist and have to function under pressures and constraints unknown in other countries, particularly in regard to uncontrolled population growth (arising from political, economic, historical, or religious influences) and extremely high population densities. The infrastructure, financial resources and technical resources generally fall far short of requirements for planning and sustaining an effective waste management service.

In the developing countries, standards of hygiene and cleanliness, which may be high in the domestic environment, are too often disregarded in public places. Established habits and customs are not easy to change. Indeed, an attempt to introduce radical changes quickly is likely to fail unless there is

---

* Past President of the Institute of Solid Wastes Management, and General Manager, Waste Dispoal Branch, Greater London Council.

493

very strong governmental direction and ability to enforce changes. Success is more likely if existing customs can be adapted to meet objectives in waste storage, collection, and disposal.

These remarks are generalities. Each situation is unique. The following paragraphs describe in outline a project in which the author was engaged for 2 years in Istanbul. The project was set up by WHO/UNDP in conjunction with the Turkish government, the object being to prepare a master plan for solid waste management in the Istanbul metropolitan area. The World Bank was also associated with the studies.

ISTANBUL METROPOLITAN AREA

At the time the project started (May 1979) Istanbul Metropolitan Area consisted of 34 separate municipalities plus some village areas, covering a population of some 4.5 million. Istanbul municipality was by far the largest of the group, with a population of 2.8 million. Istanbul is unique in being a city located on two continents, separated by the Bosphorus waterway; approximately 75% of the population of the municipality is on the European side, the remainder on the Asian (Anatolian) side of the Bosphorus. The European side itself has a natural waterway—the Golden Horn (Halic) dividing this area. The city as a whole is a mixture of new and old: modern apartment blocks and motorways; historical areas dominated by great mosques and palaces; 'gececondus' (illegal developments) where development ranges from slum 'shanty' areas to substantially built brick or concrete-block dwellings. Istanbul, by its variety, vitality, and rich history, indeed justifies a designation of one of the world's great cities.

PRESENT SITUATION IN WASTE MANAGEMENT

As far as could be ascertained (no waste is weighed), some 4,000 tons of household and commercial waste per working day are collected by the various municipalities. Collection presents many problems. Although modern compaction-type vehicles are extensively used, the municipalities lack the resources to maintain a reasonable level of serviceability. The vehicles are purchased almost entirely from abroad, although Turkey has an embryo industry producing compaction-type collection vehicles on imported chassis. Spares have to be purchased from abroad and as the country is seriously short of foreign currency, the municipalities are often unable to buy spare parts. Lack of standardization aggravates the situation; vehicles are generally bought in batches when money is available and orders are placed with whoever offers the best deal at the time. Some vehicles have been provided by foreign countries as gifts or as part of an aid programme. Workshop facilities are generally minimal, but the Turkish artisans do a remarkable job in improvisation in efforts to keep the wheels turning.

Waste collection in the Istanbul municipalities involves heavy physical work. Because low-grade solid fuel is the main source of domestic heat household waste has a very high ash content; this—plus the fact that waste receptacles of a variety of shapes and sizes (generally uncovered) are used—imposes a heavy work load; 3 tons per man per day is a typical work output for a refuse collector. Because collections are irregular in many districts, much of the waste is dumped in streets and alleys and has to be hand-shovelled into vehicles. The physical nature of Istanbul—steep hills and many narrow streets, with congested traffic—imposes further problems for the collection services.

In the metropolitan area some 4,000 tons of waste per day are collected; more than 500 vehicles comprise the municipalities' fleets but non-availability reached as high as 40% at times.

Industrial wastes are not collected by the municipalities. A vast amount of industry exists in the metropolitan area and the collection and disposal of wastes is dealt with either by the industries themselves or by contractors hired to remove the waste.

Waste disposal is entirely by landfill. Some small incinerators were locally built some years ago, but none is now in operation. The characteristics of the waste—high moisture content in summer, high ash content in winter, make it unsuitable for incineration. The landfill sites in use are mainly natural valleys, where waste is deposited in great depths (50–60 ft in some cases). The main site for Istanbul municipality receives more than 2,000 tons/day. Unfortunately, the sites is badly located in relation to water resources and pollution of wells has occurred. Some municipalities allow salvaging to be carried out at the tips; contractors are given salvaging rights in return for payment to the municipality. In such cases, the operation is under some measure of control; however, uncontrolled scavenging takes place at some major sites, where scores of men, women, and children (and animals) infest the tipping area. Even though the activity is not authorized and is forbidden, the authorities seem unable to stop it.

SOLID WASTE ANALYSIS

An extensive programme of waste sampling and analysis was carried out. Samples were collected from seven zones of the metropolitan area over a period of 10 months. The waste was classified in size and characteristics with the aid of wire-mesh screens made locally; samples of the waste as collected were analysed at Istanbul Technical University to determine CV and compostability.

The characteristics of Istanbul solid waste are, if not unique, unusual in the extremes of composition in different seasons, as can be seen from Table 1. Summer waste is very high in organic and moisture content, while in winter ash content predominates. As mentioned earlier, domestic heating is mainly produced from burning of solid fuel—a low-grade coal. Because of the high cost of oil in Turkey—nearly all has to be imported—the use of coal for this purpose will increase, as new properties are now required to be equipped with solid-fuel boilers and many existing properties are being converted from oil to solid-fuel heating. The result will be even more ash in the domestic wastes. This situation is a good example of the influence of national economic factors on waste characteristics.

The results of these analyses had, of course, a strong bearing on the methods of waste disposal which could be recommended for future use, taking account not only of present conditions but of future economic trends.

Table 1.   Solid waste analysis—middle-class property
(30% of total premises)

	Winter	Spring	Summer	Autumn
Organics (%)	36.25	39.50	61.00	56.70
Paper (%)	7.36	11.00	18.30	15.00
Textiles (%)	1.45	2.85	3.10	4.30
Plastics (%)	2.45	5.40	3.90	2.10
Glass (%)	1.12	3.37	3.00	2.40
Metals (%)	0.37	2.88	1.50	1.00
Ashes/dust (%)	51.00	33.60	9.20	18.50
Density (kg/m^3)	364	344	337	395

ECONOMIC AND SOCIOLOGICAL CONSTRAINTS

Although transfer of technology is an important aspect of aid to developing countries, it is essential, if feasibility studies are to be implemented, to propose measures which are within the economic and technical abilities of the country concerned, even if they do not measure up to hygienic and environmental standards expected in developed countries. This may seem a statement of the obvious, but is not always applied. The adage 'The best is the enemy of the good' is particularly relevant to attitudes to waste management projects in developing countries. One of the difficulties of 'master plans', aimed at some speculative and undeterminable future, is to make assumptions on long-term conditions in countries where the political and economic development is fraught with uncertainty. For example, it might seem advantageous to propose a plant or equipment of high capital cost because the investment cost will be financed by a long-term low-interest loan provided by one of the international funding agencies. But the key question is whether the municipality or other operating authority will have the financial and technical resources to operate and maintain the facility through its lifetime. Great metropolitan authorities, naturally enough, tend to want and expect modern technological solutions to their problems. These hopes and expectations are not always realizable within the realities of economic constraints. For instance, low wage rates may be the most significant factor in choosing a technology or system.

Established social customs are a further important consideration. In Istanbul, for example, there is a well-organized system of salvaging ('totting' in UK parlance) whereby very large quantities of material are recovered and sold to industry. The system operates through a number of levels of entrepreneur activity. At base level, scavengers scour the waste put out in bins and boxes for collection. These men and women go round the streets early in the morning before the collection vehicles arrive. Material thus collected is sold to merchants, who carry out some preliminary sorting before selling it to other merchants, who do further sorting and cleaning prior to sale to industry. The whole range of activity, though deficient in many aspects of hygiene and public amenity, gives employment and a living for hundreds of people; provides industry with low-cost secondary raw material, and saves the national economy considerable sums of foreign currency for imported materials. It is easy enough to say that the pracice of 'totting' and manual sorting in open yards is undesirable and should be stopped, but what is to be put in its place? A mechanized sorting plant under municipal control would be a tidier and more environmentally acceptable alternative, but the consequential reduction in employment and higher costs to industry must be weighed in the balance.

MASTER PLAN STUDIES

The feasibility studies covered the whole field of solid waste management:

    waste collection;
    waste disposal;
    transport;
    materials and energy recovery;
    industrial wastes;
    hospital wastes;
    training requirements;
    legislation;
    organization and administration.

A project office was set up in Istanbul under the management of the WHO senior technical adviser with a Turkish counterpart. Local technical and administrative staff were recruited and WHO Regional Office, Copenhagen, engaged expert short-term consultants to assist the study as required.

An important part of the study was a survey of industrial wastes arising in the metropolitan area: 297 industrial premises, employing 91,784 people, were visited. From the survey it was estimated that nearly 200,000 tonnes of wastes per year were produced. The amount of material recycled within industry and externally was found to be commendably high—some 45% of the total wastes. Prices obtained for recycled materials in Istanbul are remarkably high by Western standards, for example:

wastepaper ($US per ton)	55
plastics	33
metal	61
glass	28
textiles	44.

The handling and disposal of hospital wastes gave rise to concern, as facilities at the hospitals generally are wholly inadequate. The medical staff are aware of the problem, of course, but lack the financial resources to make improvements. There was also found to be a lack of defined management responsibility for waste handling in many cases. Unpleasant and contaminated wastes (dressings, etc.) were often mixed with domestic-type wastes and put in a yard for collection in uncovered receptacles. The study involved visits to 80 hospitals and clinics.

The main problem in the study lay in finding a long-term solution to the waste disposal problem. Incineration was considered to be impracticable as a major contribution to disposal; the C.V. of the waste had a range of 200–1900 kcal/kg. Only waste from selected areas could support combustion without the use of supplementary fuel, and the progressive use of solid fuel for domestic heating, with the attendant high ash content of the waste, made incineration a difficult and uneconomic proposition. The construction of materials recovery plants was not considered viable because of the intensive salvage activity by the private sector before the waste was collected by the municipality.

The principal method of disposal suitable for the greater part of waste arisings, therefore, had to be by landfill. Existing sites were either unsuitable or did not have long-term capacity. The cost of remedial measures for groundwater protection at the largest and principal site would have been extremely high, and the long-term availability of the site was uncertain, so an extensive search was carried out to find alternative sites. The difficulties of finding suitable and acceptable sites for landfill in developing countries are often no less than in western Europe, despite the apparent availability of great areas of non-urban land. Village dwellers are no more willing than their town or city counterparts to have waste disposal activity in their neighbourhood. Water supply for rural areas is often from wells, so groundwater protection is important. Road access to distant sites is generally poor and sometimes non-existent, and the cost of long-distance transport may overstrain the municipalities' resources. All these factors limited the area of search in the Istanbul area. Eight potential sites were eventually found; their use would involve the construction of transfer stations in the collection areas, so outline plans for three transfer stations were prepared. It was proposed to use 'roll-on/off' type of bulk containers in conjunction with eight-wheel rigid-chassis vehicles for transport from the transfer stations. Istanbul municipality had some articulated transfer vehicles in use from a small transfer station, but these proved unusable on the main landfill site in winter, when heavy rainfall made traction conditions very difficult.

Reference has been made to the recovery of materials by the private sector. Other aspects of resource recovery were also investigated. A survey of agricultural areas in the metropolitan area and its hinterland, to determine the possible markets for compost, was carried out for the project by Istanbul Academy for Economic and Commercial Sciences. Analyses of waste made by the Technical University had shown that domestic waste for 8 or 9 months of the year had good compostability. The survey showed that there was a high potential use for compost, though only experience would show whether the quality of the product would be acceptable and the price one which the market could sustain. It was decided that the construction of one plant of 300 tonnes/day input capacity was justified. This was to be designed on the simplest handling and processing methods, using the 'windrow' method of composting. Materials recovery by manual sorting would be included in the flow line. The plant would be located at one of the proposed new landfill sites.

Consideration was also given to the use of selected wastes for the manufacture of animal feed, as much of this essential material had to be imported. It was felt that an operation of this kind would best be undertaken by private enterprise so it was proposed that the waste disposal authority (yet to be established) for the metropolitan area should invite offers from international companies with experience in this field to construct and operate a plant.

Istanbul municipality owned and operated a concrete-manufacturing plant, which also produced concrete building blocks. It was arranged that tests should be carried out to determine the suitability of ash from domestic waste to be used as a supplementary aggregate in the manufacture of blocks. Boiler ash from domestic heating installations in blocks of flats could easily be kept separate from other waste for collection.

Finally, as a long-term measure, it was proposed that a detailed study should be made of the use of solid waste as a fill material for land reclamation from the sea. A separate study was already under way of the feasibility of creating new land, mainly for recreational purposes, on sections of the northern coast of the Sea of Marmara. It was considered that, in view of the very large volume of fill for which solid waste might be used (some 2 million m^3) this offered a potentially practical and economic outlet for much of the high ash content waste in winter months.

Overall, therefore, long-term solutions to Istanbul's waste disposal problems could be foreseen, but their implementation would require a strong engineering and organizational capacity, which at present was lacking.

ORGANIZATION AND MANAGEMENT

The project brief called for recommendations to be made on the most appropriate organization for solid waste management in the Istanbul metropolitan area. This posed quite a problem, with more than 20 municipalities involved and an existing complex legal structure. The situation was further complicated (but paradoxically, in the event, simplified) by the take-over of the government by the military while the study was in progress. The constitution of municipalities changed (almost overnight). The military authorities replaced mayors and senior staff by army personnel, so many previous contacts were lost and organizations such as the Union of Istanbul Municipalities were abolished. However, it is probably true to say that proposals for administrative reorganization, involving political considerations, had a better chance of being implemented than hitherto, as decisions at both government and municipal level could be more readily made. A factor which simplified the project's task was the decision of the new government to create a single municipal authority by amalgamating Istanbul municipality with 23 others, which was to take effect from March 1980. Previously, the metropolitan area for project purposes had not been clearly defined and opinions differed as to which authorities should be included. The new decision removed any doubts or arguments. The population of the new municipality is approximately 4.3 million and is expected to rise to 5.8 million by 1995.

The project was fortunate in having the assistance of Mr Alan Norton of Aston University, Birmingham, in the organizational studies, as he had previously participated in an OECD study of municipal management requirements for Istanbul municipality and was familiar with the legal and administrative usage inherent in Turkish public administration. Three options were considered and put forward regarding an organizational and management structure for the new Istanbul Metropolitan municipality:

(1) creation of a semi-autonomous solid waste management authority; this would control and operate all solid waste management functions—collection, transport, and disposal;

(2) establishment of a separate solid waste disposal authority, collection to remain with the municipality;

(3) placing all solid waste management functions under control of an enlarged municipal cleansing department.

The project recommendation was for adoption of option (1), followed by (2) as a second choice. It was felt that the municipality did not have, and was unlikely to be able to acquire, the managerial and technical resources adequate to operate solid waste management services over an area of 4.3 million population which could well rise to 7 or 8 million by the end of the century. Whichever option was chosen, it was proposed that the area of the new Istanbul authority should be divided geographically into three zones with a strong degree of decentralization of administration. Traditionally, Turkish administration tends to be strongly centralized.

FINANCIAL IMPLICATIONS

Municipalities in Turkey have few sources of revenue generation. By far the bulk of their revenue comes from direct grants or payments from central government. Budgeting is therefore not primarily related to service needs, but to apportioning what is likely to be handed out by central government.

It would not be practicable to operate a large-scale organization such as that envisaged on such an uncertain financial basis. It was therefore recommended that the solid wastes management services should be self-financing by means of direct charges for waste collection and disposal. Various methods of collecting the charges were discussed and put forward for the government's consideration.

CONCLUSION

In developing countries—and Turkey was no exception—it is often found that institutional factors give rise to more difficulties than technical problems. Success in solving the latter can only be achieved by devising systems acceptable to the population at large and within the political will and ability of the government to implement recommended measures. Master plans are often prepared more in a spirit of hope than of expectation and while long-term planning is essential to assist governments in their long-range economic forecasting, priority in solid waste management projects may well be given to measures which will bring about environmental and health improvements to populations sorely in need of both.

25

A case study in solid waste generation and characteristics in Iran

ADRIAN COAD

WEDC Unit, University of Technology, Loughborough

ABSTRACT

Taking a look at turbulent Iran before the recent political changes the author sets out the methods to be used in preparing a case study for the solution of solid waste problems in an Arab State. Common practices, the constituents of solid waste, socio economic factors and recommended solutions are considered. The text contains useful information on waste characteristics in rich and poor parts of the world.

INTRODUCTION

The influence of lifestyle on the quantity and nature of domestic solid waste was investigated by the author in Shiraz, Iran, in 1973. This chapter describes the city, the public attitude towards refuse, and the collection system used, and presents some of the data gathered on solid waste quantities and composition. The relationship between refuse characteristics and standard of living is examined and recommendations on improving the efficiency of the collection system are made. It is hoped that the information in this chapter will be valuable to any who are concerned with solid waste management in cities similar to Shiraz.

Large variations in solid waste quantities have been observed across the world. These differences are linked to differences in prosperity, climate, industrialization, community size, and consumption habits. The design of an efficient solid waste collection and disposal service requires data on refuse weights, composition, and volumes, and an understanding of local customs, and so it is necessary to obtain information from the area concerned; global

Table 1   Refuse variations worldwide

	Poona, India	Benghazi, Libya	Britain	Arizona, USA
Reference	2	3	4	5
Weight per capita g/cap./day	300	640	510	1000
Percentage vegetable	68	42	14	23
Percentage paper	9	16	22	43
Percentage inerts	—	14	19	11
Density (kg/m^3)	300	167	229	128 (ref.5)

values cannot be assumed. Table 1 shows results of refuse analyses in four countries to illustrate the wide range of quantities that has been observed.

Variations with time are also required for planning and budgeting, and so projections of quantities must be made. These forecasts are best made with the aid of data for several of the preceding years, but if these are not available, estimates can be made from a study of the relationships between refuse generation rates and standard of living—as the proportion of the community in each socioeconomic stratum changes, so the refuse from that community will change. For this reason it is often necessary to analyse solid waste from households in different income groups and make comparisons.

THE TIME AND THE PLACE

The refuse analyses were performed during February, March, and April of 1973. At that time some sectors of the Iranian economy were planning for rapid growth, funded by oil revenues. There was a very large difference in income and lifestyle between the professional rich and the unskilled poor. The apparent stability of the country was attracting foreign investment of capital and expertise.

Shiraz is located in the south of the central plateau of Iran, and is a provincial capital with a university and international airport. At that time it was an important tourist centre. The more western-orientated citizens lived in luxurious villas near the university (Figure 1), whilst the traditional merchants, artisans, and labourers lived in houses built with sun-baked clay around a courtyard, entered from narrow alleys, near the sun-baked clay around a courtyard, entered from narrow alleys, near the bazaar (Figure 2). In the latter area the population density was estimated to be 400–500/hectare. Some large blocks of flats were being built on the outskirts.

The climate is seasonal, varying from a January mean minimum of 0 °C to a

Figure 1   High-cost housing

Figure 2   Traditional housing

July mean maximum of 37 °C. An annual rainfall of 330 mm is typical, and it falls mostly between November and April.

Pit latrines and soakage pits were the most common means of sanitation. Diarrhoea and gastro-enteritis were common, being diagnosed in about 50% of the children seen in the outpatient departments. The streets (but not the alleys) were drained by means of open channelss which were concrete-lined, about 400 mm wide and of varying depths. These were also used for irrigating roadside trees and for washing clothes.

An estimated 35% of the population had completed primary education.

This information is included here because of its relevance to solid waste problems. A more detailed account of the city, and the project as a whole, is given in a dissertation written by the author for the University of Shiraz.

COMMON PRACTICES AFFECTING THE GENERATION OF SOLID WASTE

The incomes of many families were very low and there was no unemployment benefit available. Labour-intensive activities that brought very small returns were therefore common. Broken or damaged items, that would be discarded in more affluent conditions, were repaired and reused. Bottles carrying a deposit (i.e. all milk and soft-drink bottles) were always returned. Servants in wealthy homes would intercept discarded items, and newspapers and magazines were sold to shopkeepers for wrapping purchases rather than being thrown away.

Packaging wastes were low because tinned food was not popular and many groceries were sold loose, to be weighed into paper bags or the customer's own container. A large part of a woman's day would be devoted to preparing meals and so food was brought fresh and unprocessed.

All except the poorest used paraffin or gas as fuels; ash was therefore not a major constituent of the refuse. Grass was rare and private urban gardens were small, so there was little garden waste. (Grass clippings were valued as fodder.) The typical yard had a small, shallow pool, the purpose of which may have been ornamental, but such pools were often used for washing dishes and watering flowers. These pools also produced bountiful crops of algae—large quantities of wet, odorous, slimy waste were therefore periodically left in the street or drainage channel for the refuse collector to remove.

SOLID WASTE COLLECTION AND DISPOSAL

The city employed approximately 600 labourers and 21 supervisors to collect refuse on a daily basis from each house. Collectors would knock on each door and have the wastes handed to them, or pick up the refuse from just inside an open door. Since a relatively large time was thereby devoted to a small

Figure 3   Modified wheelbarrow

quantity of waste, thorough examination was possible and so the worker would separate glass and any item of value, and even return cutlery inadvertently discarded with food waste.

Most of the collectors were equipped with modified wheelbarrows to carry the refuse to the nearest transfer area. These wheelbarrows had ball or roller bearings welded to the two rear legs so that the barrows could be moved along smooth roads without any vertical force being required. (One such barrow is shown in Figure 3.) However, they needed lifting on rough ground and were very unstable—a road strewn with refuse from an overturned barrow was not an uncommon sight. Some labourers complained of back pains as a result of lifting loaded barrows. Fifteen wheelbarrow loads were weighed and the weights ranged from 60 to 168 kg. To contain this amount of material it was necessary to extend the sides with metal sheeting.

Some men used four-wheeled handcarts which were more stable, quieter and able to carry greater loads; an example is shown in Figure 4. They had been designed to carry three bins each, but were invariably used with loose steel sheets to contain the refuse. As with the wheelbarrows the load was deposited by removing the sheets. Measured loads varied between 123 and 265 kg.

The number of households visited during one round trip varied between 35 and 113 (11 routes were studied). Refuse from offices, shops, and schools was collected together with domestic wastes.

The refuse collectors were also responsible for keeping the streets clean and irrigating the trees in their own areas. This was a commendable

Figure 4   Four-wheeled hand-cart

arrangement because one man had the full responsibility for the public environment of his area. There are cases in other countries where one department is responsible for cleaning the drains and another for sweeping the roads, with the result that one man cleans the drains and deposits the debris on the road, and then another man sweeps the litter and debris into the drains. Usually in Shiraz, men would collect refuse in the morning, and irrigate or sweep in the afternoon.

The transfer areas were not so commendable. As shown in Figures 5 and 6 they were either part of a wide street or else a vacant plot of land. Solid waste brought in by handcart was deposited on the ground and then laboriously loaded into trucks by shovel, necessitating much work and long vehicle-waiting times. Refuse was often left at the end of the day, thereby encouraging the breeding of rats and flies, causing complaints, and spoiling the environment. Scavenging also took place in these areas; glass was separated to be sold for 1–3 rials/kg (in 1973 1 rial was equivalent to about 0.6p sterling), and water melon rinds were set aside for feeding to animals. Scraps of bread were also saved, but the motivation for this appeared to be more religious than economic. Wood was taken for fuel.

Once loaded, the wastes were hauled outside the town in open lorry, tractor trailer, or compactor truck. The municipality had records showing that each load was purchased by farmers for 30 rials (£0.18) to improve the organic content of the soil; the result was also a great ugliness and a severe hazard to bare feet, since no effort was made to remove non-degradable material.

Not all of the town's solid waste reached this destination—there was a

Figure 5   Refuse transfer at road junction

Figure 6   Enclosed transfer area

noticeable degree of indiscriminate dumping and burning both by collectors and by the general public, on vacant lots, waste ground, and the dry river bed.

METHOD OF SOLID WASTE ANALYSIS

The analysis was performed by accompanying a refuse collector on one complete round trip, recording the number of houses he visited and the number of residents in each, and at the transfer area sorting the refuse into standard categories. The procedures were developed at several randomly chosen places before the statistically important phase was begun.

The choice of areas that satisfactorily represented different lifestyles was made with the help of a local university sociologist and a municipality housing official. Within these areas beats were chosen that included few shops. The refuse from each beat was sorted for at least three successive days. Five such beats were chosen and designated as A, B, C, D, and E.

On each occasion the author met the collector at the beginning of his round and noted the number of houses and residents (by asking the number of families and number of adults and children at each door, after the refuse had been handed over). Solid waste from schools and shops was put into a plastic sack to keep it separate, since the study was concerned only with domestic refuse. Piles of rubbish left in alleys were included because it was supposed that they had come from adjacent houses.

This technique resulted in samples of between 60 kg and 170 kg. American research[1] indicated that 90 kg was the minimum satisfactory sample size. But if the number of households contributing is regarded as more significant than the sample weight, it is clear that 60 kg was a satisfactory sample for Shiraz, since this quantity came from at least 35 households whereas 90 kg from a twice-weekly collection in the USA could come from as few as 10 households.

Once safely delivered to the transfer area, the refuse was deposited on a plastic sheet 2 m square, which served to keep the chosen sample separate from the rest, and conserve the dust and moisture. The material was then sorted into its components, each type of waste having its own plastic sheet—or bag, if the material was light enough to be blown away. Anything small enough to pass through 20 mm openings in a sieve was classed as 'fines'.

The categories chosen were: (1) food wastes and foliage; (2) paper; (3) miscellaneous combustible material; (4) plastics and rubber; (5) glass; (6) metal; (7) inert material; (8) fines.

The weight of material in each category was measured using a spring balance reading up to 8.5 kg (greater loads could not easily be held steady for reading). Weights and observations on the nature of the material were recorded. The date, time, weather, and the date of the previous collection were also written down. Two people could sort 100 kg of refuse in this way in an hour.

The equipment used for this analysis could be wrapped up into a bundle that could easily be carried by one person. This was necessary because the sorting was performed at several different transfer areas. Density measurements were made separately by loosely loading a 200 litre drum and weighing it with a set of bathroom scales. A measure of the moisture contents was obtained by drying the paper fractions in the sun and noting the loss of weight. These simple procedures were sufficient to give reasonable accuracy at very low cost.

SOCIOECONOMIC DATA

Five areas had been chosen to represent different lifestyles, but it was necessary to determine the degree to which they differed, and a ranking. It was assumed that four factors influence solid waste quantity and composition: family size; status and wealth; cosmopolitanism (meaning the extent to which traditional ways had been replaced by a more international style); and the location of the house. The first three were determined by means of a questionnaire survey that was conducted by a local university student. The enquiry forms were filled in anonymously by those able to read, and filled in by the student for the others. The response rate varied from 83% in the poorest area to 47% in the wealthiest.

The questionnaire was designed to allow cross-checking; several questions were asked such that they provided similar information in different ways.

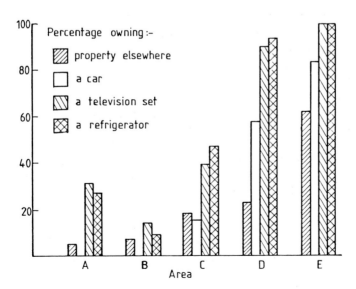

Figure 7   Socioeconomic data—wealth

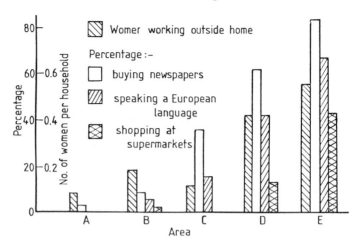

Figure 8   Socioeconomic data—cosmopolitanism

Questions relating to status and wealth concerned house size, ownership of cars, television sets, refrigerators and property, income, fraction of income spent on food, employment, use of servants, and number of wage-earners in the household. All criteria, except the last, showed that areas A and B were similar, but that there were clear differences between all the other groups, as Figure 7 shows. Questions relating to cosmopolitanism considered the extent to which women worked, newspaper readership, ability to speak a foreign language, and use of supermarkets. Figure 8 shows the results. Questions about habits related to refuse production concerned employment of servants and shopping frequency (and other questions about use of refrigerators and newspapers also had some relevance).

There was no indication of a preference for shopping on one day rather than another; most people went shopping at least three times a week and so it was reasonable to suppose that only the rest day, Friday, would produce refuse different from the other days.

RESULTS

Table 2 shows the results for areas A and B (together), and area E—the poorest and wealthiest respectively—giving both the component weights and the percentage by weight. The most important trend that is indicated is the increase of food and garden waste with income level. When the food waste component of the fines is added, the difference is even more pronounced. Since the amount of garden waste was always small, the most likely conclusion from these results is that the wealthier households are more wasteful. There was no sign of the use of the more expensive packaged foods reducing the quantities of organic waste. Much more inert material and dust

Table 2   Refuse weights and compositions from socioeconomic extremes

	Weight, g/cap./day		Percentage by weight	
	A and B	E	A and B	E
Total	250	400		
Food and garden waste	90	220	36	55
Paper				
moist	20	70	8	17
dry	18	33	—	—
Other combustible	18	16	7	4
Plastics and rubber	4	9	1	2
Glass	3	6	1	1.5
Metal	5	9	2	2
Inert material	17	14	7	4

came from the more traditional area, probably because of the widespread use of mud plaster as a construction material.

The moisture content of the paper was 16% for areas A and B and 52% for area E. Possible reasons for this difference include the large amount of dust (which would absorb water) and the lack of drains in the poorer areas, and the fact that water supplies were metered—so that changes of water in the yard pools would be less frequent in areas A and B.

Solid waste densities decreased from 260 kg/m^3 in the lower income areas to 170 kg/m^3 in the wealthier regions. Corresponding refuse volumes were 1.0 litre/cap./day to 2.4 litres/cap./day respectively. These density differences are in spite of the moisture contents and the higher degree of compaction expected from the bigger loads in the prosperous area.

PROJECTIONS

An attempt was made to forecast the way in which the quantity and composition of the city's refuse would change in the following years. In the absence of historical data three assumptions were made:

(1) that consumption habits of the wealthiest group would become more widespread;
(2) that solid waste trends in a smaller city follow those of a bigger city;
(3) that refuse generation can be estimated from production figures or imports of certain items.

One of the difficulties in the use of the third assumption is the determination of the fraction of the production that is going to the urban centres, as opposed to rural areas or for export.

The rate of growth of refuse quantities in other countries was also considered as a guide for forecasts.

The estimated increase in per capita weights over 20 years was about 100%, and the total weight of refuse to be collected in Shiraz in 1993 was projected to be over four times the 1973 figure.

It was thought that no sudden changes in the nature of the waste would take place, but only a gradual increase in the food waste and paper fractions. The amount of dust would decline as the use of traditional building methods diminished and as concrete or asphalt paving became more widespread.

RECOMMENDATIONS

The use of handcarts was regarded as a wise practice, providing a high level of service and also employment. The converted wheelbarrow type was inferior to the four-wheeled cart; the replacement of the former was suggested.

The transfer areas were inefficient and unsanitary. Simple, satisfactory transfer facilities could have been built with a split-level format to facilitate loading of the vehicles, and a wall around three sides to improve the appearance of the neighbourhood. The areas should have been paved for wet-weather working and to enable thorough cleaning of the site at the end of each day.

The wider use of tractors with trailers—each tractor responsible for several trailers—was regarded as appropriate.

The best disposal route for most of the refuse was composting because of the high percentage of compostable material, the need of the local soil for organic material, and the low cost of labour. The feasibility of separate collection should have been investigated; the personal contact between the householder and the refuse collector could have been used to encourage and explain proper separation, and the slow rate of accumulation of refuse in the handcart would have enabled frequent checking of the material destined for composting. Alternatively, if a small price were paid for material that should not enter the composting process, entrepreneurial spirit would ensure a useful degree of separation.

SEQUEL

Since this study it was rumoured that the city council was planning to buy two-stroke tricycle vehicles for house-to-house collection. Such a change would have had the following disadvantages:

(1) Cost; the money spent on these vehicles could have been used more effectively elsewhere—for example improving the transfer areas and preparing a simple composting facility;

(2) Manpower changes; the vehicle driving test included a written examination so no illiterate man could drive. Therefore the refuse collectors would have lost their jobs when the vehicles were introduced and there would have been a scarcity of licensed drivers who would consider the menial task of refuse collection. A number of skilled mechanics would also have been needed.

(3) Pollution; the noise and exhaust fumes in the narrow alleys would cause considerable nuisance.

The author is not aware of what actually happened regarding this issue.

The other changes, of which the whole world is aware, have been the revolution and war with Iraq. The resulting changes in lifestyle and industrial output have rendered the projections of 1973 totally invalid, and the new legislation on wages and employment conditions have undoubtedly affected the cost and availability of labour. The author hopes that this experience will not deter others from attempting to make forecasts!

REFERENCES

1. Britton, Paul. 'Improving manual solid waste separation studies'. *Journal of the Sanitary Engineering Division, ASCE,* **98** (October 1972), 717–30.
2. Refuse disposal studies at Calcutta; *Technical Digest of the Central Public Health Engineering Research Institute,* Nagpur, India, No. 15, March 1971.
3. Flintoff, Frank, and Hobby, Ronald. *Assignment Report, Waste Management and Disposal,* Libya, 0035/R; WHO, EM/ES/186, December 1971.
4. Stirrup, F. L. *Public Cleansing: Refuse Disposal.* Pergamon Press, London, 1965.
5. American Public Works Association. *Municipal Refuse Disposal,* 2nd edn. Public Administration Service, Chicago, Ill., 1966.
6. Gilbertson, Wesley E. Present and future trends in municipal disposal of solid wastes; a contribution in *Problems in Community Wastes Management,* Public Health Paper No. 38, WHO, 1969.

Practical Waste Management
Edited by J.R. Holmes
©1983 John Wiley & Sons Ltd

26

Solid waste aspects of public health engineering in Java, Indonesia

R. J. OWENS

ABSTRACT

Based on an Urban Development and Sanitation project in West Java, Indonesia, this paper elaborates the methods to be used in establishing a master plan in a Far East city. The text describes the geographic features of the area, the types of housing, the nature of wastes generated and the current inefficient methods of collection and disposal. Recommendations for future good practice are set out as are waste volumes and densities for the area. Design parameters and a guide to decision making paths are set out in the text.

INTRODUCTION

Bandung Urban Development and Sanitation Project

The study, over a period of 15 months, involved a multidisciplinary team of 26 expatriate and 15 Indonesian specialists from several consulting firms based in Indonesia, the Philippines, Australia, and England. The Final Report, presented in 1979, consisted of three volumes:

(1) The Main Report;
(2) Development Strategy;
(3) Feasibility Study.

In addition, twelve technical papers prepared during the study on specific study subjects, including three on solid waste, were published:

(1) Surveys and Forecasts;
(2) Alternative Disposal Methods;

517

(3) A Pilot Study of Composting.

The main objective of the study was the preparation of a long-term development plan and within this framework the formulation of first-stage projects for implementation over a 5-year period. These projects included the provision of infrastructure for new housing development, improved services to existing urban communities (kampungs), phased development of sewerage, drainage, and solid waste management. Implementation of first-stage projects commenced in 1981.

Indonesia

This archipelago consists of over 13,000 islands, extending over a distance of some 3,000 miles, with a total land area of over 1.9 million km^2 and a population of over 150 million. The official language is Bahasa Indonesia and Islam is the predominant religion. Jakarta, the capital and largest city, with a population of 6.0 million, is located on Java, one of the larger islands, as is Bandung, with a population of 1.3 million.

Agriculture is the dominant sector of the economy, employing over 60% of the labour force. Main crops included rice, rubber, palm oil, coconut oil, and, more recently, sugar. The main industries include mining, fisheries, forestry, textiles, and manufacturing. Natural resources include gas, oil and minerals. The GNP per capita is $US390 (1980).

Bandung

The study area, defined as the Bandung Region, covers an area of 3,250 km^2 with a population of approximately 3.5 million (1976 census). Within this region is the City of Bandung with an area of 8,098 hectares and a population of 1,293,000 (1976 census). The City with adjacent districts influenced by the medium-term growth form the Bandung Metropolitan area covering 540 km^2 with a population of 2,150,000 (1976 census) of who 75% live in urbanized areas. The Metropolitan population is expected to increae to 2.6 million by 2001 AD.

Key productive areas of the regional economy are agriculture and manufacturing, employing approximately 60% and 7.5% of the work force respectively.

Bandung lies at the northern edge of a flat plateau at an elevation of 700 metres, surrounded on three sides by a volcanic mountain range which rises to more than 2,000 metres. The southern half of the city, and the area south of it, lies on a flat wide plain. The plateau, drained by the Citarum River, is traversed from east to west by the main National road and rail routes, which provide the principal axis for urban development as the City spreads beyond its boundary. The radial road network from the City provides secondary spurs for development.

The climate is tropical, with temperature varying between 24 and 32 °C. Rainfall occurs mainly during the months of January and February with average annual rainfall for the Bandung Region of 1500 mm, with an average figure for January of 200 mm.

TECHNICAL BACKGROUND AND COMPONENTS OF THE SANITATION STUDY

Topography

Urban development to the south-east and west of the City and south of Citarum River is on lake deposits consisting mainly of impervious clay. Most of the City is on more permeable material, with younger volcanic deposits to the north. Land adjacent to the Citarum River is flat but ground slopes in the Metropolitan area are between 0.5 and 2.0%. An extensive system of channels and rivers drain the area with significant seasonal variations in flow.

Health

There are no reliable statistics on existing health conditions, diseases, and causes of death; but it is known that during the early 1970s life expectancy at birth was 45 with crude death rate about 19 per 1,000, and infant mortality 160 per 1,000 live births. During the same period in the UK comparable figures were 72 years, 12 per 1,000, and 18 per 1,000 respectively. Child mortality in Bandung is high; over 20% die before they reach the age of 5.

Housing

There are, generally speaking, two categories: formal and informal housing. Formal housing, which conforms to planning and building regulations, houses about 23% of the City population. It occupies 31% of the developed land with population densities varying from 100 to 400 per hectare. Informal housing has been built without assistance from formal institutions and building enterprises and has developed independently of official standards and regulations. About 77% of the City's population is housed in this way, occupying 44% of the developed area with population densities varying from 400 to 600 per hectare. Informal development consists of permanent, semi-permanent, and temporary housing in the ratio 40 : 30 : 30. The average size of households in the Metropolitan area in 1976 was 5.36 persons. Over a quarter of all households were assessed to be in shared accommodation.

Roads

The theoretical capacity of the road system is adequate for the volume of traffic. In practice—for various reasons such as lack of pavements, laybys, car parking facilities, and an overall road hierarchy—there is congestion. Non-motorized traffic and numerous street vendors effectively reduce the carrying capacity of the road system. The regional road system is routed through Bandung and the radial road structure, with the absence of completed ring roads, leads to the use of narrow streets in the city centre by through traffic. In 1977 there were 55,000 passenger cars, 1,600 buses, 21,000 trucks, and 124,000 motor cycles based in Bandung.

Water supply

The public water supply is available to 20–30% of the City population and is of doubtful quality. The daily supply amounts to 90,000 m^3, with 85% of this from a river source, 10% from springs, and 5% from artesian wells. Because the public supply is inadequate the majority of the population make alternative arrangements; surface water is polluted and unsuitable for domestic use and numerous shallow wells are used providing an estimated 60,000 m^3 per day. The quality of this groundwater is poor and there is evidence of faecal pollution.

Sewerage

Only about 7% of the City population is served by a municipal trunk sewer system which is underutilized. A small sewage treatment works gives inadequate treatment to the flow. In the City some 15–20% of the population is served by septic tanks. Part of the suburban area is served by a combined system consisting of concrete-lined open channels. This works well in areas of higher-quality housing but other areas are served by poorly constructed and unlined channels with inadequate gradients, which result in blockages and sections containing septic sewage. Few industries treat their wastewater before discharge to water courses and there is no effective control.

Drainage

As the urban areas of Bandung have expanded, the natural drainage system has been developed by constructing channels and pipelines. Canals were also constructed to divert flow from the upper catchment so that the full flow did not pass through the City but this approach is no longer practical due to urban expansion. The drainage system is adequate and major expansion is unnecessary. However, encroachment on the waterway by buildings and other developments, plus silting, dumping of solid waste, and illegal weirs have reduced the effective capacity of the system.

Solid wastes

Less than 10% of the urban area within the City is served by a collection system managed by the 'Municipal Sanitation and Beautification Office' (DKKK). This system is supplemented by a service provided by a local community organization (RW) headed by a community leader. This organization is responsible for 200 to 350 households in about seven neighbourhood units. There are some 233 such organizations which have instituted a collection system serving about 436,000 people. The principal means of collection is by handcarts, which convey the solid waste to temporary collecting places. From these transfer points waste is transported by trucks to the disposal sites. The transfer depots are unsatisfactory in design and location. They are open to weather and to scavengers; the wastes have to be picked up by hand for loading; the pick-up vehicles obstruct the traffic. The 54 trucks used for transferring wastes to disposal sites are fitted with fixed bodies and must be unloaded by hand.

At the disposal sites open dumping is used, the waste is allowed to remain uncovered, and no mechanical compaction occurs. At all stages of the system, from home storage to final disposal, material that can be reused is removed from the waste.

Summary

The review in this section of the various engineering components of the project may provide some surprising statistics to those unfamiliar with large urban conurbations in developing countries in Southeast Asia. The purpose of the project was, of course, to identify, amongst other things, those areas where action was required to improve sanitary conditions and public health in particular.

Although the existing conditions described in this section leave much to be desired, the residents in Bandung are a happy and naturally friendly people and this, combined with a favourable climate and beautiful countryside, resulted in a pleasant working environment for the study team.

The study included a review of agencies relevant to urban development, including central, provincial, and local government levels, as well as private organizations. Within the Metropolitan Area, below central government level, there are numerous agencies responsible for management of urban services. The study included recommendations regarding the provision of a structured institutional framework which, although most important, related to specific existing political conditions and have not, therefore, been covered in this review.

SOLID WASTE

Objectives

With an existing collection system covering some 30% of the urban area, the principal objective of the study was to improve various aspects of the system and extend it to cover the City area. The damage to the environment of the uncontrolled disposal of solid waste can be clearly seen. The waste is dumped in the streets awaiting transport to disposal sites and in the numerous rivers in areas where for several years residents and industry have had no alternative methods. At the disposal sites unrestricted access creates health hazards and smoke from fires is a common occurrence. The main objectives of the study can be summarized as follows:

(1) the protection of public health and environment at a cost which can be met locally;
(2) the development of a system based on local climate and physical, social, and economic factors;
(3) the selection of appropriate technology in the design of equipment and methods;
(4) eduation of the public to ensure widespread understanding of the value of the system and its operation;
(5) implementation of manpower policies and training opportunities for supervisory and management staff to develop personnel with the incentive and capacity to operate the system.

The existing system

As with most solid waste systems the operation in Bandung involved storage, collection, transfer, transportaton, and disposal. The main sources of waste were domestic, markets, and commercial premises/streets. As described in the section on solid wastes, the collection system is provided by two organizations, DKKK (municipal) and RW (local community).

Storage

Storage containers used by households varied from 10–litre buckets to fixed containers in concrete, brick, or steel either in the front garden or the street.

Collection services

DKKK is responsible for the solid waste management in the City. It is organized on a unit basis with the City divided into eight units operating from small, local inadequate depots and each unit is divided into a number of sections. These serve premises mainly in the commercial areas but also some residential areas.

Table 1 Staffing and equipment—DKKK solid waste collection—April 1978

Wilayah	Unit Number	Number of groups	Handcarts	Workers
Bojonagara	I	7	8	26
	II	10	15	60
Cibeunying	III	13	18	107
	IV	10	19	75
Karees	V	10	15	67
	VI	12	17	79
Tegallega	VII	6	8	41
	VIII	4	6	33
Totals			106	408

Source: DKKK

Details of staffing and equipment issued by DKKK in April 1978 for solid waste collection are given in Table 1. The area served is estimated to be 470 hectares or 9.5% of the area of developed land (4.931 hectares in 1976). RW organizations are established in every Kampung (urban village) within the City to provide a number of community services. Out of a total of 720 RWs in the City, 233 have instituted a collection system serving an estimated population in 1978 of 436,000. These services were established with encouragement of the City Government, although no standard is set and little financial assistance is provided. The quality of service varies, with collections daily in some cases, others on alternate days or twice weekly. Handcarts are used and although DKKK provide these in some cases there is considerable ingenuity in design of others. For example some boxes are mounted on pedal tricycles, a favoured form of transport for passengers and/or goods available for hire (with driver). The total area of the city served by the two systems is about 1,400 hectares or about 29% of the estimated urban area of the City.

Equipment

The handcart is the principal item of equipment for collection of solid waste from premises. This consists of a wooden box usually about 1 m^3 in volume mounted on a single axle fitted with solid rubber-type wheels. Empty, the cart weighs about 200 and 300 kg and when fully loaded can weigh 700 kg.

Transfer depots

These usually consist of a three-sided concrete structure, up to 10 m^3 capacity, but can be as small as 2 m^3, located at the sides of roads. Material dumped from the carts is hand-loaded into trucks for transfer to dump sites. The arrangements are unhygienic and restrict traffic movement, and the location of transfer depots is somewhat arbitrary.

Tranport

A fleet of 54 fixed-body motor trucks, either municipal or on contract, are used to transport the waste to dump sites. These are organized in four political divisions (Wilayah) of the City, as shown in Table 2. In April 1978 the transport section advised that they employed 144 men and transported an average of about 1,670 m^3 of waste each day to three dump sites. Three front-end loaders were hired either from the City or the Provincial Public Works Department, but these, as with many of the trucks, were not fully operational due to repairs and maintenance. A survey of the transport operations was made over one week in June 1978 at the three disposal sites covering zones shown in Figure 1. Truck movements and volumes of waste delivered were measured each day and the results for one week are included in Table 3.

Table 2   Transport allocation—solid waste system—Bandung, 1978

	Number of trucks		
Wilayah	Municipal	Contract	Total
Bojonagara	9	8	17
Cibeunying	8	6	14
Karees	6	7	13
Tegallega	3	7	10
Total	26	28	54

Source: DKK Transport Section.

Table 3   Transport of solid waste—Bandung, 1978

Item		Dago	Cicabe	Cieunteung
		Disposal area		
1	Volume delivered (m^3)	2,328	1,554	2,273
2	Number of vehicle trips	426	253	383
3	Man-hours	7,658	2,547	6,143
4	Vehicle-hours	845	353	632
5	Volume per man-hour	0.30	0.61	0.37
6	Volume per vehicle-hour	2.75	4.4	3.6
7	Number of temporary collecting places served	31	29	23

Source: BUDS Survey, 26 June–1 July 1978.

Disposal

The operation of the three disposal sites is similar; waste is unloaded from the trucks and either falls or is pushed over the working face and there is no mechanical compaction. On one site, which is flat, material is mounded by a front-end loader. The estimated life of the three tips based on assumed improvements and increased compaction to 400 kg/m^3 were 2, 5, and 6 years. These periods would be reduced with improved collection systems.

Recovery of materials

Useful materials are removed informally for reuse at all stages from house-holders' waste containers up to and including the tipping site. There is little of salvageable value, therefore, remaining in the waste at the disposal site which would justify recovery. Whilst this process creates a health hazard, it provides a useful service as well as income for a significant number of people. Some control is desirable but prevention of scavenging would be diffiult and a compromise solution would be appropriate in the circumstances.

Investigations and predictions

Sources

The three principal sources considered were residential areas, markets, and commercial areas. As smaller home industrial sources are scattered through-out the residential areas and some residences were used for commercial purposes the waste from these sources was included as residential areas. Street cleaning involves mainly collecting commercial waste, and quantities generated are therefore included under the commercial heading. Most of the waste generated from major industries is recycled or reused; none finds its way into the collection system, and no change is anticipated in future. Waste from City parks and gardens is assumed to be disposed of directly by the appropriate municipal department.

Some changes in the source of material can be expected as the solid waste management system and services in the city generally are developed.

Forecasts of solid waste generation

(1) *Domestic.*   With limited information available locally it was decided to carry out a solid waste survey to include the various sources of waste on a sample basis using the existing collecting system for sampling. Key parameters for the survey included:

(a) population;
(b) income levels;
(c) volume of waste;
(d) density of waste;
(e) composition of waste.

Of these, income levels and consequent spending and living habits proved to be the most difficult.

Eight areas were selected for the survey, which was conducted in two phases. These were considered to be representative and included both formal and informal housing with a reasonable geographic spread in the four Wilayahs in the City. The areas were selected taking into account factors such as:

(a) those with adequate existing collection system at community level;
(b) those where disposal to rivers or other dumping areas was difficult;
(c) those where there was a reasonable representaton of various 'domestic' sources such as cottage industry and incidental commercial activity but avoiding major industries, markets, or main road commercial activities;
(d) those where co-operation and interest was shown by the community leader.

Collection during Phase 1 survey was monitored for 8 days; the first day being a test run, the results were not used. Three of the eight areas were re-surveyed (during Phase 2) as it was found difficult to interpret results due to the irregular pattern of the gangs using the existing system of collection.

The characteristics of the area surveyed are given in Table 4 and the results

Table 4   Characteristics of areas surveyed in domestic solid waste generation survey, Bandung, March 1978

Lingkungan/RW	Average daily population surveyed	Average daily no. houses surveyed	Average number people per house	Percentage of house type		
				I and II	III	IV
1 Pasirkaliki RW 02	563	76	7.5	99	1	—
2 Cihapit RW 03	475	65	7.4	100	—	—
3 Cijagra RW 01	1,356	193	7.0	96	4	—
4 Pasirkaliki RW 04	286	43	6.7	72	24	4
5 Nyengseret RW 05	477	74	6.5	71	28	1
6 Cihaurgeulis RW 13	924	100	8.5*	62	36	2
7 Cicadas RW 14	976	131	7.5	49	40	11
8 Pungkur RW 12	692	91	7.6	85	6	9
Total survey	5,749	773	7.4	80	16	4

* Excluding two institutions with 131 people between them.

Source: BUDS Domestic Solid Waste Survey. Dates of survey, 13–20 March 1978. Only last 7 days' results used; first day's results disregarded.

in Table 5. The results in terms of volume and weight generation rates show a larger than expected range.

Analysis of the results raised two main problems: measurement of contributing population and classification of house types (based on definitions used in the 1971 census) for different levels of waste generation. These were researched more thoroughly to eliminate anomalies arising from irregular collection patterns and classification of house types in the 1971 census in terms of building construction in Bandung.

A second survey using improved techniques in areas with more reliable collection systems and improved supervision was arranged to provide additional data from three areas. The results showed an increase in generation rates in all three areas, reflecting the view that a regular systematic system will record higher generation rates than an irregular one. Planning for an

Table 5   Results of first phase of domestic solid waste generation survey, Bandung,
March 1978

Lingkungan/RW	Waste generation per day			Density (kg/litre)	
	Volume per head (litres)	Volume per house (litres)	Weight per head (kg)	Average	Range
1 Pasirkaliki RW 02	3.74	27.9	1.10	0.29	0.16–0.40
2 Cihapit RW 03	2.89	21.3	0.51	0.18	0.11–0.29
3 Cijagra RW 01	0.84	5.89	0.24	0.28	0.16–0.40
4 Pasirkaliki RW 04	1.7*			0.14	0.07–0.22
5 Nyengseret RW 05	2.23	14.4	0.53	0.24	0.14–0.32
6 Cihaurgeulis RW 13	0.63	6.09	0.11	0.18	0.15–0.21
7 Cicadas RW 14	0.52	3.86	0.15	0.29	0.28–0.30
8 Pungkur RW 12	0.92†	6.96	0.20	0.25	0.20–0.29

* Population served was difficult to estimate due to pattern of collection; therefore this estimate is less reliable.

† For first 3 days of survey, generation rate was 0.92 litres per head per day. This is regarded as the most reliable measure for this survey area.

Source: BUDS Domestic Solid Waste Survey, Phase 1.

improved solid waste collection system should therefore allow for the higher generation rates. The following basis was therefore used:

(a) 80% of the area, mainly informal housing, an average of 1.6 litres per person per day, the average in the Phase 2 survey (range 1.3–2.4);
(b) 20% of the area, mainly formal housing, an average of 3.3 litres per person per day.

(2) Markets.   These are one of the four main sources of solid waste generation. Some 52 markets were in operaton in 1977 varying from 1,200 to 61,870 m^2; a total of 153,840 m^2. Some 16 new (including replacements) markets were under construction in 1978 with areas varying from 1,400 to 14,850 m^2, a total of over 51,000 m^2. With significant volumes from each site a separate collection system can be justified. Generation forecasts were based

Table 6   Estimate of market waste generation in Bandung, 1986

Kecamatan	Market area (m^2)	Volume generated per day (m^3)	Population ('000)	Volume per head of population per day (litres)
1 Sukasari	2,725	13.6		
2 Sukajadi	8,350	41.7		
3 Cicendo	5,489	27.4		
4 Andir	78,271	341.3		
Wilayah Bojongloa	94,835	474.0	328.8	1.44
5 Cidadap	2,004	10.0		
6 Coblong	1,917	9.6		
7 Bandung Wetan	27,533	137.7		
8 Cibeunying	17,935	89.7		
Wilayah Cibeunying	49,389	247.0	399.9	0.62
9 Bandung Kulon	2,600	13.0		
10 Astanaanyar	10,924	54.6		
11 Babakan Ciparay	8,606	43.0		
12 Bojongloa	8,560	42.8		
Wilayah Tegallega	30,690	153.4	341.0	0.45
13 Regol	7,390	36.9		
14 Lengkong	8,236	41.2		
15 Batununggal	3,369	16.8		
16 Kiaracondong	12,255	61.3		
Wilayah Karees	31,250	156.2	397.7	0.39
Kotamadya Bandung	206,164	1,030.6	1,467.4	0.70

Source: BUDS

on a survey of four markets representing the range and size and type. The survey of the four markets included occupancy ratios, number of stalls, percentage of total stalls by type, solid waste generation in litres/m^2/day for various users. There are, of course, a number of variables such as type of market, hours of operation, and seasonal supply of some goods. Floor area was adopted as the most important variable and, from records of market size and location, forecasts of market waste generation (mainly organic matter) prepared for the City.

A detailed analysis of market facilities was prepared for 16 administrative areas in the City including two regional markets as well as local ones. The market areas per 1,000 population varies considerably, but for existing markets in the City the average area per 1,000 population is 73 (local) and 119 (all markets), the corresponding anticipated figures for 1986 were 84 and 140.

The total quantities of solid waste from markets at the end of the first-stage project (1986) and the average volume per head of population in various areas in the City is given in Table 6. The solid waste study not only provided

Table 7   Survey of solid waste collection from commercial premises—Bandung, April 1978

Unit/ group	Trip 1 Vol. Wt (m³)(kg)	Vol. Wt	Length	Days since swept	Trip 2 km/ day	Tot vol.	Generation (m³/km/day)	Date (1978
III B3*1.35 506	1.01 102	2,515	1	2.5	2.36	0.94	10 April
1.22 242	0.95 178	2,515	1	2.5	2.17	0.81	11 April
1.35 430		2,515	1	2.5	1.35	0.54	12 April
	0.54 262	1,125	0.5	0.6	0.54	0.9	12 April
				8.1	6.42	0.82	(Average)
III B4†1.09 218	0.89 160	1,605	1	1.6	1.98	1.23	10 April
1.30 346		1,605	1	1.6	1.3	0.81	11 April
	0.54 210	872	0.5	0.4	0.54	1.35	11 April
1.09 150		1,605	1	1.6	1.09	0.68	12 April
	0.83 164	460	3	1.4	0.83	0.59	12 April
	0.43 34	872	0.5	0.4	0.43	1.07	12 April
				7.0	6.17	0.96	(Average)
III C1‡1.35 316		1,762	1	1.8	1.35	0.75	10 April
1.48 236		1,762	1	1.8	1.48	0.82	11 April
	1.28 462	825	1	0.8	1.28	1.6	11 April
1.55 318		1,762	1	1.8	1.55	0.86	12 April
	0.95 114	825	1	0.8	0.95	1.2	12 April
				7.0	6.61	1.05	(Average)

Average over these groups = 0.94 m³/km/day.

  * Jl Asia Afrika, Tamblong, Naripan, Sunda.
  † Jl Braga, Asia Afrika, Suniaraja, Merdeka, Lembong.
  ‡ Jl Asia Afrika, Alkateri, ABC, Banceuy.

Table 8   Solid waste generation forecasts for Bandung, 1978–2001

Year	Source	Unit	Quantity	Rate	Total waste generated (m³)
1978	Residential				
	Formal	population	258,600	3.3 l/h/day	853
	Informal	population	1,034,200	1.6 l/h/day	1,655
	Markets	population	1,292,800	0.42 l/h/day	543
	Commercial	kilometres of street	47.2	0.94 m³/km/day	44
	Estimated total solid waste generated				3,095
1986	Residential				
	Formal	population	292,400	3.3 l/h/day	965
	Informal	population	1,175,000	2.0 l/h/day	2,350
	Markets	population	1,467,400	0.69 l/h/day	1,013
	Commercial	kilometres of street	60	1.76 m³/km/day	106
	Estimated total solid waste generated				4,434
2001	Residential				
	Formal High density	population	300,000	3.8 l/h/day	1,140
	(formal)	population	163,000	3.1 l/h/day	505
	Informal	population	1,200,000	3.1 l/h/day	3,720
	Markets	population	1,693,100	1.1 l/h/day	1,862
	Commercial	kilometres of street	93.6	2.74 m³/km/day	256
	Estimated total solid waste generated				7,483

Source: BUDS, Background Paper No. 10.

Table 9   Chemical composition of solid waste—Bandung

Parameter	Percentage
Moisture content	83.5
pH	6.6
Organic matter	66.5
Carbon	38.6
Nitrogen	0.5
Phosphorus as P_2O_5	0.4
Potassium as K_2O	11.5
Calorific value	231.7
Carbon/nitrogen ratio	77.2

Source: BUDS

valuable original forecasting data for feasibility studies but also provided guidelines for the urban planning component of the project.

(3) Commercial premises   Significant quantities of solid waste are generated daily, and generally this is deposited on footpaths and streets to be collected daily. Street cleaners using brooms and handcarts collect waste from the street and also baskets used for storing waste in the premises. A solid waste management system would reduce the amount of work undertaken by the street cleaners.

A survey was arranged of one group in each of these administrative units of DKKK in the commercial areas of the City in April 1978. The generation rate was calculated on a street frontage basis, DKKK providing a list of streets swept in the whole of the City, measured and classified between residential and commercial/industrial. The total length of commercial streets was estimated to be 47,225 km and without details of use of premises the length of streets was used as the best basis for waste generation and collection efficiency. The survey covered a total length of 5,880 m and serviced 379 shops, 89 offices, 121 houses, and 32 other users. A total of 19.2 m^3 was collected from these premises in three days but not all streets were cleared each day and one was cleared twice daily. The results of the survey are given in Table 8 and an average rate for commercial premises of 0.94 m^3/km/day was established.

To provide forecasts of solid waste generation from commercial premises for the first-stage project (1986) allowance was made for population growth, changes in disposal income, and improvement in solid waste management. The sum of these three factors was assessed to be 8.15% per annum and a figure of 1.76 m^3/km/day adopted for 1986, a total of 106 m^3/day.

Summary.   Forecasts of solid waste generation for 1978, 1986, and 2001 from various sources based on detailed study and surveys are included in Table 8.

Solid waste in Bandung has a very high percentage of vegetable matter, as can be seen from Table 9. This results in high densities and rapid decomposition in the tropical climate. Collection and disposal of waste must therefore be frequent and reliable to avoid offensive conditions.

The high moisture content and low calorific values indicate that incineration is an unsuitable disposal method. The high vegetable content suggests that composting should be considered, and a pilot project was recommended as part of the first-stage project.

The carbon content was established by dividing the percentage of organic matter by 1.724 as apparatus for determining total carbon (electric combustion method) was not available in Bandung. This method appeared to produce a high carbon to nitrogen (C/N) ratio for the type of waste found in countries like Indonesia. The tests were repeated over one week on composite samples collected each day and the results are incorporated in Table 10. The C/N ratio of 77.2 is on the higher end of the range for composting but should be acceptable if the method is financially sound.

Density. The loose density of solid waste was about 140 kg/m^3; this increased when measured in handcarts to a range from 250 to 300 kg/m^3. On transfer to open trucks density was found to be in the range between 180 and 250 kg/m^3.

A Dutch-sponsored solid waste pilot study at one market over several weeks (October 1977 to January 1978) used a DAF compactor truck for transporting wastes. Densities in this truck were found to be in the range of 220–320 kg/m^3.

Future trends in composition.   Changes in the composition of solid waste in Bandung are difficult to predict, as there are no historical data available in Bandung or other cities in Indonesia. Trends in other countries can be used as a guide; these indicate increases in paper, plastics, metals, and glass. However, no significant change can be foreseen in the vegetable content, which is the predominant component of the present waste in Bandung, and the informal sorting of useful materials during the disposal process can be expected to continue for some time. Routine analysis of waste should provide more reliable data for long-term strategic planning.

RECOMMENDED SOLID WASTE MANAGEMENT SYSTEM

Introduction

The recommended solid waste management system is briefly described in this section. The procedures, investigations, and socioeconomic factors leading to the selection of the recommended scheme are those generally used and are not, therefore, detailed.

Decision areas on the generation, storage, and collection of solid waste are shown in the form of a flowchart in Figure 1. The generation of solid waste, the physical and chemical characteristics, and reduction in volume by uncontrolled salvaging have been covered in previous sections. A description of the recommended system for storage, collection, transport and disposal is given in summary form in Table 10, and brief details of the various components are given later in this section.

Collection Frequency

Bin Sizes   Formal
                                         Informal
                                         Markets
                                         Industries

Handcarts

Formal Sector   Size
                                         Capacity
                                         Performance/day

Handcarts

Informal Sector   Size
                                         Capacity
                                         Performance/day

Transfer Depots   Size
                                         Location
                                         Range Served

Transport Vehicles   Capacity
                                         Payload
                                         Performance/day

Disposal Areas   Daily Generation
                                         Volume Required
                                         Cover Required
                                         Plant Required

Other Solutions   Composting
                                         Materials Recovery
                                         Incineration

Some base ratios used in feasibility study

Household size	4.86 persons
Collection frequency	3 times per week
Bin sizes, formal	40 litres
Bin size, informal	30 litres
Bin size, markets	160 litres

Generation rate, formal	3.3 litres/day/house
Generation rate, informal	2.0 litres/day/house
Handcart capacity	6 × 160 litres Bins
Mean handcart performance	160 houses/day
Refuse compactor vehicle capacity	4.1 tonnes
Number of loads from transfer points	5 per day
Daily capacity per vehicle	20.5 tonnes
Number of handcarts serving one vehicle	20
Population served per transfer depot	15,550
Households served per transfer depot	3,200

Storage bins

Three sizes of storage bin were proposed: 30, 40, and 180 litres capacity. The smaller sizes were to be used in residential areas, the smallest in informal development, the larger in formal development, reflecting the different generation rates, access, and storage problems. The 180-litre capacity containers are for market use on a communal basis and are sized so that six of them can be accommodated on a standard hand-cart. The material, size, shape, covering, lifting arrangements, and method of identification of the containers were specified.

Hand-carts

A standard hand-cart was designed for collection throughout the City, capacity about 1 m^3 to limit gross weight. This consisted of a lightweight tubular frame fitted with pneumatic-tyred wheels carried on a single axle with tapered roller bearings. A hinged tailgate provides a ramp for loading purposes.

Transfer depots

These buildings provided access ramps for hand-carts; containers (160 litres) are off-loaded under cover, emptied into the truck, and washed for return to hand-carts. Storage space is provided for full and empty clean containers. A washroom is provided for operatives. Depots are located to limit the operation of hand-carts to a maximum of 1 km and a substantial increase in depots is needed to cover the City; a total of 95 is required. However, it must be appreciated that 80% of the population is located in informal housing areas where access for larger vehicles is not possible and hand-powered carts are more economic than those using electric motors or internal combustion engines.

Transport

The transport of solid waste from transfer depot to disposal area will be by motor truck. This will be fitted with a special purpose, medium-size body (11.5 m^3) for use in solid waste transport. This vehicle would be powered by a diesel engine (at least 120 brake horsepower) and have a nominal carrying capacity of 7 tonnes (payload of 4.7 tonnes). The body is non-tipping for stability at the disposal sites with compaction and emptying by means of a single packing plate, hydraulically operated.

Disposal

The disposal method adopted, after considering various alternatives, was sanitary landfill; after evaluation of 11 sites, including the three existing ones, four sites were selected, two existing and two new. Evaluation of each site included reviewing 17 principal factors such as area available, cost, access, topography, geology, hydrology, environmental factors, and cost penalties in operation. A transport plan was prepared to ensure most efficient routeing of trucks from the transfer depots to the disposal sites. The method of operation of each disposal site was also specified.

Other factors

The feasibility study included proposals for phasing, estimates of capital cost, operation and maintenance costs, financial analysis, investment requirements, sources of finance, financial feasibility, economic benefits, revenue, and organizational structure for implementation. These factors are of particular relevance when, as in this case, the project was financed by a Development Bank.

ACKNOWLEDGEMENTS

This chapter is a brief synopsis of the solid waste component of the Bandung Urban Development and Sanitation Study Report. The project was funded through the Asian Development Bank and the implementing agency for the Government of Indonesia was the Directorate General of Housing, Building, Planning and Urban Development, Ministry of Public Works, whose permission to publish this chapter is gratefully acknowledged.

Practical Waste Management
Edited by J.R. Holmes
©1983 John Wiley & Sons Ltd

27

Assessment of solid waste management problems in China and Africa

P. A. Oluwande

Professor of Civil Engineering, Faculty of Technology, University of Ibadan, Nigeria

ABSTRACT

A contribution from a prominent academic in West Africa and a consultant to the World Health Organisation, this chapter looks at very basic solid waste management problems in the author's experience in China and Africa. The peculiar nature of urban solid wastes, the rate at which the waste is generated, its storage, collection and disposal are considered. Some basic low cost but effective systems of collection and disposal are recommended as are guidelines in the selection of refuse vehicles and plant. The shortage of funds, a common problem in much of the developing world, is viewed as to its impact on solid wastes management.

The problems of solid waste management in the developing countries are discussed in this chapter from the point of view of experiences in China, Nigeria, Ethiopia, and Kenya. The most important aspects of solid waste management in the developing countries are related to problems of (1) effective storage in generating premises, (2) collection, and (3) efficient transportaton of the waste to disposal sites. The different factors which aggravate these problems are discussed together with the various approaches being adopted for their solutions. Suggestions are also made for more effective and efficient approaches.

INTRODUCTION

The problem of sanitary disposal of refuse is becoming very serious in many developing countries. In a few countries, especially those which are enjoying some levels of economic boom (like Nigeria), the management of disposal of solid waste seems to be so formidable as to defy the existing financial, technical, and administrative potentials and capabilities of the authorities in charge.[1] In spite of the regular publicity in the mass media and the constant political statements and actions made by highly placed personalities in authority, refuse remains a permanent feature along urban roads. On many occasions the important streets are blocked completely by refuse[2] (see Figure 1).

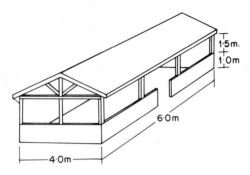

Figure 1   A rectangular enclosure with roof (shed type)

Culturally, many people in the developing countries have taboos and beliefs which make household refuse items odious objects to live with. Among many tribes, one of the first duties in the morning, of the housewife or the youngster in a household is to sweep the floor. This is illustrated by this quotation translated from Odunjo[3]:

Wash your face in the morning when you wake up,
Brush your teeth with chewing stick,
The plates are all to be washed and
The floor to be swept clean

It is forbidden among the Yoruba of Western parts of Nigeria to perform any activity in the 'stale' floor of the household until it had been swept. The 'stale' floor means the floor as it remains overnight. It is the tradition of the people to select special locations for dumping the refuse and other solid wastes produced in the communities. The people would never deposit refuse in any other place except the approved and recognized dumping areas. This is still the practice among all the villages and other rural communities, with the result that in the rural areas of the developing countries one does not normally find refuse and other solid wastes deposited indiscriminately about the environment.[4]

In Nigeria, during the pre-independence period, the towns were kept clean by strict enforcement of the public health laws and regulations against littering. The 'sanitary inspectors', as the public health personnel in charge of environmental health used to be known, were very respectable and fearsome. They were more dreaded than the local authority 'policemen'. However, since independence the emphasis on the enforcement of the sanitary laws and regulations has been slackened. In many places the laws have either been totally removed from the statute books or the public health personnel have been prevented from enforcing them. This, and other reasons—like the rapid urbanization rate, the heterogeneous nature of the cities, the poor finances of the local authorities, and the low executive and managerial capabilities—are the main causes of the refuse disposal crisis which exists in all the urban centres of Nigeria and other developing countries today.

This chapter will review the topic of solid waste management in the developing countries under the following headings:

(1) the peculiar nature of the solid waste;
(2) household storage of refuse;
(3) peculiar collection problems;
(4) formidable transportation logistics;
(5) the common disposal methods;
(6) the financial aspects and;
(7) the administrative and managerial problems.

Under each heading the existing situations will be outlined and suggestions for improvement will be made.

THE PECULIAR NATURE OF THE SOLID WASTE

It is universally accepted that the quantity, characteristics, and the composition of refuse vary mainly according to the socioeconomic status, the food habits, local customs, geographical locations, occupations, and climatic conditions.[5] However, the variations within most of the communities in the developing countries are more dramatic.[2,7,8] The gap between the rich and the poor varies more widely in the developing countries. These wide variations are very much reflected in the refuse quantity and quality. Table 1 gives the composition and characteristics of refuse from three areas of Ibadan, Nigeria[2], (Oluwande 1974). Table 2 gives the picture in another community in Nigeria. In Addis Ababa, in Ethiopia, although there is no marked division of the city into low-income and high-income residential areas, the quantity and characteristics of refuse still vary according to the proportions of poor and rich people in different areas of the city. In China, as would be expected, the variations are not as much.

Seasonal variations in the quantity and quality of refuse in most of the developing countries are also very wide. There are distinct dry and wet seasons. The quantity and the types of foodstuffs for the two seasons are

Table 1   Composition and characteristics of refuse from three residential areas of Ibadan[2]

| Components | Mean percentage by weight | | |
	GRA (Government Reserved Areas)	Private layout areas	Traditional old parts
Leaves	13.7	33.7	81.3
Paper	12.6	11.3	2.5
Garbage	65.3	41.6	8.2
Tin	4.6	6.2	3.5
Glass	2.1	2.5	0
Rag	1.6	3.4	4.3
Dust	0.6	1.3	0.2
Density	256 kg/m^3	280 kg/m^3	296 kg/m^3
Moisture content	64.8%	61.4%	49.7%

different. In Nigerian urban centres the refuse disposal authorities have more difficult tasks during the wet seasons in trying to collect and transport refuse (which is very bulky because of the high proportions of remnants of green vegetables, plus maize cobs and husks). During the periods when maize is not available the refuse men are very happy because the volume of refuse they have to handle is very much reduced.

The rates at which the characteristics of refuse change in most of the developing countries are higher than those normally experienced in the developed countries. A country such as Nigeria, which has little difficulty in paying for imported luxury consumable goods but very low technical capacity for efficient maintenance of such goods, has her urban environment littered with carcasses of automobiles, gas cookers, mattresses, beds, fans, and other metallic objects. Although these items are not normally found in the household refuse they form a considerable proportion of the solid waste which has to be catered for in all the urban centres.

HOUSEHOLD STORAGE OF REFUSE

One of the prerequisites in the chain of activities that must be properly carried out in order to have an efficient and sanitary management of solid waste is proper storage in the generating premises. In the developed countries the standard household dustbins are common and familiar to everybody. Such

Table 2 Comparison of Refuse from two areas of a community in Ibarapa, Nigeria[2]

| Components | Mean percentage by weight | |
	Local university campus	Local people
Leaves	28.1	49.9
Paper	8.1	0.7
Garbage	48.0	4.4
Tin	3.7	0
Glass	4.3	0
Rag	2.0	0.5
Dust	10.0	44.5

dustbins comply with certain standards which enable them to function efficiently.[9] In most of the developing countries the majority of the people do not have dustbins. The few people that have, make use of all sorts of containers which do not ensure sanitary storage of refuse in the premises of the households. In many areas there are no regulations or bye-laws on household dustbins. Where such laws exist, they are never enforced.

As a first step towards sanitary and efficient management of solid waste in the developing countries, all the household units must have standard dustbins. The dustbins must be watertight, durable, and resistant to corrosion. They must have tight-fitting lids. The capacity must depend on the average household population and the frequencies with which they are to be emptied. It is important to distinguish between houses and households in many communities of the developing countries in order to have sanitary storage of refuse. Normally, a house contains many households—that is many family units live in each house. In urban centres in Nigeria a house with six rooms may contain six households. The number of people living in such a house may be well over fifty. Some aspects of results of a survey carried out in 15 urban centres of Nigeria between 1981 and 1982 are given in Table 3. It is customary for the landlords of such houses to provide a container for each house. The 200-litre diesel drums, without lids, are the commonest containers. Since two-storey buildings often have well over 100 occupants and the refuse disposal authorities may not visit such houses over a 4-week period, the insanitary conditions are not difficult to imagine.

It is therefore very important to study the living pattern in every community in the developing countries before the capacities of the dustbins are fixed. As much as people, household units and not houses should be encouraged to have dustbins. Where this is not possible, landlords must be made to provide special dustbins designed to meet the local situations.

Another aspect of refuse storage problems in the developing countries is sanitary storage of solid waste generated in the shops and market stalls—that is, the commercial refuse. Normally the traders do not have dustbins. They and their customers litter the markets and the shopping centres, thereby making sanitary refuse collection difficult for the disposal authority. In Nairobi, Kenya, it is obligatory for shop-owners to have dustbins. In Jos, Nigeria, it is also illegal for stall and shop-owners not to have dustbins. The bye-law is effectively enforced in Jos. This, coupled with regular health education programmes, helps to make the city the neatest in Nigeria.

To summarize, in order to have sanitary and effective storage for refuse in domestic premises in the developing countries, the following steps must be taken:

(1) surveys must be carried out to establish the house population density;
(2) the capacities of the dustbins must be related to the house population density and the frequencies at which the dustbins are likely to be emptied;
(3) the refuse disposal authorities must provide the dustbins and sell to those who may wish to buy from them;
(4) there must be regular health education of the people on the proper use of the dustbins;
(5) there must be bye-laws which make it obligatory for householders and traders to have dustbins.

REFUSE COLLECTION

Even in the developed countries, where towns and cities are well planned and the people are well informed on environmental health matters, refuse collection is the most difficult and most expensive aspect of solid waste management.[10,11] In such countries the cost of collection and transportation of refuse takes well over 70% of the entire budget for solid waste management.[12-14] In most parts of the developing countries efficient refuse collection is more complicated because house-to-house collection is not possible. This is because houses are not accessible to motor vehicles. In the traditional core areas of Ibadan, Nigeria, houses are so built closely together and arranged in such orderless manner that it is impossible to ride bicycles between them.[15] The same situation exists in parts of Addis-Ababa, Ethiopia, and in major cities in China. Even in those areas where house-to-house collection is practicable, the refuse disposal authorities are not able to provide efficient services. In Nigeria only very few houses enjoy a form of house-to-house refuse collection, which is very irregular and unreliable. The drivers and the labourers who take the refuse vehicles to the houses in Ibadan,

Table 3   Household refuse survey in Nigeria

(a) Variations in No. of people and quantity of refuse in the medium–high-income area in Lagos

No. of occupants	Annual household income (₦)	No. of rooms in the house	Average refuse quantity (kg/h/day)
11	NA	5	0.2
10	2,000	6	0.2
11	15,000	6	0.23
40	NA	10	0.10
24	NA	6	0.15
6	10,000	4	0.30
10	13,000	5	0.30
7	19,000	5	0.90
8	NA	6	0.17
5	4,200	6	0.14
5	NA	5	0.60
10	NA	6	0.33
13	NA	6	0.30
13	NA	6	0.30
16	NA	8	0.20
21	NA	8	0.15
14	NA	NA	0.25

NA = not available.

(b) Variations in number of people and quantity of refuse in medium–low-income area in Lagos

No. of occupants	Annual income (₦)	No. of rooms in the house	Annual refuse quantity (kg/h/day)
45	NA	12	0.1
20	NA	6	0.07
74	NA	16	0.15
96	NA	16	0.11
28	NA	8	0.08
53	NA	12	0.06
65	NA	12	0.12

Nigeria, will never visit any house unless the landlord or the tenants in the house are prepared to give them special tips, in spite of the fact that appropriate payment of ₦ 4(about £3.20 sterling) per flat per month had been paid to the authority in charge of disposal. Therefore, in addition to the constraint of poor urban planning, efficient house-to-house refuse collection in many developing countries is made impossible by bad administration, poor management, and non-availability of funds. All these factors are interrelated.

SOME PRACTICAL APPROACHES TO EFFICIENT REFUSE COLLECTION IN DEVELOPING COUNTRIES

> Get mobilized, pay attention to hygiene, reduce diseases, improve the health conditions.
>
> Mao Tse-Tung

The above statement of the great Chinese leader is relevant when discussing how to improve solid waste collection in the developing countries. Whatever approach is adopted, it is necessary to mobilize the people for maximum co-operation and participation. In many developing countries refuse collection is so poorly organized and executed that less than 25% of the refuse produced is actually collected. The remaining 75% is allowed to remain, and to cause nuisance and pollution of the urban environment.

In order to have an efficient refuse collection programme in the developing countries, the following four approaches must be efficiently employed in different situations.

(1) effective house-to-house collection;
(2) appropriate depot method;
(3) the Chinese practice; and
(4) modified garchey technique.

Effective house-to-house collection

House-to-house refuse collection method makes refuse collection a straight-forward municipal engineering task in the developed countries. It is the most convenient method for refuse collection. If properly organized it will enable disposal authorities to derive revenue direct from solid waste management. All houses with suitable access roads must have standard dustbins. All the dustbins must be visited periodically by the refuse vehicles and the collection crews. The frequency of the visits must be determined after taking into consideration the refuse generation.

However, the frequency of collection should not be less than fortnightly.

The frequency of visits, as much as possible, should be regular. The landlords or the tenants should pay for this house service if this has not been included in the property rating system in the community. The levy for refuse disposal should be related to what it costs the disposal authority to provide the service. Over-subsidization by the authorities often prevents efficient services in the developing countries. In Ibadan, Nigeria, the disposal authority charges ₦ 2 (about £1.6 sterling) per dustbin per month. The people are often reluctant to pay this amount because the refuse crews may never visit the premises over a period of 3 months unless they are induced by tips. In the same city there are many private firms which collect refuse and charge monthly levies per dustbin ranging between ₦ 5 and ₦ 10 (£4–8 sterling). The people are happy to pay because the services are regular and efficient.

Market stalls, shops, and other commercial and business premises should also have their own dustbins. The monthly payment by the owners of such premises should depend on the sizes of the dustbins. As already stated, the dustbin size in any premises should be related to the refuse generation rate in the premises.

For house-to-house refuse collection programme to succeed in the developing countries there must be regular and effective health education of the public. The authorities must have an adequate number of suitable refuse vehicles together with efficient maintenance workshops and a regular supply of spare parts.[16] The collection crews must be adequate in number. They must be adequately motivated, remunerated, and supervised to prevent situations where the crews serve only those premises where they collect illegal payments. The crews should also be provided with protective clothing, gloves, nose-masks, and rubber boots. In areas where traffic hold-up during the day is likely to reduce the collection efficiency, night collection should be carried out. This has been practised in Addis-Ababa, Ibadan, Dar es Salam, and parts of China. In such situations the crews should be adequately compensated for this.[17]

Depot method

For those areas where the houses are not accessible by roads, depots located at strategic places along the main roads must be employed. Depots are also necessary in market places and other commercial centres.

Different forms of depots are being employed in various communities of the developing countries (Figure 2). In selecting the type of depots to employ in a particular area the most important factor to consider is the method of loading the refuse vehicles, or the method of emptying the depots. Normally, up to 75% of the time spent on refuse collection may be taken up during loading of refuse into the vehicles.[7,13] For this reason the most efficient loading device practicable under a given situation must be investigated and

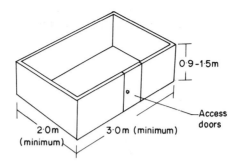

Figure 2   Un-roofed rectangular enclosure

employed. For manual loading the type shown in Figure 2, which consists of a shed with a low wall and roof, is suitable. The roof is an advantage in regions with high rainfall. In urban centres where mechanical loading may be necessary, simple but strong-walled rectangular enclosures illustrated in Figure 2b are suitable. In different parts of Nigeria, where this type is being employed, the depots are often damaged by the mechanical equipment employed in loading the refuse vehicles. Overhead depots illustrated in Figure 2c were recommended for easy manual loading in parts of Ibadan.[7] Unfortunately the introduction was not backed up with adequate health education and regular servicing; this led to the discontinuation of the system. The experts who introduced the system intended that the refuse could be easily 'pushed' manually with shovels from the overhead depots into the refuse vehicles. This was why the discharge entrance of the depot was set at the height of the refuse vehicle. Movable depots, usually made of metal or sometimes with wooden bodies, are also very common in the developing countries. They take different forms, shapes, and capacities. In Nigeria some are provided with wheels and towing gear, others have handles for mechanical lifting, while others are mere containers. The commonest type of depot found in Nigerian cities is the ordinary dumping site along the main roads. These depots are normally created by the people themselves.

Whatever type of depot is recommended, the following must be borne in mind:

(1) refuse in the depots is liable to being set ablaze no matter how good the effectiveness of the emptying programme;
(2) nuisance is common from full and overflowing depots—it is therefore important to employ depots of the correct capacity and to ensure regular emptying;

(3) depots must not be located too far from the road or be placed anywhere which is not easily accessible to prevent the refuse being deposited outside the depots—it may be necessary to employ labourers to police the depots.

The Chinese collection practice

In places like Shanghai, Canton, and other big cities, the Chinese employ the principle of mass participation for both refuse and night-soil collection. This reduces collection costs and time. The refuse vehicles visit residential areas at specific times of the day (normally early in the morning or in the evening). A familiar noise is made, and the people bring out their dustbins or the night-soil containers to be emptied into the vehicles. This approch has been adopted in parts of Addis-Ababa in Ethiopia and some local government areas in Nigeria. If backed up with proper health education of the public, it will work in most of the developing countries. It is highly recommended for markets and other commercial centres.

The modified garchey technique

In the real garchey method the normal household refuse is flushed with water through 20 cm diameter pipes into the tanks provided in the ground of multi-storey blocks of flats. This is too sophisticated and expensive for most of the developing countries where water supply is grossly inadequate and operation–maintenance is often not efficient.

A modified form of the garchey method is recommended for multi-storey blocks of flats in the developing countries. In place of the 20 cm pipes, chutes 60 cm in diameter (or square), may be provided. Common household refuse is then thrown, to drop through the chutes into collecting depots located at ground level. For proper operation the people must be educated so that they do not block the chutes with bulky refuse like mattresses. The chutes must also be readily accessible for clearing of blockages, and for other maintenance.

In a few multi-storey blocks of flats in Lagos, Nigeria, a system similar to that described above was introduced some years ago. The passages through which the refuse is sent to the containers on the ground are made of fencing wire. The main complaint against this system is that refuse particles are easily blown about, since the openings are not airtight.

REFUSE TRANSPORTATION

The transportation of refuse to the disposal sites is a very important aspect of refuse collection activities. In the urban areas refuse transportation may be very difficult. Unless it is properly programmed, it will adversely affect the sanitary disposal of refuse.[12,18] In the developing countries the importance of regular and efficient transportation cannot be over-emphasized because the majority of the people normally empty their dustbins into the roadside depots. If these depots are not emptied regularly, because vehicles are not available to transport refuse to the disposal sites, the failure of the disposal programme is easily perceived. Refuse then blocks the main roads. Many factors make efficient refuse transportation difficult in many developing countries. Some of these factors are:

(1) the nature of refuse;
(2) the traffic situation
(3) inadequate number of refuse vehicles;
(4) shortage of funds; and
(5) haulage distances.

The nature of refuse

Refuse in most of the developing countries like Nigeria (Table 1), China, Ethiopia, and Uganda tends to have high organic (garbage) content.[2,7,8] Therefore the refuse cannot be compacted, to take up a small space in the vehicles, as is normally the case with less dense refuse from the developed countries. This means that, in many areas, vehicles which compress the refuse cannot be employed to noticeable advantage.

The traffic situation

In urban centres of many developing countries the traffic situation is so chaotic that refuse collection and transportation are seriously affected. In such situations the best solution is to effect collection and transportation during the period of minimum traffic, especially in the night. The period between 9 p.m. and 6 a.m. can be ideal. In such cases the need for adequate financial inducement for the workers cannot be over-emphasized. Night collection and transportation has been introduced in cities such as Dar es Salam in Tanzania, Beirut in Lebanon, Addis-Ababa in Ethiopia, and a few others.

Inadequate numbers of refuse vehicles

This is the most serious of all the factors which affect efficient transportation of refuse in the developing countries. No refuse handling authority can ever have the number of refuse vehicles it requires at any given time because the vehicles are very expensive and the wear and tear on them are very great.

Many authorities in the developing countries do not have more than 20% of the number of vehicles they need at a given time. Authorities with 50% of the number needed are very lucky indeed. Many local authorities also employ many types of vehicles imported from different suppliers. This complicates maintenance and provision of spare parts.

In order to minimize problems caused by shortage of vehicles, the following points are important[19]:

(1) Mixed fleet of petrol and diesel vehicles should be avoided. The diesel engine vehicles are preferable.
(2) Where the road conditions allow, heavy-duty vehicles which have high payload capacity are preferable because the number of trips to disposal sites will be reduced.
(3) It is not advisable for a local authority to have more than two types of vehicles. This ensures that the maintenance staff can be familiar with the vehicles, thus promoting high maintenance efficiency. It also facilitates easy stocking of spare parts and local fabrication of those spare parts which are not supplied by the manufacturers.
(4) The vehicles should preferably be equipped with tilting gear, for off-loading, but it should also be possible to off-load manually. This can be ensured if the vehicle body can be opened up.

Other important points to consider on refuse vehicles in the developing countries are those relating to the use of tippers and other non-conventional refuse vehicles. Tippers are commonly employed in many local authorities. The use of tippers should be discouraged unless special covers can be provided to ensure that refuse does not fly off to litter the routes. Hand-carts and wheelbarrows are very useful for collecting the refuse from those places which are not accessible to conventional vehicles. It is important that refuse collection crews have a separate apartment on the vehicles for themselves and their equipment. The crew should never sit on the refuse.

Shortage of funds

In most of the developing countries the refuse disposal authorities are not able to provide efficient services because of scarcity of funds. When there are no funds, collection and transportation are affected most. Scarcity of funds for refuse disposal is often due to the following:

(1) inefficient revenue collecting system;
(2) poor budgeting;
(3) mismanagement of the meagre available funds—more will be said on the financial aspects later.

Distance of the disposal sites

The commonest refuse disposal method practised in the developing countries is tipping (controlled but often only partly controlled). Urban centres in these countries are expanding very rapidly, with the result that existing tipping sites are quickly rendered inadequate and the new ones are often found in locations which are very far from the areas where the bulk of the refuse is generated. In the developed countries this may not pose serious problems for refuse transportation because the disposal methods such as incineration and composting, which may be readily employed, do not have to be sited very far from the generating centres.

In order to minimize the effects of distant disposal sites on refuse transportation it will be necessary to investigate every situation carefully to decide on the most appropriate solutions.[17] In certain places it may be economical to pretreat the refuse by pulverization or compaction into blocks to reduce the volume to be transported. In other areas it may be reasonable to employ incineration in spite of the cost and the high technology involved. The railway, the water barges, and other transportation methods may have to be employed.

THE COMMON DISPOSAL METHODS

In most of the developing countries of Asia and Africa the commonest method of solid waste disposal is crude controlled tipping or semi-controlled tipping. In China, and to certain extent in India, composting is the main method of disposal.[20]

Crude or semi-controlled tipping is so called because although the refuse is tipped in selected areas, other controlling aspects, such as the limitation of the layer thickness and the covering of the tipped refuse, are not normally performed. The waste is usually employed to reclaim and level depressed areas like ravines, old quarries, and valleys. In Nairobi, Kenya, old quarries are being reclaimed for residential buildings. It has been reported that such reclaimed quarry sites are ready for building blocks of flats 10 years after tipping has ceased. This is because the authority in Nairobi control the tipping by ensuring adequate compaction for each layer.

It should always be borne in mind that the tipping sites can pollute underground water.[21,22] Hand-dug wells are very common, and are important sources of water in both urban and rural areas of the developing countries. Those who live near tipping sites must be warned about this danger.

Composting

As already mentioned, composting is not as popular as it should be in many developing countries. This is because most of these countries have more than

80% of their working populations engaged in agriculture.[23] Some of these countries spend huge sums of money to import chemical fertilizers. Although the compost is no substitute for chemical fertilizers, such countries can still benefit from the soil-conditioning nature of organic composts. In China between 90 and 95% of the total wastes (sewage and refuse) produced is employed to make composts to promote agricultural production. Simple and labour-intensive methods of composting are employed to reduce cost.[20] In Lagos, Nigeria, a multi-million naira composting plant completed in 1979 has not been commissioned for use because the present administration feels rightly that the running and operation cost of nearly ₦ 20 million per annum is unreasonable for the stage of the country's development.

Incineration

As for composting, modern incineration is not common in the developing countries. The refuse in most of the communities has very low calorific values. Moreover, to operate the huge urban incinerators without coupling them up with heating and power generation is too expensive for most developing countries.[24] However, simple incinerators similar to the one illustrated in Figure 3 are very efficient in rural and semi-urban communities where the refuse is mainly combustible materials.[2] This type of incinerator can be built locally with laterite, cement blocks, and metal. Care should always be taken to ensure that the refuse is first deposited on the ground near the incinerators. Towards the evenings, after the refuse has been exposed to the sun, the refuse is placed in the incinerators for combustion. In this way the simple incinerators, when strategically located at the rate of about one incinerator per 200–300 people, has been found to be adequate for semi-urban communities with populations up to 50,000. It is pertinent to add that the rates of refuse production in such communities are between 0.1 and 0.25 kg/head/day.

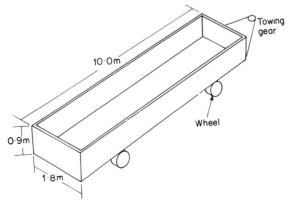

Figure 3   Typical movable depot on wheels

FINANCIAL ASPECTS OF SOLID WASTE MANAGEMENT

The main factor which makes efficient solid waste management difficult in many developing countries is shortage of funds. That is, the authorities in charge of refuse disposal do not have the necessary financial resources to operate efficient services. In Nigeria the annual budgets for refuse disposal of some local government authorities are given in Table 4. Since the bulk of the refuse generated in these areas is collected from roadside depots, it is not possible to charge the people separately for refuse disposal. There is no effective tenement or property rating system in any of the cities either; therefore, the local authorities depend on government grants for nearly all their programmes. Since refuse disposal does not generate any revenue, and since environmental health programmes are generally low among the priorities of public authorities in the developing countries (and refuse disposal being rated far below water supply and housing) it is not surprising that refuse disposal is often poorly funded.

Table 4   Annual budgets for refuse disposal of some local authorities in Nigeria

Local authority	Estimated population	Annual budgets for solid waste management (₦)
Ibadan	3,000,000	Not Available
Warri	700,000	300,000
Onisha	2,000,000	297,730
Kaduna	1,300,000	392,000
Maiduguri	600,000	3,000,000
Sokoto	600,000	2,750,000
Kano	6,000,000	3,000,000

In order to ensure that adequate funds are provided for refuse disposal, the revenues of the local authorities must improve. The best way to improve the revenues of the local authorities is through efficient and effective tenement or property rating programmes. Landlords should pay rates commensurate with the environmental facilities and services provided for their property.[15] When facilities and services are efficiently provided, investigations have revealed that the people will be willing to pay. If the property rating cannot be made to take care of the operation and maintenance costs of all the services such as water supply, sewage, and refuse disposal, then revenue should be generated from refuse disposal. The only way of doing this is to introduce house-to-house refuse collection for all those who have reasonable access to their houses.

ADMINISTRATIVE AND MANAGEMENT PROBLEMS

In many developing countries the local authorities are responsible for refuse disposal. In those countries where municipal authorities have city engineers, the engineering services (including solid waste management) are often properly managed. In Nigeria the argument continues as to which organization should be in charge of refuse disposal. The constitution says the local authorities should be in charge. The local authorities cannot do the job efficiently because they lack financial, technical, and managerial capabilities. Also, in various local government authorities, in-fighting goes on as to which department should handle refuse disposal. The Medical Officers of Health (MOH) want to do it and the city engineers also want to do it. Whenever the MOH departments are charged with the responsibility, the sanitarians quarrel with the medical officers for interfering with what they regard as purely the responsibility of the environmental officers. Partly because of this confusion, and partly because the local authorities cannot manage refuse disposal efficiently, separate institutions have been created in Nigeria. These institutions are called different names in various states. In Lagos there is the Waste Disposal Board; in Owerri there is Imo State Environmental Sanitation Authority (previously the Ibadan Waste Disposal Board); in Kaduna it is the Kaduna Capital Development Board; while in Jos it is the Plateau Urban Development Board, to cite few examples.

It is the view of this writer that solid waste management in urban areas is mainly an engineering programme. The departments or institutions charged with the responsibility must approach it with the skill of engineering management and adequate funding.

REFERENCES

1. Oluwande, P. A. (1979) Sanitation appliances in Nigeria. In *Proceedings of Cairo International Regional Seminar in Sanitary Engineering*, pp. 31–38. Building Research Establishment, Garston, UK.
2. Oluwande, P. A. (1974) Investigations of certain aspects of refuse in Western State of Nigeria. *Journal of Solid Wastes Management*, **64**, 22–32.
3. Odunjo, J. F. (1945) *Alawiye*, Book I, 1st edn. St Paul Press, Lagos.
4. Oluyemi, I. O. (1972) Refuse and Sewage Disposal. Paper presented at annual conference of Nigerian Society of Engineers, Benin City, 6–8 December 1972. NSE, Lagos.
5. *New Civil Engineer* (1973) Wastes: Special Feature (Magazine of the Institution of Civil Engineers), 30 August 1973, London.
6. Clark, R. M., Sweeten, J. M., and Greathouse, D. G. (1972) Basic data for solid wastes pilot study. *J. San. Eng. Div. ASCE*, **98** (SA6), 897–907.
7. Maclaren International Ltd. (1971) *Immediate Measure Report*: Master Plan for Waste Disposal and Drainage in Ibadan UNDP/WHO Project. Western State Secretariat, Ibadan.

8. Clarke, F.A.J. (1970) Solid waste disposal in the Tropics. Personal communications on Uganda refuse. Kampala, October 1970.
9. British Standard Institution, 1960 The storage and collection of refuse from residential buildings. London, *BS code of Practice CP 306*.
10. Clark, R. M. (1973) Measures of efficiency in solid waste collection. *J. Env. Eng. Div. of ASCE,* 99 (NEE4), 447–459.
11. Malina, J. F., and Morgan, W. T. (1972) Refuse reclamation and recycle. *J. San Eng. Div. of ASCE,* **98** (SA3), 213–223.
12. HMSO (1967) Her Majesty's Stationery Office. Refuse storage and collection (Report of the Working Party on Refuse Collection). Great Britain Ministry of Housing and Local Government, London.
13. Lawrence, J. P., and Harrington, J. J. (1974) Multi-variate study of refuse collection efficiency. *J. Env. Eng. Div. of ASCE,* **100** (EE4), 963–978.
14. Bodner, R. M., Cassel, E. A., and Andros, P. J. (1970) Optimal routing of refuse collection vehicles. *J. San. Engr. Div. ASCE.* **96** (SA4), 893–904.
15. Oluwande, P. A. (1978) Some new approaches to urban renewal in developing countries. *ITCC Review* **VII** (4), 36–41. Tel Avivi, Israel.
16. Clark, R. M. and Helms, B. P. (1972) Fleet selection for solid waste collection systems. *J. San. Eng. Div. ASCE,* **98** (SA1), 71–78.
17. Houseberg, E. (1972) Facility selection and haul optimization model. *J. San. Eng. Div. of ASCE,* **98** (SA6), 1005–1019.
18. Grossma, D., Hudson, J. F., and Marks, D. A. (1974) Waste generation models for solid waste collection. *J. Env. Eng. Div. of ASCE,* **100** (EE6), 1219–1230.
19. Acharya, R. J. (1970) Collection, Transportation and Disposal of city refuse. Lecture delivered during training course on city refuse disposal, 23 to 26 November 1970. CPHERI, Nagpur, India.
20.. McGarry, M. G., and Stainforth, J. (1978) *Compost, Fertilizer and Biogas Production. Human and Farm Wastes in the People's Republic of China.* IDRC, Ottawa, Canada.
21. Holmes, R. (1980) The water balance method of estimating leachate production from landfill sites. *Solid Wastes,* **70** (1), 20–33.
22. Snoxell, J. (1979) A tip leachate problem. *Water Pollution Control,* **7** (3), 362–363.
23. Bassam, E. El, and Thorman, A. (1979) Potentials and limits of organic wastes in crop production. *Compost and Land utilization,* **20** (6), 30–35.
24. Kampert, W. (1967) Refuse incineration with heat recovery. In *Report of the 9th International Conference of the International Association of Public Cleansing,* p. 166. London.

Index